T0122551

Studies in Big Data

Volume 124

The series "Studies in Big Data" (SBD) publishes new developments and advances in the various areas of Big Data- quickly and with a high quality. The intent is to cover the theory, research, development, and applications of Big Data, as embedded in the fields of engineering, computer science, physics, economics and life sciences. The books of the series refer to the analysis and understanding of large, complex, and/or distributed data sets generated from recent digital sources coming from sensors or other physical instruments as well as simulations, crowd sourcing, social networks or other internet transactions, such as emails or video click streams and other. The series contains monographs, lecture notes and edited volumes in Big Data spanning the areas of computational intelligence including neural networks, evolutionary computation, soft computing, fuzzy systems, as well as artificial intelligence, data mining, modern statistics and Operations research, as well as self-organizing systems. Of particular value to both the contributors and the readership are the short publication timeframe and the world-wide distribution, which enable both wide and rapid dissemination of research output.

The books of this series are reviewed in a single blind peer review process.

Indexed by SCOPUS, EI Compendex, SCIMAGO and zbMATH.

All books published in the series are submitted for consideration in Web of Science.

Aleksei V. Bogoviz
Editor

Big Data in Information Society and Digital Economy

Editor
Aleksei V. Bogoviz
Moscow, Russia

ISSN 2197-6503 ISSN 2197-6511 (electronic)
Studies in Big Data
ISBN 978-3-031-29491-4 ISBN 978-3-031-29489-1 (eBook)
https://doi.org/10.1007/978-3-031-29489-1

This Springer imprint is published by the registered company Springer Nature Switzerland AG
The registered company address is: Gewerbestrasse 11, 6330 Cham, Switzerland

Basics of Applying Big Data in the Information Society and Digital Economy (Introduction)

Aleksei V. Bogoviz

Doctor of Economics, Professor, Independent Researcher, Moscow, Russia

Big data is an advanced Industry 4.0 technology that plays an important role in building cyber-physical systems. The first peculiarity of big data is an incredibly large volume of information that is systematized (collected in a special way, clearly structured, with its logical relationships marked) and ready for further processing and analysis. The second peculiarity of big data lies in the involvement of the digital representation, storage, transmission, and processing of information. The third peculiarity is that big data is collected automatically.

Due to the above features, big data is an important link in cyber-physical systems that can completely exclude human participation in almost all phases. Thus, digital devices (e.g., ubiquitous computing, the Internet of Things, or machine vision) can collect big data to be processed by artificial intelligence. Nevertheless, cyber-physical systems are created for people as their users and beneficiaries. Therefore, big data plays an important role in the formation, functioning, and development of the information society and digital economy.

The information society is a progressive social environment that emerges from a developed knowledge society (a high level of education, the value of knowledge), evolving under the influence of scientific and technological progress (the intensive creation and distribution of high technology). As the embodiment of knowledge and technology, information is the highest value (i.e., social reference point, priority) of the information society. The digital economy is also referred to as the data economy because data is at its core—a key resource and a source of economic growth and development. Thus, information is a systemic and central component of the information society (social structure) and digital economy (economic structure) in Industry 4.0.

When studying big data as part of the information society and digital economy, it is advisable to consider it as an institution. Therefore, it is necessary to consider four institutional aspects of big data. The first aspect has to do with the need for technological infrastructure and a legal framework for the distribution and use of big data. Since big data is only one element of cyber-physical systems, its use in practice requires an overall high level of technological development of the information society and digital economy. On the one hand, technologies are required to collect big data and, on the other hand, to process it.

The legal framework should regulate and encourage the active application of big data as part of cyber-physical systems in e-government practices, electronic payments and communications, etc. Together, the technological infrastructure and the legal framework trigger the process of institutionalizing the use of big data in the information society and digital economy. Government regulation involves the creation of common "rules of the game" in the application of big data, along with other technologies of Industry 4.0.

The second aspect is the need for favorable market conditions to introduce and apply big data. On the one hand, there must be sufficient, solvent, and sustainable demand for Industry 4.0 products. To form this demand in the information society, digital competencies must be massively mastered, and consumer preferences for high-tech products must be formed. On the other hand, there must be a supply, the basis of which is a highly skilled digital workforce ready to use big data.

The third aspect lies in the importance of highly effective organization and management of big data in business. Big data can be collected and used in different ways by businesses depending on the industry. For example, in agriculture, big data is collected through phyto- and zoomonitoring—gentle on plants and animals. Industry requires environmental monitoring and application of big data to protect the environment and develop green industries. In the service industry, big data is perspective to create and manage social networks, which makes the ethical issues of big data important.

The fourth aspect is the importance of a favorable social environment and effective social interactions for the dissemination and application of big data in the information society. The transition to Industry 4.0 must be systemic and involve all businesses in the overall value chain because the introduction of big data only in one enterprise or a single link in the value chain will not unlock their potential to improve the efficiency of the digital economy and information society. Electronic document management and electronic state and public monitoring of the entire value chain are required to realize this potential.

Cooperation is a promising mechanism for developing big data, which allows for meeting this condition of consistency. It can take the form of business cooperation, allowing the pooling of resources to accelerate digital modernization. Consumer cooperation is also possible to increase demand for Industry 4.0 products. The cooperation of the government and society to jointly control society and the economy using big data is also possible.

This book aims to systemically explore and reveal the basics of using big data in the information society and digital economy. The first part of the book outlines the technological and institutional framework for the dissemination and use of big data. The second part focuses on the use of big data in the digital economy. The third section explores the application of big data in digital business by sectors, including industry, construction, agro-industrial complex, energy sector, and other sectors of the digital economy. The fourth section examines the cooperative mechanism for the dissemination and application of big data in the information society. The fifth section reveals a case study on the use of big data in the digital economy of the Eurasian Economic Union (EAEU).

The novelty of this book lies in the institutional approach to the study of big data, which allows us to explore it more deeply and reveal its socio-economic nature. Using an institutional approach, this book considers big data as a technology for the information society and digital economy. The advantage of studying big data as an institution is that it bridges the gap between theory and practice and provides the applied focus to this book.

The primary target audience of this book is academics involved in studying Industry 4.0, the information society, and the digital economy. This book provides them with a conceptual presentation and theoretical and methodological research on developing the information society and digital economy based on big data. An additional target audience for this book is expert practitioners in Industry 4.0. In this book, government regulators will find applied recommendations for regulating the dissemination and application of big data in the information society and digital economy. For managers of high-tech companies in the digital economy, the book provides the author's recommendations to improve the efficiency of applying big data in business. The book is of particular interest to readers from the EAEU because it contains many case studies and examples, as well as highly detailed recommendations for the EAEU.

Contents

Technological and Institutional Framework for the Dissemination and Use of Big Data

Big Data: A System-Forming Role in the Development of the Information Society and the Digital Economy for the Transition to Industry 5.0

Elena G. Popkova

Abstract The purpose of the work is to determine the role of Big Data in the development of the information society and the digital economy during the transition to Industry 4.0. Using the regression analysis method, the impact of Big Data and analytics on the information society is determined—agility of companies, e-participation and e-government based on IMD statistics. A sample of 10 developed and developing countries leading in the field of digital economy in 2021 was formed for the study. As a result, it is proved that Big Data can and already now play an important system-forming role in the integrated development of the information society and the digital economy, contributing to the transition to Industry 5.0. The theoretical significance of this conclusion is that it has expanded the range of applications of Big Data, as well as filled a gap in the technological support of the transition to industry 5.0, supplementing artificial intelligence with Big Data. The practical significance of the obtained results lies in the fact that they opened up the possibility of a practical transition to Industry 5.0 based on the use of Big Data, thereby supporting and accelerating the Fifth Industrial Revolution.

Keywords Big data · Information society · Digital economy · Socialization of intelligent machines · Artificial intelligence · Industry 5.0

JEL Codes C31 · C55 · C82 · O31 · O32 · O33 · O35 · O38

1 Introduction

Big data is an advanced digital technology of Industry 4.0, which emerged and became widespread under the influence of the Fourth Industrial Revolution. The advantage of Big Data is that it allows you to form especially large collections of data and its highly efficient analysis, thereby providing high-accuracy characteristics of the processes and systems under study. The technical component of Big Data and

---

E. G. Popkova (✉)
Peoples' Friendship University of Russia (RUDN University), Moscow, Russia
e-mail: elenapopkova@yahoo.com

A. V. Bogoviz (ed.), *Big Data in Information Society and Digital Economy*,
Studies in Big Data 124, https://doi.org/10.1007/978-3-031-29489-1_1

its importance in the functioning of cyber-physical systems have been studied in sufficient detail and described in the existing literature [5, 7, 11].

Rapid scientific and technological progress drives the imminent onset of the Fifth Industrial Revolution and, accordingly, the transition to Industry 5.0. The problem lies in the uncertainty of the role of Big Data (along with other advanced technologies) of Industry 5.0, as well as in ensuring the transition to it. In this regard, the study of the social component of Big Data, which has not been practically explored, becomes relevant. The purpose of this article is to determine the role of Big Data in the development of the information society and the digital economy in the transition to Industry 4.0.

2 Literature Review

Industry 4.0 is the current technological order, which is dominated by cyber-physical systems. The peculiarity of this mode is the autonomy (independence) of intelligent machines controlled by artificial intelligence, which does not involve human participation in their work. This feature has led to a breakthrough in the development of the digital economy, allowing, for example, the creation of "smart" industrial enterprises that operate continuously and demonstrate increased productivity [8, 9, 14, 17].

Industry 5.0 is a future technological order in which cyber-social systems will prevail. This mode is characterized by the fact that the synthesis of human and artificial intelligence, the balanced development of technologies and societies will be achieved in it. Artificial intelligence is cited in the existing literature as a technology contributing to the transition to Industry 5.0. It really makes it possible to adapt intelligent machines to the needs of people [3, 13, 15].

But it is obvious from the accumulated practical experience that artificial intelligence alone is not enough to create harmonious cyber-physical systems. For example, a person can become a hostage of a "smart" home if his settings are not flexibly adjusted as human needs change (i.e. they will continue to work according to a standard, pre-programmed scheme). "Smart" traffic cameras are already sending false parking fines for illegal parking to drivers who are actually caught in traffic jam.

"Smart" (unmanned) transport can travel only in ideal conditions, which can only be created artificially, but do not occur in real life. In human–machine communications, in most cases, the linearity of artificial intelligence is immediately noticeable, its work according to a strictly defined algorithm, while human intelligence is much more flexible. And there are many such examples of the limitations of artificial intelligence. Taken together, they indicate that artificial intelligence alone is not enough for the transition to Industry 5.0—other advanced technologies are needed.

In this regard, Big Data deserves special attention. After all, if they provide machine communications, they have the potential to support human–machine communications [2, 6, 10, 12, 18]. The results of the literature review and gap analysis are clearly demonstrated in Fig. 1.

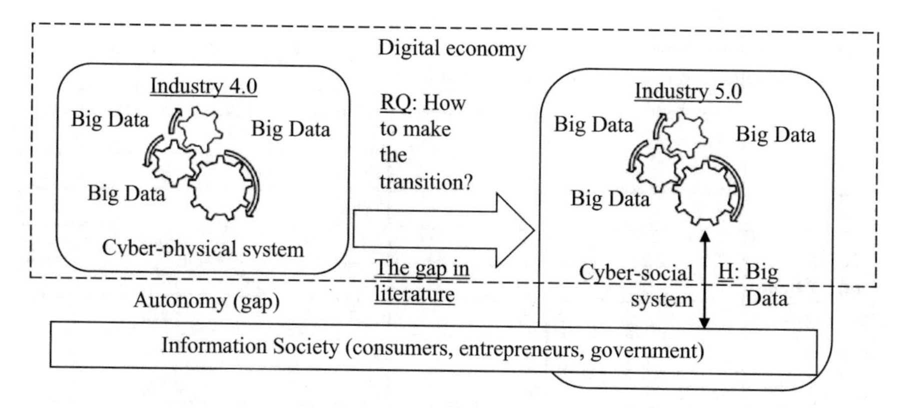

Fig. 1 Scientific knowledge about Industry 4.0 and Industry 5.0, gap in literature, research question (RQ) and research hypothesis (H). *Source* Developed and compiled by the author

As shown in Fig. 1, the conducted literature review revealed a gap associated with the uncertainty of the prospects for the transition to Industry 4.5. This raises a research question (RQ) about how to make the transition to Industry 5.0. The essence of this RQ is how to make the socialization of intelligent machines. In this paper, hypothesis H is put forward that Big Data plays a system-forming role in the development of the information society and the digital economy for the transition to Industry 5.0.

3 Methodology

To test the hypothesis (H), a regression analysis method was chosen. This method is used to determine the impact of use of Big Data and analytics on the information society—agility of companies, e-participation and e-government based on IMD statistics (2022) [4]. For the study, a sample was formed consisting of 10 developed and developing countries leading in the field of digital economy in 2021 (included in the rating of the 64 best digital economies in the world according to IMD, 2022 [4]) and, in particular, in the field of Big Data development (included in the ratings based on [1, 16]).

4 Results

To determine the contribution of Big Data to the development of the information society, a regression analysis of statistics from Table 1 was performed, as a result of which the following models of paired linear regression were obtained.

Model 1: $y1 = 6.02 + 0.51x$. According to the obtained model, with an increase in the activity of use of Big Data and analytics by 1 place, the agility of companies

Table 1 The level of development of Big Data and information society in developed and developing countries leading in the field of digital economy in 2021

Country	Use of big data and analytics	E-Participation	Agility of companies	E-Government
	x	y_1	y_2	y_3
USA	5	1	7	9
Canada	8	16	21	26
Great Britain	18	6	20	7
Switzerland	23	18	6	16
Russia	31	26	57	33
South Africa	40	45	55	56
Mexico	49	35	34	50
Argentina	46	28	60	29
Chile	41	28	26	31
Brazil	56	18	44	47

Source Compiled by the author based on [4]

increases by 0.51 places. The close relationship of the indicators is confirmed by their high cross-correlation: 68.01%. The significance F = 0.03045 indicates that the model corresponds to a significance level of 0.05. At a given significance level, the tabular F is 5.32. The calculated F was 6.88, exceeding the tabular distribution, therefore, the Fisher F-test was passed, and the model is reliable at a given level of significance.

Model 2: $y2 = 8.41 + 0.77x$. According to the obtained model, with an increase in the activity of use of Big Data and analytics by 1 place, e-participation increases by 0.51 places. The close relationship of the indicators is confirmed by their high cross-correlation: 67.55%. The significance F = 0.03207 indicates that the model corresponds to a significance level of 0.05. At a given significance level, the tabular F is 5.32. The calculated F was 6.71, exceeding the tabular distribution, therefore, the Fisher F-test was passed, and the model is reliable at a given level of significance.

Model 3: $y3 = 6.98 + 0.74x$. According to the obtained model, with an increase in the activity of use of Big Data and analytics by 1 place, e-government increases by 0.51 places. The close relationship of the indicators is confirmed by their high cross-correlation: 77.17%. The significance F = 0.00893 indicates that the model corresponds to a significance level of 0.01. At a given significance level, the tabular F is 11.26. The calculated F was 11.78, exceeding the tabular distribution, therefore, the Fisher F-test was passed, and the model is reliable at a given level of significance.

In accordance with the obtained econometric models (through substitution), the prospects for the development of the information society based on Big Data are determined, which are demonstrated in Fig. 2.

As shown in Fig. 2, at the maximum level of use of Big Data and analytics (1st place, +96.85%), the agility of companies increases by 70.47% (up to 6–7 places),

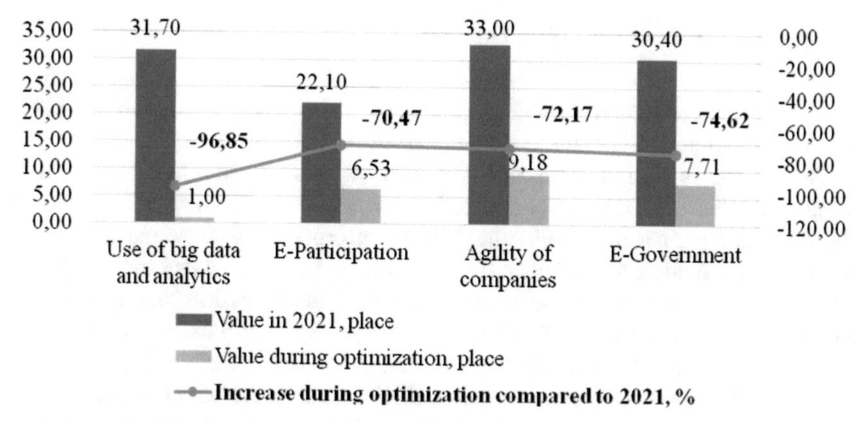

* The increase in indicators has a negative sign due to the fact that indicators are measured in places, and the smaller their values, the better. From a qualitative point of view, an improvement in indicators and a positive increase in their values are achieved

Fig. 2 Prospects for the development of the information society based on Big Data *Source* Calculated and constructed by the author

e-participation—by 72.17% (up to 9th place) and e-government—by 74.62% (up to 7–8 places). To clarify the quantitative results obtained, they were supplemented with a qualitative analysis, which revealed the following potential system-forming role of Big Data in the development of the information society and the digital economy, supporting the transition to Industry 5.0:

- Socially-oriented business management aimed at studying consumer preferences and identifying their changes (in the formation of Big Data) to better meet consumer demand. In addition to this, taking into account the individual characteristics of employees (when generating Big Data) to optimize team building with the participation of human and artificial intelligence (people and intelligent machines). The advantage of performing this role is associated with the support of agility of companies;
- Marketing support of product selection taking into account the individual consumer preferences continuously studied (integrated into Big Data) and analyzed by artificial intelligence. The advantage of performing this role is related to the support of e-participation;
- Personification of the e-government system through the selection of in-demand services and their adaptation to the unique needs of each household and business. The advantage of performing this role is associated with the support of e-government.

5 Discussion and Conclusion

The contribution of the article to the literature consists in clarifying the prospects for the application of Big Data. Unlike the researchers [5, 7, 11], who attribute Big Data to Industry 4.0 (limit the scope of their application to the digital economy), it is justified that Big Data can and already now play an important system-forming role in the comprehensive development of the information society and the digital economy, contributing to the transition to Industry 5.0. This is the answer to RQ and confirmation of the hypothesis H, which was put forward in this article. The theoretical significance of the conclusion lies in the fact that it expanded the range of applications of Big Data, as well as filled the gap in technological support for the transition to Industry 5.0, supplementing artificial intelligence with Big Data.

The practical significance of the obtained results lies in the fact that they opened up the possibility of a practical transition to Industry 5.0 based on the use of Big Data, thereby supporting and accelerating the Fifth Industrial Revolution. The social significance of the results of the study is related to the fact that it provided scientific support for the socialization of intelligent machines for their "friendliness" towards people. As this study has shown, Big Data can serve the interests of the development of the information society through increasing the flexibility of the digital economy and the creation of cyber-social systems.

This study is limited to considering only the role of Big Data. The revealed significant potential of their contribution to the transition to Industry 4.0 is the basis for conducting similar studies on the example of other advanced digital technologies related to Industry 4.0—there is the probability that many of them can also serve as a transition to Industry 5.0. In future studies, it is also advisable to deepen the examination of the prospects for the new role of Big Data related to the systemic development of the information society and the digital economy, as well as to offer management recommendations for this.

References

1. Chakravorti, B., Bhalla, A., Chaturvedi, R. S. (2019). Which Countries Are Leading the Data Economy? Harward Business Review. Retrieved April 14, 2022, from https://hbr.org/2019/01/which-countries-are-leading-the-data-economy.
2. Charles, V., Emrouznejad, A., & Gherman, T. (2022). Strategy formulation and service operations in the big data age: The essentialness of technology, people, and ethics. *Studies in Big Data, 98*, 19–48. https://doi.org/10.1007/978-3-030-87304-2_2
3. Devi, B. S., & Muthu Selvam, M. (2022). SoloDB for social media's big data using deep natural language with AI applications and Industry 5.0. *Smart Innovation, Systems and Technologies, 243*, 279–294. https://doi.org/10.1007/978-981-16-3675-2_21
4. IMD. (2022). World Digital Competitiveness Ranking–2021. Retrieved April 14, 2022, from https://www.imd.org/centers/world-competitiveness-center/rankings/world-digital-competitiveness/.

5. Keshk, M., Moustafa, N., Sitnikova, E., & Turnbull, B. (2022). Privacy-preserving big data analytics for cyber-physical systems. *Wireless Networks, 28*(3), 1241–1249. https://doi.org/10.1007/s11276-018-01912-5
6. Nasrollahi, M., & Fathi, M. R. (2022). Modeling big data enablers for service operations management. *Studies in Big Data, 98*, 49–94. https://doi.org/10.1007/978-3-030-87304-2_3
7. Niu, C., & Wang, L. (2022). Big data-driven scheduling optimization algorithm for Cyber-Physical Systems based on a cloud platform. *Computer Communications, 181*, 173–181. https://doi.org/10.1016/j.comcom.2021.10.020
8. Popkova, E., Bogoviz, A. V., Sergi, B. S. (2021). Towards digital society management and 'capitalism 4.0' in contemporary Russia. *Humanities and Social Sciences Communications, 8*(1), 77. https://doi.org/10.1057/s41599-021-00743-8.
9. Popkova, E. G., Inshakova, A. O., Sergi, B. S. (2021). Venture capital and Industry 4.0: The G7's versus BRICS' experience. *Thunderbird International Business Review, 63*(6), 765–777. https://doi.org/10.1002/tie.22235.
10. Ray, S. K., Alani, M. M., & Ahmad, A. (2022). Big data for educational service management. *Studies in Big Data, 98*, 139–161. https://doi.org/10.1007/978-3-030-87304-2_5
11. Sabbagh, R., Živković, S., Gawlik, B., Sreenivasan, S. V., Stothert, A., Majstorovic, V., & Djurdjanovic, D. (2022). Organization of big metrology data within the Cyber-Physical Manufacturing Metrology Model (CPM3). *CIRP Journal of Manufacturing Science and Technology, 36*, 90–99.https://doi.org/10.1016/j.cirpj.2021.10.009.
12. Sangwan, S. R., & Bhatia, M. P. S. (2021). Soft computing for abuse detection using cyber-physical and social big data in cognitive smart cities. *Expert Systems*. https://doi.org/10.1111/exsy.12766
13. Saniuk, S., Grabowska, S., Straka, M. (2022). Identification of social and economic expectations: Contextual reasons for the transformation process of Industry 4.0 into the Industry 5.0 concept. *Sustainability (Switzerland), 14*(3), 1391. https://doi.org/10.3390/su14031391.
14. Sergi, B. S., Popkova, E. G. (2022). Towards a 'wide' role for venture capital in OECD countries' Industry 4.0. *Heliyon, 8*(1), e08700. https://doi.org/10.1016/j.heliyon.2021.e08700.
15. Sindhwani, R., Afridi, S., Kumar, A., Banaitis, A., Luthra, S., & Singh, P. L. (2022). Can Industry 5.0 revolutionize the wave of resilience and social value creation? A multi-criteria framework to analyze enablers. *Technology in Society, 68*, 101887. https://doi.org/10.1016/j.techsoc.2022.101887.
16. Srivastava, S. (2019). Top 10 countries & regions leading the Big data adoption in 2019. Retrieved April 14, 2022, from https://www.analyticsinsight.net/top-10-countries-regions-leading-the-big-data-adoption-in-2019/.
17. Yankovskaya, V. V., Kelina, K. G., Chutcheva, Y. V., Alekseev, A. N. (2021). The consumer economy and the pleasure economy: Similarities and differences in developing countries in Industry 4.0. *International Journal of Trade and Global Markets, 14*(4–5), 507–515. https://doi.org/10.1504/IJTGM.2021.116737.
18. Zhang, S., Yang, L. T., Feng, J., Wei, W., Cui, Z., Xie, X., & Yan, P. (2021). A tensor-network-based big data fusion framework for Cyber–Physical–Social Systems (CPSS). *Information Fusion, 76*, 337–354.https://doi.org/10.1016/j.comcom.2021.10.020.

Digitalization in Business Management: Issues of Legal Regulation and Protection

Mikhail G. Ivanov, Olesya M. Ivanova, Alexander I. Kalinichenko, Mikhail S. Galiev, and Marina A. Semenova

Abstract Digitalization, informatization and robotization penetrate into all spheres of socio-economic, industrial relations, and act as an integral component in the management structure of complex economic processes. Advanced ideas on the development and application of high technologies in the field of business management create additional opportunities to improve the efficiency of economic indicators. A sufficiently high level of development of socio-economic relations, achievements of science and technology create the necessary conditions for this. The reasonable application of technical standards, symbols and the systematic use of legal norms, the combination of new and traditional legal regulation regimes to neutralize contradictions between the interests of the digital, information and communication industry and other socio-economic institutions increase the efficiency of business management. Legal support for the normal development of business is a real problem that requires a doctrinal study of social practice and an explanation of the strategy for improving the mechanisms of legal regulation and protection of service and management institutions in modern conditions of active development of the digital, information and communication industry (Ivanov in Bulletin of the Russian University of Cooperation 3:105–108, 2016 [3]).

M. G. Ivanov (✉) · M. A. Semenova
Cheboksary Cooperative Institute (branch) of the Russian University of Cooperation, Cheboksary, Russia
e-mail: imkafedra54@mail.ru

O. M. Ivanova
Moscow University of the Ministry of Internal Affairs of Russia Named After V. Ya. Kikot, Moscow, Russia

A. I. Kalinichenko
Department of the Internal Affairs Directorate for the North-Eastern Administrative District of the Main Directorate of the Ministry of Internal Affairs of the Russian Federation for the City of Moscow, Moscow, Russia

M. S. Galiev
Kamchatka Branch of the Russian University of Cooperation, Petropavlovsk-Kamchatsky, Russia
e-mail: galiev87@inbox.ru

A. V. Bogoviz (ed.), *Big Data in Information Society and Digital Economy*, Studies in Big Data 124, https://doi.org/10.1007/978-3-031-29489-1_2

Keywords Business · Law · Management · Engineering · Digitalization · Economics · Performance management · Regulation

JEL Classifications L84 · P37 · O17

1 Introduction

It is undoubtedly, in order to overcome risks and exclude the development of chaotic events in business management, it is necessary to form legal principles, norms and standards of a new format. In this regard, the problem of taking into account the criminal risks in the system of service and management relations, determining the degree of criminal Internet impact on the economy, on its managerial component, is particularly relevant. The evidence of the presented problems is explained by the fact that both fraudulent actions and other types of cybercrime, along with "old" property crimes [6], are not only widespread, but also aggressive and offensive. For example, the number of cases of cryptocurrency thefts from e-wallets of owners, investment funds or other companies owning individual or collective e-wallets is not decreasing [8]. They, in particular, reflect the degree of risk, the possibility of causing significant harm to official and managerial activities in the field of business [5].

Consequently, the illegal use of digital information and communication technologies in the field of business management and other circumstances affect the level of misappropriation of the results of economic activity, has a detrimental effect on the nature, pace and sources of development of the entire economy. Criminally obtained products occupy a considerable share among the products produced by the business, which a priori determines the place of individual regions in the all-Russian division of labor.

2 Methodology

The scientific validity of the main conclusions and judgments is confirmed by the study of fundamental research works of scholars in the field of legal regulation and legal protection of business in modern conditions of development of the digital, information and communication industry.

The degree of reliability was achieved by studying social practice, statistical data on economic abuse; judicial and investigative practice related to the implementation of legal responsibility for malfeasance and economic abuse.

An integrated approach to the study allowed us to rethink and turn to qualitatively new level of legal assessment of the use of digital, information technologies in the field of business management and the development of conceptual provisions for its regulation and protection from criminal influence.

3 Results

The authors' desire to study the problem under consideration is explained by the fact that the progress of economic culture, the development of information technology changes the ratio of "human" and "technical" in business management. The authors' concerns are due to the fact that science and practice are following the path of forming a kind of human intelligence, but much more effective. At the same time, this process is not limited only to digitization of information, it is supposed to develop artificial intelligence built on a specific system and database. Of course, this will require "the study of methods and means of extracting, presenting, structuring and using knowledge. This technological breakthrough is undoubtedly a significant factor in the development of modern society" [7].

Thus, it is the introduction of digital, information technologies in combination with the robotization of production that puts before the modern doctrine of law the problem of a comprehensive understanding of the processes of legal regulation of official and managerial activities in the field of business and the protection of national economic and financial interests. Therefore, research is needed that would make it possible to comprehend the institutions of a private–public nature, the importance of the human factor in the field of high technology, various causes and conditions of deviant human behavior in the context of new realities.

At the same time, we should note that recently a lot of publications have been devoted to the legal aspects of digitalization. Undoubtedly, those researchers are right who, not without reason, pay attention to the fact that the large-scale digitalization of society and the introduction of artificial intelligence have so penetrated into the public consciousness that the fastest resolution of legal regulation problems is required. Otherwise, scientists believe, the whole process of introducing and using digital technologies can turn into chaos.

In addition, when solving the problems of legal regulation of official and economic activities, it is impossible to ignore the issues of interdependence of conceptual provisions of both international law and national legislation. This issue is particularly relevant in the context of the mass introduction of high technologies in the field of management of socio-economic processes. Foreign scientists have spoken more categorically about this. They assess the regulatory role of law not only as a characteristic of the national legal system, but consider it as a category of a transnational nature [10].

Therefore, in this regard, it will be necessary to address the problem of realizing and understanding the intersectoral nature of legal regulation. It seems that the methodology of forming an integrated approach in this case will allow to resolve existing contradictions and expected disagreements. Of course, it is very problematic to combine such elements as norms, legal relations, legal acts of civil, administrative, criminal systems, nevertheless, the expected effectiveness of the process of legal regulation of high technologies in the field of business management, as well as its dynamic, working regime, incline the authors to strengthen the position of an integrative approach to solving problems. The proposed approach will allow not only to develop legal regulation in the field of business and management relations, but also to bring them into a stable and orderly state. In this case, the toolbox contents, that is, the special legal content of the mechanism of legal regulation can be considered as an effective means of legal influence, which in its development would form a single consistent chain of implementation of legal norms.

Thus, we can talk about the methods of legal influence (permission and prohibition) both for management activities and in relation to the powers of its subjects. This will allow to control the actual behavior of the subjects involved in business management and to exclude the manifestation of deviance [9]. For example, it is possible to trace how positive legal prescriptions of a private-legal, public-legal nature (obligation, permission, prohibition) affect the mental activity of subjects, the formation of motives for their deviant behavior. The formation of positive motives among the subjects of official and economic activity, the rejection of deviant behavior, in our view, are an important indicators of the effectiveness of the implementation of legal norms. This judgment should be applied to the results of this activity, and it is fair. It follows that the process of legal influence can be considered as the fulfillment of the requirements of legal norms in order to ensure a single economic legal space. Consequently, in this place, it is possible to focus attention on the fact that the norms of criminal law, for example, without interfering in a person's personal space, ensure the stability of economic relations, protecting them from deviant behavior of subjects. This is a conceptual positionIt is important to understand that criminal legislation punishes only those acts of a deviant nature that violate the norms of common law or the economic freedom of others. This fact, as a phenomenon of social practice, cannot be questioned. Thus, the right combination of permits and prohibitions, to one degree or another, contributes to the formation of socially justified motivations, freedom of development of material, spiritual and other non-legal, but socially significant incentives.

The social aspect of the integrative approach consists, first of all, in the indirect impact on those institutions of society that are outside the legal mechanisms of regulation and protection of business and management relations, but, nevertheless, somehow determine the deviant behavior of business management subjects. Considering that the process of informatization and digitalization penetrates into all spheres of a person's socio-economic life, the problem under consideration is very relevant and acquires the level of a national security problem [4].

Consequently, the process of digitalization makes it necessary to analyze not only and not so much the technical, but also the social and legal aspects of its regulation. At the same time, we believe it is important to understand that the use of high technologies carries certain risks of harming the existing economic order, ensuring the integrity of the economic security system, which in turn can be preserved only within the legal framework. In addition to all that, when forming electronic images and meanings, it should be taken into account that they are the result of human intellectual activity. Consequently, the moral and ethical side, the social significance of a particular type of entrepreneurial and other economic activity and its specifics are also an important part of the framework of legal regulation and legal protection of the process of digitalization and informatization of business management.

Therefore, the study of the legal aspects of scientific and practical development should take into account the level of use of digitalization tools. At the same time, the motivation for improving legislation in the field of ownership and use of Internet resources may be the desire to ensure strict compliance with the rule of law in the field of legislative regulation of business management. Unification of national legislation in the field of digitalization will help to avoid the difficulties and uncertainties that are generated by gaps and conflicts of laws on legal regulation and protection of business management.

Unfortunately, scientific thought encounters certain obstacles and difficulties. Resistance comes, first of all, from within, i.e. every representative of the branch scientific school is very jealous of his "sovereignty". Scientific public opinion is more or less characterized by distrust of the norms of other branches of law. Nevertheless, we believe that it is possible to develop a doctrinal model of a strategy for the development of legal policy, taking into account international legal regulations and the guidelines of the current national legislation.

The strategy of forming the legal space of digital information and communication technologies is not a one-day job. If the development strategy is presented as a comprehensive methodological model, then it must necessarily also include a program of legal regulation of the actions of both administrators (authors) and users of Internet resources in the field of business management and their legal protection.

Clearly, the choice of legal mechanisms for their regulation is associated with the variety of tasks and functions performed. Constitutional law and administrative law may be involved in this case. Constitutional law, for example, is the basis and guarantee for the formation and development of digital information and communication technologies for managing economic relations in the form of norms on economic rights and freedoms of citizens.

Administrative law modifies and supplements the competence of public bodies, establishes appropriate control and supervision regimes. Providing for possible violations of technology regimes, the law establishes responsibility for their non-compliance.

A special place in the national legal system is occupied by the norms of civil law, which regulate both general legal relations related to scientific and technical developments and specific relationships between economic entities. In addition, the jurisdiction of civil law includes legal certificates of various facts, circumstances and actions using digital, information and communication technologies.

It is noteworthy that subjective and real rights of the new format have received an appropriate legal characteristic in the current civil legislation [1, 6]. For example, according to the norms of the Civil Code of the Russian Federation, the law of obligations and others are recognized as digital rights, the content and conditions of which are determined in accordance with the rules of the information system. The specifics of legal regulations are manifested in the fact that the exercise of the right, the disposal of the right, including the transfer, pledge, encumbrance of digital rights in other ways or restriction of the disposal of digital rights are possible only in the information system.

The rules of criminal law play an equally important role. Federal Law No. 207-FZ of November 29, 2012 introduced into the Criminal Code of the Russian Federation a whole group of norms on special types of fraudulent acts provided for in Articles 159.1–159.6 [2, 6].

The above suggests that almost all branches of law are involved in the mechanism of legal support of digital, information and communication technologies in the field of business management, although each industry has its own subject and, accordingly, methods of regulation.

At the same time, for law enforcement, it is important to correctly understand the terminology that the legislator uses when forming the disposition of legal norms. This is also due to the fact that scientific and technological progress is developing in several areas of economic activity. Somewhere digitalization is needed, in another place robotization prevails, the third direction is provided by informatization. In addition, some legal categories can contribute to the digitalization process as their own part of socio-economic development programs. Other concepts contain technical characteristics and introduce them into legal circulation as equivalents of traditional actions.

Consequently, the process of forming a legal form is complicated by the fact that the development and functioning of digital information and communication technologies, computer and other technical systems is a closed system that exists according to a given program of a technical mechanism. In this regard, the direct application of general legal norms is hardly acceptable. On the other hand, those general, conceptual provisions explaining, for example, the existence of competencies of business management entities, the emergence of threats and risks can be preserved in the specifics of legal regulation of the process of introduction and use of new technologies. The development of special legal regulations on self-regulation according to a given program, on liability, etc. can be considered as special conditions.

Thus, the legislative activity on the formation of legal mechanisms for regulating the digitalization of business management has a rather complex and multidimensional nature. Obviously, it was not sufficient to simply consider the norms of only one branch of law, as well as to use only traditional legal concepts, categories and institutions.

4 Discussion

The conceptual and categorical apparatus of legal support for the process of digitalization of business management should be considered as a complex mechanism of relationships between business entities and should serve as a starting point for their legal concept. The complexity of the structure of the conceptual apparatus is explained by the combination of legal provisions and technical regulators, the description of the modes of creation and use of symbols, standards, programs, their influence on legal mechanisms and management institutions. The content of the conceptual apparatus will influence both legal forecasts and the diagnosis of risks, errors, as well as the determination of the immediate and long-term prospects for digitalization of management.

In modern Russian legal science, the problems of strategic development of legal policy are quite complex. The sphere in question is viewed by scientists as one conceptual problem that requires its speedy resolution.

5 Conclusion

The norms of law, legal regulations in the system of digital, information and communication technologies are aimed at improving the efficiency and security of business management. The Russian reality requires the definition of fundamental directions for the development of mechanisms of legal regulation and legal protection of digital, information and communication technologies.

The development and resolution of organizational and legal, social problems based on the reasonable application of standards, symbols and the systematic use of norms of different branches of law, the combination of new and traditional legal regulation regimes to neutralize contradictions between the interests of the digital, information and communication industry and other socio-economic institutions, which, in principle, increases the efficiency of business management, is in demand.

These doctrinal provisions can play a key role in the development of the Russian legal policy of digital, information and communication technologies in the field of business management. The hypotheses presented are derived from those fundamental requirements that are necessary for doing business in the conditions of the normal existence and functioning of the rule of law, and to ensure the realization of basic economic rights and freedoms of man and citizen.

References

1. Bezverkhov, A. G., Krivokapich, B. (2020). The counteraction to legalization (laundering) of criminal incomes: a comprehensive legal approach. *Courier of Kutafin Voscow State University (MSAL), 10*(74), 194–204. https://doi.org/10.17803/2311-5998.2020.74.10.194-204.
2. Coleman, J. W., & Cressey, D. R. (1990). *Social problems* (4th ed.). Longman Higher Education.
3. Ivanov, M. G. (2016). Prospects of criminal responsibility optimization for the service and economic crimes. *Bulletin of the Russian University of Cooperation, 3*(25), 105–108.
4. Ivanov, M. G., Andreev, V. V., Kuznetsov, A. P., Bezverkhov, A. G., & Ivanova, O. M. (2021). Official economic abuse in economic management as a criminological and legal category. *Studies in Systems, Decision and Control, 316*, 1001–1008.
5. Ivanova, O. M. (2017). Current problems of precautionary impact on economic offenses in the sphere of business management. *Fundamental and Applied Research of the Cooperative Sector of the Economy, 6*, 147–151.
6. Ivanova, O. M. (2019). *Crimes against property: Theoretical and applied research (history, timeliness, trends)*. RUK.
7. Khabrieva, T. Y., & Chernogor, N. N. (2018). Law in the conditions of digital reality. *Journal of Russian Law, 1*(86), 85–102.

8. Sidorenko, E. L. (2014). The OECD anti-corruption standards and their implementation in national criminal law (the experience of countries passing the third assessment phase). *Journal of Foreign Legislation and Comparative Jurisprudence, 1*, 85–88.
9. Sutherland, E. H. (1983). *White collar crime*. Greenwood Press.
10. Taylor, V. (2017). Regulatory rule of law. In P. Drahos (Ed.), *Regulatory theory: Foundations and applications* (pp. 393–413). ANU Press.

Analysis of the Features of Sustainable Development of the Mercosur Countries in the Context of the Covid-19 Pandemic

Rail R. Khussamov, Elena A. Galiy, Elena V. Ivanova, Tatyana V. Bodrova, and Elena V. Zubareva

Abstract *Purpose*: The purpose of this article is to identify the features of sustainable development of the MERCOSUR countries in the context of the COVID-19 pandemic. *Design/methodology/approach*: The authors use comparative and retrospective analysis to identify the distinguishing characteristics of countries meeting the sustainable development goals. The objects of research are the MERCOSUR countries. *Findings*: It has been established that Uruguay is the undisputed leader in sustainable development in the context of COVID-19. Two countries: Brazil and Venezuela slowed down the pace of implementation of national sustainable development strategies due to the pandemic and other reasons. *Originality/value*: According to the results of the analysis, it was revealed that countries that have long-term national strategies for sustainable development are more stable in achieving sustainable development goals. The size of a national economy does not guarantee that it can successfully overcome an external shock such as the lockdown caused by the COVID-19 pandemic.

Keywords MERCOSUR · Sustainable development goals · Government policy · National economy · COVID-19

JEL Classifications Q01 · Q58 · F15 · F64 · O57

R. R. Khussamov
Financial University Under the Government of the Russian Federation, Moscow, Russia

E. A. Galiy
Russian Presidential Academy of National Economy and Public Administration, Moscow, Russia

E. V. Ivanova (✉)
The Military University of the Ministry of Defense of the Russian Federation, Moscow, Russia
e-mail: evi77@yandex.ru

T. V. Bodrova · E. V. Zubareva
Russian University of Cooperation, Mytishchi, Russia
e-mail: ezubareva@ruc.su

A. V. Bogoviz (ed.), *Big Data in Information Society and Digital Economy*,
Studies in Big Data 124, https://doi.org/10.1007/978-3-031-29489-1_3

1 Introduction

Currently, the countries of the world are going through a difficult, dramatic moment in their history—overcoming the negative consequences of the COVID-19 pandemic in economic and social processes.

The implementation of the ambitious Sustainable Development Goals (SDGs), which were adopted in 2015, is in jeopardy, or at least the time for their implementation is delayed due to the impact of the COVID-19 pandemic.

Let us consider how the countries of the MERCOSUR union carried out the implementation of the SDGs in modern conditions.

MERCOSUR includes five countries: Argentina, Brazil, Paraguay, Uruguay and Venezuela, whose membership has been suspended. Bolivia has been in the process of joining this union since 2015.

MERCOSUR is the fifth economy in the world, its population is slightly less than 300 million people, the total area of the countries of this union is 14.8 million km^2 [8].

Each state forms its own unique model for the implementation of the SDGs for the period up to 2030, depending on the strategic development goals of national economies, available material and human resources.

2 Methodology

Let us analyze what measures the MERCOSUR countries have taken to achieve sustainable development goals in a situation of overcoming the consequences of COVID-19.

Argentina is the first object of our analysis.

Currently, the body responsible for coordinating the actions of all state and public organizations of the country on the implementation of the sustainable development strategy is the National Interdepartmental Commission for the Implementation and Monitoring of the SDGs, which was established in 2016 [2]. Before that, there was the National Council for Sustainable Development of Argentina created in 1999.

The Ministry of the Environment and Sustainable Development, created in 2019, and the National Council for the Coordination of Social Policy of Argentina play an important role in the implementation of the SDGs.

Argentina's first national sustainable development strategy, the "Environmental Action Strategy for Sustainable Development", was developed in 2001 [12]. It is also necessary to note such national strategies as the National Strategy and Action Plan for Biodiversity Conservation 2016–2020; National Strategy for the Integral Management of Urban Solid Waste (2005); National Strategy for Environmental Education (2006, 2020), etc.

Argentina's active position in the international discussion and implementation of the SDGs should be noted. Argentina has prepared voluntary national reports on the implementation of the SDGs in 2017, 2020 and is preparing such a report in 2022 [3].

In response to the COVID-19 pandemic, the Argentine government developed the National Action Plan "ACCIONAR" in December 2020, the main goal of which was to achieve an integrated territorial approach in social policy to the most vulnerable communities in the country. This plan included 15 social tasks, the solution of which was aimed at a comprehensive improvement in the quality of life of the country's population [11].

The second object of research is Brazil. Brazil actively participated in the international debate on sustainable development goals in 2015. In 2016, the government of this country established the National Commission on Sustainable Development Goals, the main tasks of which were: (1) a comprehensive study of sustainable development goals; (2) dissemination and popularization of the concept of sustainable development; (3) ensuring transparency in the implementation of sustainable development goals in the country [10].

In 2019, the Special Secretariat for Social Affairs of Brazil was created to replace the National Commission on Sustainable Development Goals to coordinate the actions of various government agencies. The functions of this secretariat were established as follows: (1) assistance to the government in the field of sustainable development; (2) development of government decisions on the implementation of sustainable development goals; (3) analysis, generalization and generation of reports on the implementation of sustainable development goals.

The analysis of the implementation of sustainable development goals, carried out by the authors, revealed the features of the Brazilian national model of sustainable development:

(1) the country has developed the traditions of multi-year state planning (Plano Pluri Anual) throughout the country, which were recorded in article 165 of the Brazilian Constitution [6]. All multi-year plans began to be checked for compliance with the sustainable development goals starting in 2016. 17 sustainable development goals were transformed into 54 national programs and further detailed into smaller goals [5];
(2) the country has implemented an initiative called "Atlas of Vulnerability of Municipalities"—a platform that reflects the level of social vulnerability in all municipalities and major metropolitan regions of the country. The social vulnerability index, which includes three equivalent components: (1) the level of urban infrastructure development; (2) the level of development of human capital and (3) the level reflecting the income and labor opportunities of the population of the municipality was developed [4].

According to a group of 106 experts, despite significant government efforts to address sustainable development challenges, Brazil has not been able to achieve significant progress in achieving the planned sustainable development goals [7]. According to experts, Brazil is failing in at least 9 out of 17 goals in the following areas: the environment, the creation of peaceful and inclusive societies, the fight against poverty and hunger [13].

The third country that is part of MERCOSUR is Paraguay.

To solve national problems in the field of sustainable development in the country, the Interdepartmental Coordination Commission for the implementation of the SDGs was formed in 2016. This commission was expanded to 17 people—representatives of the main ministries and departments of the country in 2020 [14].

In 2014, the country adopted the National Development Plan "Paraguay 2030", which included three main areas of action: (1) poverty reduction and social development; (2) inclusive economic growth and (3) correct inclusion of the country in world processes, expansion of economic integration [9].

The analysis of the implementation of sustainable development goals, carried out by the authors, revealed the features of the Paraguayan national model of sustainable development:

(1) the implementation of the universal principle of "leaving no one behind", which means the use of policies based on the facts of inclusive involvement of all citizens in the process of implementing national priorities in the field of sustainable development;
(2) creation of a new National Statistical Institute, which will participate in the creation of a multidimensional poverty index;
(3) positive experience of multisectoral project management aimed at the implementation of 17 sustainable development goals.

The fourth object of our study will be Uruguay.

In 2021, the Government of this country presented the fourth voluntary report on the achievement of the Sustainable Development Goals, which reflected the progress made by the country [18].

The legal framework of the sustainable development sphere of Uruguay is more representative than other MERCOSUR countries: (1) the national development strategy—Uruguay 2050, which was adopted in 2019 after 5 years of intensive work on it; (2) National Strategy for Gender Equality (2021); (3) national environmental plan (2021), etc.

In our opinion, the main features of sustainable development in Uruguay are:

(1) no special state body responsible for the implementation of sustainable development goals has been identified. In this case, the country's government itself is responsible for all decisions and results in this area;
(2) the national regulatory framework for the implementation of sustainable development goals, which makes it possible to unambiguously understand the tasks set, strategic guidelines and allocated resources for solving the tasks set, has been formed;
(3) unlike other MERCOSUR countries, a unique national long-term strategy until 2050, based on the paradigm of sustainable development, has been developed [16].

In conclusion, let us consider the fifth member of MERCOSUR—Venezuela.

Currently, this country is going through a difficult, critical stage in its development. There is very little relevant information necessary to determine the characteristics of sustainable development of this country, so we will use the UN information on this country in the field of sustainable development [1].

In 2016, the Bolivarian Republic of Venezuela presented its first and so far the only national voluntary report dedicated to the achievement of the sustainable development goals [17].

In our opinion, the features of sustainable development of Venezuela at the moment are:

(1) sustainable development goals are not a priority for the government of this country at the present time;
(2) the new UN strategy to support sustainable development goals in Venezuela, which includes three important areas: (1) restoration and further development of the national economy, sustainability of the social protection system; (2) assistance to social groups most affected by the crisis and COVID-19 has been in effect since 2018;
(3) promoting the depolarization of society and comprehensive humanitarian assistance in solving problems related to human rights.

3 Results

Table 1 presents the ranking of the MERCOSUR countries in the implementation of measures aimed at achieving the sustainable development goals.

The following conclusions can be drawn from the presented table: (1) currently Uruguay, which over the past two years has improved its position in the global ranking from 45 to 41, is the leader in the implementation of the sustainable development goals; (2) a group of stable countries has been identified—these are Argentina and Paraguay, which have slightly worsened their positions in the global ranking, respectively by 1 and 2 places; (3) a group of outsider countries. Venezuela is a country that performs poorly in achieving the Sustainable Development Goals, and its place in the global ranking has dropped by 6 positions. Brazil also showed negative dynamics in

Table 1 Ranking of MERCOSUR countries in achieving the SDGs

Country	Country's place in the global ranking (2020)	Country's place in the global ranking (2021)	Country dynamics in the global ranking (2021)
Argentina	51	52	−1
Brazil	53	61	−8
Paraguay	90	92	−2
Uruguay	45	41	+4
Venezuela	118	122	−6
Average	71,4	73,6	−2,2

Source Compiled by the authors based on data [15]

the global rankings. It is important to note the general downward trend in the average rating of the MERCOSUR countries in the period 2020–2021 by 2 positions.

4 Conclusion

Let us sum up the analysis of the national characteristics of sustainable development of the MERCOSUR countries.

1. Argentina. The following was revealed in the implementation of the sustainable development goals of this country: (1) there is extensive experience in the implementation of sustainable development strategies; (2) there is a broad legal basis for the implementation of the SDGs; (3) the national action plan as a reaction to the COVID-19 pandemic has been developed.
2. The Brazilian model of sustainable development differs from others in the following: (1) long-term planning which takes into account the achievement of the SDGs in the country is being implemented; (2) an open and accessible platform that reveals the level of social vulnerability in the municipal areas of the country is used; (3) Brazil under the influence of COVID-19 has not yet reached the set benchmarks in the implementation of the SDGs.
3. The Paraguayan model of sustainable development has the following features: (1) the inclusive principle of "leaving no one behind" is being implemented; (2) development of the system of the national statistical system in order to more accurately measure the SDGs; (3) a multisectoral approach is used to achieve the SDGs.
4. The Uruguayan model of sustainable development has the following differences: (1) the entire government of the country is responsible for achieving the SDGs; (2) the legal framework for the implementation of the SDGs has been formed; (3) the country's long-term strategy until 2050, based on the paradigm of sustainable development, has been developed.
5. Venezuela. This country is characterized by the following features of the implementation of the SDGs: (1) sustainable development goals are not currently a priority for the government of this country; (2) there is a system of support in achieving the SDGs from the UN.

References

1. About our work for the Sustainable Development Goals in Venezuela. (2021). Retrieved October 17, 2021, from https://venezuela.un.org/es/sdgs.
2. Argentina. (2021). National Coordination of Social Policies. Retrieved October 17, 2021, from https://www.argentina.gob.ar/politicassociales/ods/nacion.
3. Argentina. (2021). Sustainable Development Knowledge Platform. Retrieved October 17, 2021, from https://sustainabledevelopment.un.org/memberstates/argentina.

4. Atlas. (2021). Atlas of Social Vulnerability in Brazilian Municipalities. Retrieved October 17, 2021, from http://ivs.ipea.gov.br/index.php/pt/sobre.
5. Brazil Pluriannual Plan. (2019). PPA 2020–2023 of Brazil. Regional Observatory of Planning for Development. Retrieved October 17, 2021, from ***.
6. Constitution of the Federative Republic of Brazil. (1988). Retrieved October 17, 2021, from ***.
7. Gameiro, N. (2021). Report shows that Brazil did not advance in 2030 Agenda Goals. Retrieved October 17, 2021, from https://portal.fiocruz.br/en/news/report-shows-brazil-did-not-advance-2030-agenda-goals.
8. MERCOSUR. (2021). MERCOSUR in brief. Retrieved October 17, 2021, from https://www.mercosur.int/en/about-mercosur/mercosur-in-brief/.
9. National Development Plan: Paraguay 2030. (2014). Regional Observatory of Planning for Development. Retrieved October 17, 2021, from https://observatorioplanificacion.cepal.org/es/planes/plan-nacional-de-desarrollo-paraguay-2030.
10. National Governance for the SDGs. (2019). Secretary of the Government of the Presidency of the Republic. Retrieved October 17, 2021, from http://www4.planalto.gov.br/ods/noticias/governanca-nacional-para-os-ods.
11. National Plan "Action". (2020). Legislation and Official Notices. Official Gazette of the Argentine Republic. Retrieved October 17, 2021, from https://www.boletinoficial.gob.ar/detalleAviso/primera/238285/20201211.
12. National Sustainable Development Strategy. (2001). Retrieved October 17, 2021, from https://www.argentina.gob.ar/sites/default/files/estrategia-nacional-sayds-2001.pdf.
13. Nilo, A., Fernandes, C. (2020). The Risk of Losing the 2030 Agenda in Brazil. Retrieved October 17, 2021, from https://sdg.iisd.org/commentary/guest-articles/the-risk-of-losing-the-2030-agenda-in-brazil/.
14. Paraguay and the 2030 Agenda. (2020). Retrieved October 17, 2021, from https://www.mre.gov.py/ods/?page_id=2252.
15. Rankings. (2021). The overall performance of all 193 UN Member States. Sustainable Development Report. Retrieved October 17, 2021, from https://dashboards.sdgindex.org/rankings.
16. Uruguay National Development Strategy 2050. (2019). Regional Observatory of Planning for Development. Retrieved October 17, 2021, from https://observatorioplanificacion.cepal.org/es/planes/estrategia-nacional-de-desarrollo-uruguay-2050.
17. Venezuela, Bolivarian Republic. (2021). Voluntary National Review 2016. Sustainable Development Knowledge Platform. Retrieved October 17, 2021, from https://sustainabledevelopment.un.org/memberstates/venezuela.
18. Voluntary National Report 2021. (2021). Uruguayan Agency for International Cooperation. Retrieved October 17, 2021, from https://www.gub.uy/agencia-uruguaya-cooperacion-internacional/comunicacion/publicaciones/informe-nacional-voluntario-2021.

Transformation of Public Administration in the Digital Economy

Maria A. Ekaterinovskaya, Olga V. Orusova, Irina Y. Shvets, Yuriy Y. Shvets, and Karina V. Khaustova

Abstract The paper aims to find out how much digitalization can give a positive impetus to the involvement of citizens in public administration, increasing the effectiveness of the use of public funds to achieve the goals of socio-economic development. The authors used systemic and hierarchical approaches, as well as software tools and methods of mathematical statistics. The involvement of civil society in the process of public administration within the framework of community budgeting (participatory budgeting) allows for enhancing the participation of citizens in the online budgeting process, monitoring budget expenditures, and developing state (municipal) programs. The authors propose the best forms of participation of civil society in the practice of community budgeting: project initiation; the discussion and prioritization of proposals; selection of the best projects; the opportunity to participate in the implementation of projects; the open nature of the procedures; public control over the implementation of projects.

Keywords Public administration · Public sector · Public services · Budgeting · Efficiency · Civil society

JEL Classifications H5 · H41 · H53 · H72

M. A. Ekaterinovskaya (✉) · O. V. Orusova · K. V. Khaustova
Financial University Under the Government of the Russian Federation, Moscow, Russia
e-mail: efcos@mail.ru

O. V. Orusova
e-mail: oorusova@fa.ru

K. V. Khaustova
e-mail: khaustova.karina@mail.ru

I. Y. Shvets · Y. Y. Shvets
V.A. Trapeznikov Institute of Control Sciences of the Russian Academy of Sciences, Moscow, Russia
e-mail: i.y.shvets@gmail.com

Y. Y. Shvets
e-mail: jurijswets@yahoo.com

A. V. Bogoviz (ed.), *Big Data in Information Society and Digital Economy*,
Studies in Big Data 124, https://doi.org/10.1007/978-3-031-29489-1_4

1 Introduction

The paper aims to assess the use of digitalization to expand the involvement of citizens in public administration, increasing the efficiency of using budgetary funds to achieve the goals of socio-economic development.

For this purpose, the authors solved tasks in terms of analyzing the institutions that ensure Russia's transition to public administration in a digital format, introducing new technologies in this area at the federal and regional levels, and highlighting approaches to involving civil society in public administration.

2 Materials and Methods

Digital technologies are successfully implemented in all areas of the socio-economic system, including public administration, as part of the implementation of the national program of the Government of the Russian Federation "Digital Economy of the Russian Federation" (approved by the minutes of the meeting of the Presidium of the Council under the President of the Russian Federation for Strategic Development and National Projects dated June 4, 2019 No. 7) [1].

The effective tools for improving the efficiency of the public component of public administration are state and municipal services, appeals of citizens, etc., which are especially important for authorities at the regional and local levels.

The general theoretical foundations for the study of the public aspect of public administration were considered by Kupriyashin [3], Slobotchikov et al. [6].

It is also necessary to note the work by Barabashev and Guseletova [2] that highlights the characteristic features of public administration in the interests of civil society.

Thus, despite the theoretical study of public administration issues and justification of the advantages of the digital economy, we believe that the issues of adapting the digital format of public administration and, in particular, involving civil society in the public administration process in Russia remain rather fragmented. In this regard, the relevance of studying the features of Russian approaches to the involvement of civil society in public administration in a new digital format, using new technological resources, is increasing.

The methodological and substantive basis of the work is the systemic and hierarchical approaches. The authors also applied the methods of mathematical statistics. The theoretical basis includes the works of Russian and foreign scientists in the areas of public administration and the digital economy.

The information and empirical base of the research include the following:

- The Russian regulatory legal regarding public administration in digital format;
- Information and analytical databases on the websites of the territorial bodies of the Federal State Statistics Service of the Russian Federation;
- Works of Russian and foreign authors.

The reliability and validity of the results obtained are ensured by a comprehensive and systemic approach to research and qualitative methods for collecting and processing quantitative data of public administration in digital format on the territory.

3 Results

Digitalization provides for the use of information data and services for interdepartmental work by government bodies. For this purpose, e-government platforms and public services at all territorial levels are used.

The use of digital platforms in public administration allows for reducing subjectivity in decision-making, reduces the corruption factor, expands analytical tools, and increases the efficiency of interdepartmental interaction.

The contour of the digital government includes a single digital portal, a single array of public administration data with access, and services for interdepartmental interaction.

The essential advantages of e-government are the reduction of transaction costs of services to the people, the reduction of errors due to the overload of public administration employees in data processing, and the openness of data and operations.

Table 1 presents the provision of public services in various fields and industries.

Some shortcomings and risks require leveling, including insufficient digital training of civil servants, dependence on foreign information resources, low level of cybersecurity, heterogeneity of development across regions, and low level of public confidence in information services as a result of the "rut effect."

The advantages of e-government are the reduction of transaction costs of services to the population and transparency of all operations, which prevents corruption.

The way to improve risk management is to increase the digital education of the population and the technological security of the country, which is solved in the federal project "Digital Technologies" [4].

Table 1 Distribution of the benefits of digitalization of services

Area	The benefits of digitalization of services
Transport	Formation of optimal schedules and timetables for the movement of vehicles, ensuring transport security
Industry	Reduction of production costs, automation of production processes, reduction of injuries at work
Housing and communal services	Improving the uninterrupted operation of public utilities
Financial sphere	Reduction of corruption and criminogenic factor in the financial sector

Source Compiled by the authors

In Russia, service systems are created for use in the process of social services for the population, for example, the social navigator of the Federal Social Insurance Fund for the social services in a remote format.

As part of the regional prevalence of social services in digital format, it is necessary to note the Moscow project "Social Assistant," the social platform in the field of public services "Portal of social services" in the Khanty-Mansi Autonomous Area, etc.

As part of improving the efficiency of public administration, most regions have implemented state information systems to ensure the work of government bodies at the request of citizens. Nowadays, the level of average openness of regional platforms and open data is about 50%.

Developing the issue of publicity, the authors would like to note the possibility of participation of citizens in the process of public administration, using the potential of digital transformation through community budgeting tools.

As part of the democratization of public administration, the principles of efficiency and transparency of civil participation in politics are highlighted.

One of the effective tools for implementing the principles of efficiency and transparency of civil participation in politics is community budgeting (participatory budgeting), which allows involving civil society in the development of state (municipal) programs and monitoring of their implementation, which is why the popularization of public participation in budgeting in Europe has seen a significant increase in recent years [5].

The practice of community budgeting, implemented through information platforms and face-to-face mechanisms, is focused on the initiation of projects by civil society, the choice of priorities for financing, and the ability to control costs in terms of the best use of limited resources for socially significant purposes. The evaluation of the "votes" of the community budgeting process is governed by regulatory legal acts that confirm the participation of civil society.

Figure 1 shows a list of areas of community budgeting projects.

There are the following best procedures for the selection of projects:

- Commissions of government representatives that evaluate projects according to specified criteria (about 50 regions);
- Face-to-face meetings focused on complex systemic tasks, where ideas are first discussed, and then priority is selected (used by more than 40 regions);
- Voting and polls on social platforms and networks of regions and municipalities, websites of municipal authorities (used in more than ten regions).

The Nizhny Novgorod Region showed the best result in terms of the share of beneficiaries from community budgeting projects—about 69% in 2018.

Table 2 shows directions for improving community budgeting.

The implementation of these measures will allow spreading the practice of community budgeting at the municipal and regional levels. The result should be the synchronization of approaches to the participation of civil society in public administration, providing an end-to-end principle of work at the regional and municipal levels, as well as increasing the awareness and involvement of citizens in the field of information security through the use of advanced technologies.

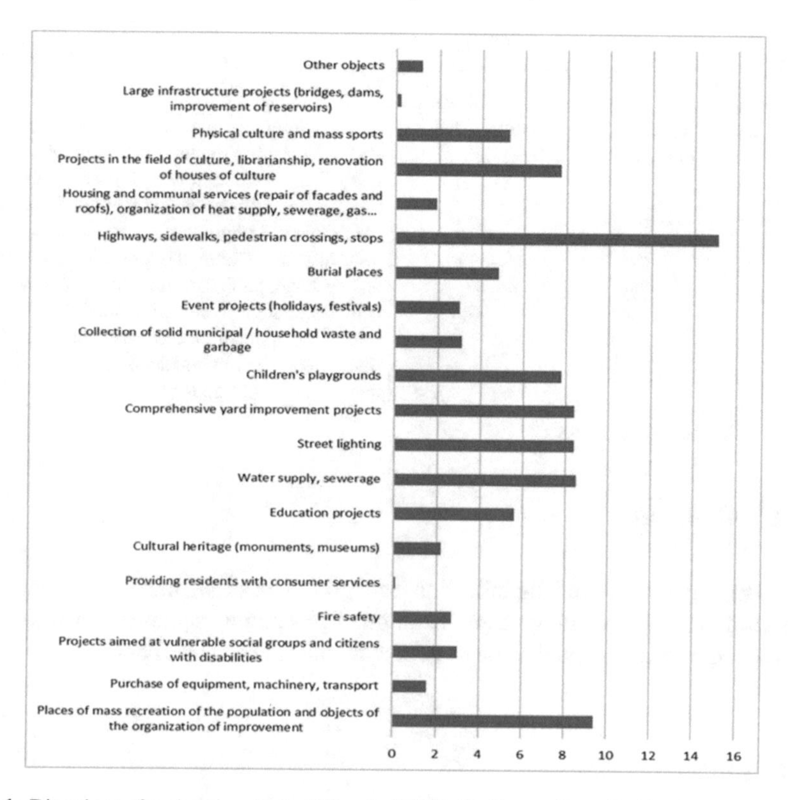

Fig. 1 Directions of regional projects of Russia (2018), %. *Source* Compiled by the authors

Table 2 Directions for improving community budgeting

No	Problem	Solution
1	Lack of necessary knowledge about community budgeting among the target groups of citizens, as well as lack of motivation to generate ideas that have scientific novelty	Develop a mechanism to encourage initiative groups of citizens (who generate ideas), for example, 50 thousand rubles. With the possibility of spending these funds for personal purposes and not for co-financing projects. Ensure promotion of the community budgeting portal "Initiative budgeting" https://budget4me.ru/
2	Lack of unified tools for identifying the best practices of proactive community budgeting at the municipal level	Conduct an annual selection of the best practices and legislate the main mandatory criteria at the regional level
3	Lack of a unified regulatory framework for the development of community budgeting at the municipal level	At the regional level, develop a strategy for the development of information security, which will be aimed at scaling the practice of information security on the territory of each municipality

(continued)

Table 2 (continued)

No	Problem	Solution
4	Lack of up-to-date community budgeting portals in many regions	Develop a mobile application and a unified Russian portal in the field of community budgeting, in which it will be possible to select a separate region and municipality, track the number of ongoing projects, the timing of their implementation, the allocated funds, the voting stage, photos of the future project. Projects could also be commented on and rated. The application can have a news feed that reflects the implemented projects and those that are put up for voting

Source Compiled by the authors

4 Conclusion

Summarizing the above, digitalization can give a positive impetus in terms of involving citizens in public administration, accelerating economic growth, and developing human potential through good governance and social policy.

References

1. Presidium of the Council under the President of the Russian Federation for Strategic Development and National Projects. (2019). The National Program "Digital Economy of the Russian Federation" (approved by the minutes of the meeting of the Presidium of the Council under the President of the Russian Federation for Strategic Development and National Projects dated June 4, 2019 No. 7). Moscow, Russia.
2. Barabashev, A. G., & Guseletova, E. L. (2010). The study of public administration in the USA: Origins, stages of development, current state, and evolution of educational programs. Public Administration Issues, 2, 66–80. Retrieved January 11, 2022, from https://ecsocman.hse.ru/data/2011/11/28/1270196859/Барабашев%2066-80.pdf.
3. Kupriyashin, G. L. (2016). Public administration. *Political Science, 2*, 131–201.
4. Popov, E., & Semyachkov, K. (2017). Analysis of trends in the development of the digital economy. *Problems of Theory and Practice of Management, 10*, 82–91.
5. Sgueo, G. (2016, January 04). Participatory budgeting: An innovative approach. Retrieved January 11, 2022, from http://www.europarl.europa.eu/RegData/etudes/BRIE/2016/573894/EPRS_BRI%282016%29573894_EN.pdf.
6. Slobotchikov, O. N., Kozlov, S. D., Shatokhin, M. V., Popova, S. A., & Goncharenko, A. N. (2020). *Number and power: Digital technologies in public administration.* Institute of World Civilizations.

Economic Security: Identifying Migration Policy as a Key Element

Nazeeh M. Elsebaie, Anastasia A. Sozinova, Elena V. Karanina, and Olesya A. Meteleva

Abstract The paper aims to define the place of migration policy in the national security system of the Russian Federation. Nowadays, the migration policy is unlisted in this structure. Some keys of migration policy are considered in the demographic, cultural, and political security, although the economic security system does not include the key features of the migration policy, which is a big sign of inadequacy. The research is based on the institutional approach. The authors analyzed, compared, and contrasted the similar institutional features and attributes of the migration policy and socio-economic security. However, the research revealed that the features coincide in terms of the object of security. According to the rest attributive criteria, the migration policy falls within the structure of socio-economic security. The study has a great theoretical benefit because it actualizes the importance of ensuring the socio-economic security of the country, considering migration as a key element.

Keywords National security · Economic security · Socio-economic security · State policy · Migration policy · Migration · Russia

JEL Classifications F52 · H55 · H56 · Z13 · F22 · O15 · R23 · D7 · F63 · F66 · F68 · F69 · J18 · J20 · J29 · J60

1 Introduction

The relevance of national security issues is associated with the end of World War II. This relevance was due to two factors: "the atmosphere of urgency generated by the unremitting stress of cold war and the emergence of a fabulous new technology of violence" [6]. The national security risks determine seven dimensions of national

---

N. M. Elsebaie · A. A. Sozinova (✉) · E. V. Karanina · O. A. Meteleva
Vyatka State University, Kirov, Russia
e-mail: aa_sozinova@vyatsu.ru

E. V. Karanina
e-mail: karanina@vyatsu.ru

A. V. Bogoviz (ed.), *Big Data in Information Society and Digital Economy*, Studies in Big Data 124, https://doi.org/10.1007/978-3-031-29489-1_5

security "economic security, energy security, physical security, environmental security, food security, border security, and cyber security" [25]. In many countries, the national security has seven dimensions, whereas the Russian national security system includes eight dimensions—political, social, economic, military, cyber, ecological, demographic, and cultural security. Additionally, there is a great difference in understanding and comprehending threats, risks, and elements of national security.

Economic security has particular importance because it additionally includes food security. The state of the country's socio-economic development ensures food independence, availability, accessibility, and adequacy [3]. Also, unlike in some other countries, social security is a unique element of the Russian national security system and has a special comprehension. Based on the analysis of documents, economic security is a fundamental principle of the national security of the Russian Federation that "delivers sustainable development of society and independence of the state" [9].

Studies on the economic security of the Russian Federation have been due to the transition to new types of administrative systems, along with the period of perestroika (1985–1991) and the subsequent most severe crisis (1992–1998) [8]. The need for a detailed study on this topic was triggered by the formation of a fundamentally new system of national security [18]. Hence, several scientific schools were formed in many universities to meet this need. One of those schools is in the Vyatka State University, which is interested in studying many topics related to risk analysis methodology and the economic security of Russian regions in terms of a formation and a diagnosis of factors—the quality of life, methods of analysis and diagnosis of risks [15], demographic and economic security threats [10, 12, 14, 23], the environmental security as an element of national security is studied by [17, 19].

The post-industrial society brings the social security domain to the fore. Thus, the decisive goal of national security is the sustainable development of the society and not only the economic system. Consequently, "the national interests of the country in the socio-economic domain are to improve the level and quality of life of the population at large, as the main condition for ensuring long-term economic growth and stability. Such an improvement can be achieved by forming a relatively large segment of the population with growing incomes that provide a decent standard of living, a steady growth in the demand, and the population's savings transformation into investments" [1].

Among challenges and threats, the economic security strategy of Russia for the period up to 2030 highlights the lack of labor resources, uneven spatial development, a significant differentiation of regions in terms of the degree of socio-economic development, and the global competition for high-skilled staff.

The school of the economic security at the institute of economics of the Russian Academy of Sciences (RAS) defines the economic security as "such a state of the economy and authority institutions, which guarantees the national interests, the socially-oriented development of the country, and appropriate defense potential" [21]. The scientists of the institute of demographic research of RAS emphasize that "these tasks can be implemented through the country's migration policy" [26]. O. D. Vorobyeva highlights the important compensating role of the migration factor in reproducing the country's potential during periods of depopulation [27]. The migration component most certainly can be considered in complex components of the national economic security in general and the socio-economic security in particular [5]. Foremost, migration is not institutionalized and has no clear place in the Russian national security system. In addition to the complexity and versatility of the migration status in the country, the ambiguous information and results obtained from migration processes for various business entities in the country lead to contradictory interpretations, which increases controversy surrounding the migration policy and its impact [20, 22].

The inclusion of the migration component in the interests of national security occurred in the 1970s in the USA and Europe [7]. In Russia, papers on this topic appeared only in the twenty-first century [24].

The leading hypothesis of the research is the assumption that the migration policy is an integral element of socio-economic security.

2 Methods and Materials

The research is based on the institutional approach, according to which the migration policy is evaluated along with other elements of Russia's national security to understand its place in this system.

The attributes of each element of the country's national security are certain institutional features, the analysis of which will determine the place of the migration policy [11]. These attributes should include the following:

- Subject and object of security,
- Threats and risks associated with it,
- Legislative tools allow this institution to provide a regulatory form [13].

It is necessary to mention the administrative regulation, which is also one of the institutional features. However, since the entire national security system is regulated by state authorities (the president, the federal assembly, and the government of the Russian Federation), the consideration of this element is inappropriate [16]. In this case, the objects of security include the constitutional order, material and spiritual values of the society, the rights and freedoms of the individual (migrants and citizens), and the sovereignty and territorial integrity of the country.

In accordance with the research aim, the authors studied the institutional features of the migration policy and compared them with the relevant attributes of socio-economic security. All data and information used in this paper are based on official sources. Thus, the methodological basis served as a general scientific descriptive method, including techniques of interpretation and comparisons and generalizations.

3 Results

All identified threats and risks of migration to the country's national security were combined in Table 1.

According to the Presidential decree "On the Concept of the State Migration Policy of the Russian Federation for 2019–2025" [2], the subjects that ensure the migration policy are the president, chambers of the federal assembly, the Government of the Russian Federation, and the administrative governmental bodies.

The legislative instruments of the migration policy are provided by the regulatory framework, consisting of Presidential decrees, Resolutions of government, Federal laws, Executive regulations of relevant departments, and agreements between Russia and other countries.

Thus, the analysis indicates that the migration policy has all institutional features that are generally inherent to the elements of the country's national security. However,

Table 1 Migration threats and risks to the national security

No	Topic	Threats and risks
1	Failing in the number of immigrants	A decline in the national labor force participation rate A decline in the tax base Low the extrabudgetary funds
2	Imbalanced migration flows	Lake of forecasting the labor market Failure to meet the economy's demands for labor resources in a timely manner
3	Unbalanced composition of immigrants	Capital outflow abroad Resettlement of undesirable categories of migrants
4	Low skilled immigrants (not relevant to the country and workforce needs)	Discrepancies between demand and supply in the labor market An increase in the labor migration costs to improving skills as well as job matching Deterioration of the security situation and high crime rates An increase in the financial support for adaptation programs
5	Uncontrolled resettlement of immigrants	Emergence of the parallel economy and ethnic cartels Changes in demographic composition and labor force participation rates
6	Illegal immigration	Unpaid taxes Negative attitude towards immigrants Violation of human rights
7	Failure of adaptation and integration	Emergence of the isolated parallel societies, ethnic enclaves, and segregation Ethno-social stratification of the population

Source Compiled by the authors

Table 2 Comparison of institutional features of the migration policy and the socio-economic security

No	Institutional feature	Migration policy Socio-economic security
1	Object of security	The same
2	Threats and risks	Threats and risks of the migration policy are included in the list of threats and risks of the socio-economic security
3	Subject of security	Security subjects of the migration policy are included in the list of subjects of the socio-economic security
4	Legislative instruments	The legislative instruments have specific characteristics, but it is included in the socio-economic security system

Source Developed and compiled by the authors

it is necessary to compare the migration policy attributes with the relevant attributes of the socio-economic security (Table 2) to more accurately determine the place of the migration policy in the national security system of the Russian Federation (whether it is an element of the socio-economic security or it should be singled out as a separate type of security). The study of the institutional features of the national security dimensions was conducted using the orientation toward identifying one's subject field described by S. Afontsev [4].

Thus, all institutional features of migration policies indicate that migration policy is an integral part of socio-economic security and a part of its composition as well.

4 Discussion

In addition to the influence on socio-economic security, there are other dimensions of the national security affected by migration policy (Fig. 1). The influence of migration policy is not limited to socio-economic security but extends to affect demographic and cultural security.

The demographic security criteria include preserving and increasing average life expectancy, improving the physical and genetic health of people, regulating the country's population, and preserving the ethnonational structure. These criteria have something in common with the criteria of socio-economic security in terms of high birth rates and average life expectancy in the country. The migration policy affects the number of the country's population and its ethnonational composition due to the volume of external migration and regulating the composition of migration.

The criteria for cultural security are associated with spiritual values, moral values, conservation and development of the all-Russian identity of all citizens, the unified cultural space, and the rising role of Russia in the global humanitarian and cultural space. The impact of the migration policy on cultural security is associated with the adaptation and integration policies on arriving migrants, preventing marginalization

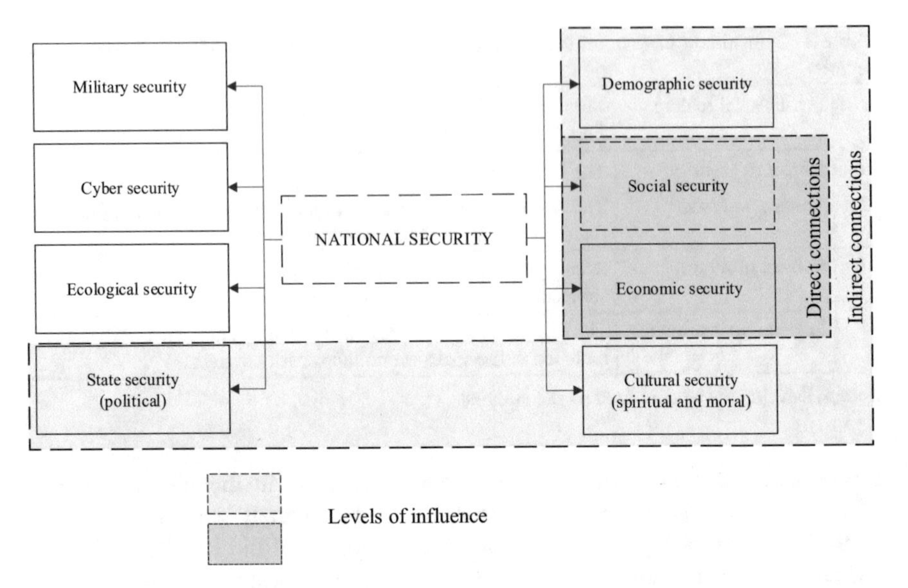

Fig. 1 The dimensions of the national security affected by the migration policy. *Source* Compiled by the author

of national groups and embedding national identities in the multicultural environment of the Russian society.

Political security is associated with ensuring the citizens' will, maintaining the democratic system of the country, and ensuring the protection and rights of all citizens while observing the norms of laws and ensuring external state sovereignty. Since migrants arriving in the Russian Federation are required to know and comply with the country's legislation, migration policy is also indirectly built to ensure the political security of the country.

5 Conclusion

It is required to consider the national security system structure of the country to meet emerging threats and potential risks. Nowadays, the migration policy is unlisted in this structure. Although some keys of migration policy are considered in the demographic, cultural, and political security, the economic security system does not include the key features of the migration policy, which is a big sign of inadequacy.

In this paper, the place of the migration policy in the structure of the national security system of Russia has been determined. The authors analyzed, compared, and contrasted the similar institutional features and attributes of the migration policy and socio-economic security. However, the research revealed that the features coincide in terms of the object of security. According to the rest attributive criteria, the migration policy falls within the structure of socio-economic security. The paper has a great theoretical benefit because it actualizes the importance of ensuring the socio-economic security of Russia, considering migration as a key element.

The research results form a pool of further research questions aimed at improving organizational and methodological aspects of ensuring the socio-economic security considering the migration policy: tools and methods for analyzing implementation results, criteria for evaluating the effectiveness, indicators for monitoring the socio-economic security, etc.

References

1. Presidential Executive Office. (2010). Order "On the approval of the food security doctrine of the Russian Federation" (January 30, 2010 No. 120, update by Order of the President of the Russian Federation on January 21, 2020 No. 20). Moscow, Russia. Retrieved February 8, 2022, from http://kremlin.ru/acts/bank/45106.
2. Presidential Executive Office. (2018). Decree "On the Concept of the State Migration Policy of the Russian Federation for 2019–2025 (October 31, 2018 No. 622). Moscow, Russia. Retrieved February 8, 2022, from http://kremlin.ru/events/president/news/58986.
3. Afontsev, S. (2020). Conceptual framework for the analysis of national and international economic security. *The Economy under Guard, 2*(13), 27–47. https://doi.org/10.36511/2588-0071-2020-2-27-47
4. Billiet, J., Meuleman, B., & de Witte, H. (2014). The relationship between ethnic threat and economic insecurity in times of economic crisis: Analysis of European Social Survey data. *Migration Studies, 2*(2), 135–161. https://doi.org/10.1093/migration/mnu023
5. Bock, P. G., & Berkowitz, M. (1966). The emerging field of national security. *Word Politics, 19*(1), 122–136. https://doi.org/10.2307/2009846
6. Bourbeau, P. (2015). Migration, resilience and security: Responses to new inflows of asylum seekers and migrants. *Journal of Ethnic and Migration Studies, 41*(12), 1958–1977. https://doi.org/10.1080/1369183X.2015.1047331
7. Dobryshina, N. L. (2011). Social and economic safety: Essence, evolution, factors. *Transport Business of Russia, 10*, 5–7.
8. Elsebaie, N. (2020). Migration policy as a component of social and economic security. *Bulletin of the Academy, 3*, 127–134.
9. Elsebaie, N., & Sozinova, A. (2021). Socio-economic security: Defining and developing indicators to assess the effectiveness of migration policy. *Journal of Economy and Entrepreneurship, 7*(132), 72–75. https://doi.org/10.34925/EIP.2021.132.7.009
10. Fufacheva, L. A., Lepeshkin, S. M., Fokina, O. V., Sozinova, A. A., & Shchinova, R. A. (2017). Improvement of the mechanism of provision of food security of Russia within management of risk system of entrepreneurship. In E. Popkova (Ed.), *Overcoming uncertainty of institutional environment as a tool of global crisis management* (pp. 9–14). Springer. https://doi.org/10.1007/978-3-319-60696-5_2.
11. Grigoreva, E., & Garifova, L. (2015). The economic security of the state: The institutional aspect. *Procedia Economics and Finance, 24*, 266–273. https://doi.org/10.1016/S2212-5671(15)00658-9
12. Ibragimovich, I. B., Bobutshah Bokhodir ogli, I., Kholikovich, P. S., & Jamol ogli, I. B. (2020). Economic security: Threats, analysis and conclusions. *International Journal of Multicultural and Multireligious Understanding, 7*(8), 64–76.https://doi.org/10.18415/ijmmu.v7i8.1806.
13. Inshakova, A. O., Sozinova, A. A., & Litvinova, T. N. (2021). Corporate fight against the COVID-19 risks based on technologies of Industry 4.0 as a new direction of social responsibility. *Risks, 9*(12), 212. https://doi.org/10.3390/risks9120212.

14. Karanina, E. V., & Ryazanova, O. A. (2021). Development of the methodology of complex diagnostics and ranking of regions' economic security for sustainable development of their digital economy. In J. V. Ragulina, A. A. Khachaturyan, A. S. Abdulkadyrov, & Z. S. Babaeva (Eds.), *Sustainable development of modern digital economy* (pp. 335–342). Springer. https://doi.org/10.1007/978-3-030-70194-9_33.
15. Karanina, E. V., & Selezneva, E. Y. (2020). Methods of analysis and diagnosis of risks of the socio-economic security of the region, taking into account factors in the development of the consumer market. *Issues of Risk Analysis, 17*(2), 10–21. https://doi.org/10.32686/1812-5220-2020-17-2-10-21
16. Mamoon, D. (2012). Economic security, well-functioning courts and a good government. *International Journal of Social Economics, 39*(8), 587–611. https://doi.org/10.1108/03068291211238446
17. Meskhi, B., Bondarenko, V., Efremenko, I., Larionov, V., Rudoy, D., & Olshevskaya, A. (2020). Technical, technological and managerial solutions in ensuring environmental safety. *IOP conference series: Materials science and engineering* (Vol. 1001, p. 012100). https://doi.org/10.1088/1757-899X/1001/1/012100.
18. Mityakov, E. S. (2018). Key elements of methodology and tools of monitoring of economic security of Russian regions (Dissertation of Doctor of Economics). Nizhny Novgorod, Russia: Nizhny Novgorod State Technical University named after R. E. Alekseev.
19. Pisarenko, P. V., Samoilik, M. S., Plaksienko, I. L., & Kolesnikova, L. A. (2019). Conceptual framework for ensuring resource and environmental safety in the region. *Theoretical and Applied Ecology, 2*, 131–136. https://doi.org/10.25750/1995-4301-2019-2-137-142
20. Ryazantsev, S. V., & Kasymov, O. K. (2021). Transformation of Russia's migration policy: Trends, priorities, and the place of Central Asian countries. *Migration Law, 2*, 16–20. https://doi.org/10.18572/2071-1182-2021-2-16-20
21. Rybakovsky, O. L., & Martynenko, S. V. (2013). Migration policy of today's Russia: Structure and directions. *Population, 2*(60), 51–62.
22. Senchagov, V. K. (2012). *The economic security of Russia: A textbook*. BINOM. Knowledge Lab.
23. Silantyeva, V. A. (2016). Regulation of migratory processes as basis of Russian national security. *Historical and Social-Educational Ideas, 8*(3/1), 22–27. https://doi.org/10.17748/2075-9908-2016-8-3/1-22-27
24. Sozinova, A., Savelyeva, N., & Alpidovskaya, M. (2021). Post-COVID marketing 2019: Launching a new cycle of digital development. In E. De La Poza, & S. E. Barykin (Eds.), *Global challenges of digital transformation of markets* (pp. 419–431). Nova Science Publishers, Inc.
25. Stepanov, A. V. (2014). Security, national security, immigration security, national migration policy: Analysis of definitions, the ratio of categories. *Bulletin of the Institute: Crime, Punishment, Correction, 2*(26), 75–78.
26. The Ammerdown group. (2016). Rethinking security: A discussion paper. Retrieved February 8, 2022, from https://rethinkingsecurityorguk.files.wordpress.com/2016/10/rethinking-security-a-discussion-paper.pdf.
27. Vorobyeva, O. D., Topilin, A. V., & Khrolenko, T. S. (2021). Reproduction of labor resources and the migration policy. *Migration Law, 2*, 21–25. https://doi.org/10.18572/2071-1182-2021-2-21-25

Linguo-Cognitive Modeling of the Legal Regulation of Digital Economy: Research Methodology Background

Olga M. Litvishko, Tatyana A. Shiryaeva, and Galina V. Stankevich

Abstract The paper aims to validate the feasibility of applying linguo-cognitive modeling as the research methodology for studying the legal regulation of the digital economy. The authors consider the digital economy and its legal regulation as a kind of professional communication that involves the identification of subjects of legal relations, electronic document management, and collection, storage, and processing of data. Thus, it possesses specific linguistic features, which are proposed to be analyzed using the method of linguo-cognitive modeling. The authors point out that systemic understanding of professional communication in the field of law and in interdisciplinary areas (e.g., the legal regulation of economic discourse mediated by the digital environment) requires a thorough study of language, including the identification of the specifics of language used in the area of legal regulation of the digital economy. Applying linguo-cognitive modeling to the analysis of legal regulation of the digital economy will allow investigating cognitive and discursive mechanisms of knowledge mediation using digital technologies and their linguistic representation. Besides, the role of language in categorizing and conceptualizing economic realities in digital dimensions will also be studied. What is more, linguo-cognitive modeling is useful in identifying and analyzing the system of frames associated with legal regulation of the digital economy, which are verbalized through lexical units that nominate objects and the reality of this area of professional communication.

Keywords Digital economy · Legal regulation · Linguistics · Linguo-cognitive modeling · Model · Professional communication · Terminology

JEL Classfications K2 · O03 · Y8

O. M. Litvishko · T. A. Shiryaeva
Pyatigorsk State University, Pyatigorsk, Russia
e-mail: olitvishko@yandex.ru

T. A. Shiryaeva
e-mail: shiryaevat@list.ru

G. V. Stankevich (✉)
North Caucasian Institute of Mining and Metallurgy (State Technological University), Vladikavkaz, Russia
e-mail: stankevich.g@yandex.ru

A. V. Bogoviz (ed.), *Big Data in Information Society and Digital Economy*, Studies in Big Data 124, https://doi.org/10.1007/978-3-031-29489-1_6

1 Introduction

The rapid development of the digital economy, which has become a leading factor in contemporary global trends, poses new challenges and generates new problems, immediately requiring the search for solutions in the field of legal regulation and public administration, economy, and public sectors. In the twentieth century, the country's economic growth depended on the technologies of management. In turn, the twenty-first century witnesses the shift in the trajectories of economic and social development, which are primarily determined by digital technologies. Russia is consistently ranked in the top 50 international digital development rankings. However, the decline in indicators for some indices, including the e-government development index and the ICT development index, demonstrates the lack of development and stability of these sectors in the digital economy system. In this regard, the imperatives of national development related to digitalization include creating a flexible system of legal regulation of the digital economy, removing priority barriers to the development of advanced technologies, and doing business in several areas closely related to the processes of digitalization [19]. In this context, the solution to such issues as identification of subjects of legal relations, electronic document management, and collection, storage, and processing of data deserves special attention.

There is no denying that these changes contribute to the intensification and acceleration of professional communication, the formation of new communicative practices, and the use of various linguistic tools, causing the transformation of the content of social processes within the professional sphere, which in recent years has become increasingly multifaceted, dynamic, changing, and which cannot but affect the formation of thinking styles of various professional groups. Professional thinking, namely, the predominant use of methods of analyzing professional situations, principles and algorithms of interaction within a specific professional group, the typology of situations of professional interaction, a specific nomenclature of communicative intentions, a typical set of communicative strategies to achieve them, a special thesaurus—is directly reflected in fundamentally new models of professional interaction characteristic of representatives of a certain social institution, in particular, law. Being the most important repository and transmitter of the bulk of socially significant information, these models often have under-investigated architectonics, fragmentary studied content, and undeclared principles of generation, dissemination, and interpretation. Recent communication studies and discourse theory have no adequate methods and methodological techniques to study these models. Although an array of scholarly publications in Russia and abroad are currently focused on analyzing various linguistic levels of professional communication, there are only scattered studies of individual aspects and stages of professional communicative interaction. A unified communicative and discursive model of professional communication has not yet been developed due to the implementation and active expansion of the digital economy into all areas of life. Cognitive and linguistic support of all aspects of legal regulation of the digital economy is particularly important today.

The totality of the facts listed above determines the high relevance and significance of the proposed study. Moreover, the development of a complex and multi-vector model of institutional communicative and discursive space of digital economy and its legal regulation is more relevant than ever because the developed model will help overcome the problems that complicate the solution of strategic tasks of state policy in the field of the development and wide application of digital economy in the Russian Federation.

2 Materials and Methods

The goal of this research is to substantiate the applicability of linguo-cognitive modeling to the development of a communicative and discursive model of legal regulation of the digital economy, which determines the use of certain general scientific and linguistic methods, such as:

- Continuous sampling method is used to collect and select the research material;
- Content analysis and discourse analysis are involved in identifying the general architectonics of communicative-discursive models of professional communication;
- Classification and categorization of the factual material are necessary for compiling an extensive corpus of data;
- Method of linguo-cognitive analysis is used to develop a cognitive model that reflects the basic concepts of the studied professional area;
- Method of cognitive discursive structuring is employed to detect the correlation of discursive, pragma-communicative, and institutional attitudes with basic cognitive structures of the researched discursive space and institutionally-dependent speech markers;
- Contextual-semantic method is relevant for analyzing language units in certain professional contexts;
- Descriptive and empiric methods are actively employed in classifying and interpreting the research material.

The research is conducted on the material of existing legislation concerning the regulation of the digital economy in English and Russian, as well as documents reflecting the existing practices of law application in this area.

3 Results

Professional communication as a kind of institutional communication implies the interaction of subjects of professional activity in accordance with relatively defined and stable models and algorithms. The main goal of professional communication is to optimize professional activity, which is achieved by carrying out private goals within the genres of this discourse determined by the forms of communication. Extralinguistic and linguistic factors are equally significant in the profound analysis of professional communication proposed by Russian and foreign scholars. Among extralinguistic factors, they point at the spread of professional realities typical of specific areas of activity across all areas of human life. It is also noted that one of the most significant features of today's society is the existence and replication of these realities by various target audiences. Experts unanimously recognize the fact that professional relations are among the key mechanisms of today's civilization, while the increase in the level of communicative competencies of all participants in the professional community is required by the new economic and social environment [5, p. 19].

From a linguistic point of view, professional communication seems to be an extremely interesting research object because the study of its functional paradigm and the system of interaction of various language levels in the process of professional interaction draws the attention of linguists representing different schools and research approaches. Many scholars point out that the diversity of approaches to the analysis of professional discourse clearly demonstrates how significant it is in the framework of current cross-disciplinary research. Every single research reveals a unique feature of professional discourse, which is undoubtedly beneficial for a deeper understanding of its nature [14, p. 23].

We must point out that, despite a sufficient number of scientific publications devoted to textual, structural, discursive, and other numerous parameters of professional communication [1, 4, 6, 9–11, 16, 17], a comprehensive analysis of various professional discursive spaces related to law, including the communicative-discursive space of professional communication in the field of the digital economy, demonstrates significant gaps, the most serious of which we consider the lack of a holistic approach to the analysis of this area of communication. In this case, it is necessary to remember the fact that an important characteristic of professional communication in the field of the digital economy is its status-oriented type, describing the communication of a specialist with a specialist, as well as a specialist with a non-specialist, which is extremely important for understanding the cognitive-linguistic essence of legal regulation mechanisms related to the digital economy.

Systemic understanding of professional communication in the field of law and in interdisciplinary areas, such as the legal regulation of economic discourse mediated by the digital environment, requires a thorough study of language, including the identification of language features present in the area of legal regulation of the digital economy. It only becomes possible if we consider the principle of anthropocentrism within the framework of research coordinates—"person-profession-text" using the tools of linguocognitology, pragmalinguistics, and sociolinguistics, on the one hand, and with the involvement of extensive empirical material in multi-system languages, on the other. A comprehensive study of the language involved in the application of principles of the digital economy in Russia may be considered insufficient if the latest methods and models of text analysis used in the theory of text and (institutional) discourse, the theory of mental representations, functional linguistics, communication theory, pragmalinguistics, legal linguistics, and legal expertise of the text are not involved in the analysis.

We believe that an effective methodological basis for studying the discourse of legal regulation of the digital economy is linguo-cognitive modeling, which allows conducting thorough and detailed research of cognitive-discursive mechanisms of knowledge mediation carried out by digital technologies and the way they receive linguistic representation. Equally important is the role of language in the process of categorizing and conceptualizing the economic realities in digital dimensions. What is more, it is our firm belief that linguo-cognitive modeling opens up opportunities for the identification and analysis of the frames system typical of a specific area of human experience and linguistically represented by means of lexemes nominating objects and the reality of the studied area while reflecting the systemic connections between the latter. In our opinion, it is the linguo-cognitive approach that allows penetrating into the depths of the nature of an internalized language and approaching the solution of the problems of reconstructing a person's linguistic ability [21, p. 121], as well as provides an explanation of the knowledge structures formed in the process of communication.

Nowadays, "model" and "modeling" have been studied within a number of linguistic schools. Their popularity is explained by their close connection to N. Chomsky's theories of syntactic structures [8], which may be viewed as models of sentence generation. Modeling is also considered to be synonymous with cognitive processes, which also help to construct and recreate real objects, phenomena, and processes.

Nowadays, it might be hard to overlook the relevance of studies in which modeling is proclaimed the main research method of linguistic and even broader—humanitarian research: a considerable number of the different scholarly publications describing models of particular fragments of linguistic and cultural reality have appeared [2, 3, 12, 13, 15, 18], pointing out that "the geography of philological problems has been noticeably detailed" [7, p. 95]. Alongside the rapid development of applied linguistics, serving as the major framework of linguistic models research, including traditional and new paradigms of linguistic studies, novel approaches are being developed aimed at conducting computer-mediated studies and the research of various professionally-oriented fields such as business linguistics, political linguistics, and legal linguistics. Models developed within these domains are of comprehensive character; they possess significant explanatory potential [4, p. 1128]. It is also important to emphasize that in contemporary linguistics, the linguistic modeling method is quite useful at structuring the knowledge verbalized by a great diversity of language means; it also determines the model's content structure.

Equally significant in the context of the present research is the fact that "social reality can be studied as a theoretical construct of various sciences" [20, p. 251], and linguistics is no exception. In terms of linguistics, many objects connected with a language, such as a language competence, language ability, language environment, language personality, language policy, language game, and language picture of the world, reflect the existing social reality. Being heard or understood within this reality, a person should be aware of conventions and values typical of the society, which, on the one hand, influences the choice of language means to express this reality and, on the other hand, is vital for shaping and structuring this society.

The immediate process of recreating the current model of the institutional communicative and discourse space of professional communication relating to the digital economy demands the study of a wide range of issues related to the communicative-discursive bilingual space of communication in the professional sphere (the material in Russian and English is involved), which requires a comprehensive socio-linguistic construction of the special (professional) knowledge represented by them, including linguo-cognitive, pragmatic-communicative, discursive-communicative, and structural-semantic aspects.

Such a development of a model of the institutional communicative and discursive space of professional communication in the area of the digital economy will contribute to the further theoretical improvement and practical testing of one of the most popular in contemporary linguistics methods of linguo-cognitive modeling of professional communication of specialists in the field of the digital economy of the Russian Federation on a regional scale. The electronic dictionaries created as a result of the conducted research and the principles underlying their structuring will help improve the basic postulates of scientific socio-lexicography and contribute to a deeper understanding of the studied field of activity—the regulatory framework for the digital economy—by scientists, specialists, and students of training areas related to the legal sector and the digital economy of the Russian Federation.

The language material collected as a result of the analysis of the regulatory framework for the digital regulation of economic activity of the Russian Federation and presented in the form of representative databases and corpora, as well as the methods used for its multidimensional analysis and structuring, will be able to lay the foundation for the creation of subsequent extensive databases based on the material of other languages. The textbook created based on the received theoretical and practical procedures and methodological developments concerning the conduct of training events, as well as the communicative-discursive, pragmatic-communicative, socio-linguo-cognitive, and professionally-oriented approaches used in their compilation, will become an impetus for further improvement of the methodology for developing materials of an educational and methodological nature in language teaching for special purposes.

4 Conclusion

Considering our main task, which is to reconstruct a complex and multi-vector model of the institutional communicative and discursive space of legal regulation of the digital economy, we think that at least three conceptual questions need answering:

- What does the discourse of legal regulation of the digital economy represent?
- What does the process of modeling this discourse space consist in (why can exactly this type of discourse be modeled this way and not the other)?
- What results can we expect while developing a model of this type of discourse?

We believe that the answers to these questions are the fundamental basis on which the universal concept of institutional communicative and discursive space of the legal regulation of the digital economy should be built, the main advantage of which is the definition and explanation of the role of language structures in the mental processes of comprehension and speech generation of members of today's business society engaged in the legal regulation of the digital economy.

It should be emphasized that the complex development of a multi-vector model of professional communicative interaction of legal professionals in the aspect of legal regulation of the digital economy has a pronounced interdisciplinary character, combining advanced developments and research in the field of communication studies, theory of text and discourse, cognitive linguistics, pragmatics and sociolinguistics, legal linguistics, and legal expertise of the text. We will not hide that the depth and scale of the new knowledge analyzed and potentially deduced within the framework of the development of such a model will contribute to the development of advanced areas of contemporary linguistics, namely: the study of institutional types of discourse, discourse analysis, theory and practice of cognitive modeling of mental representations, and theoretical and applied terminology and terminography, on the one hand. On the other hand, this research will be extremely useful to various researchers working within the framework of contemporary cognitive science because it has a high level of fundamentality, transparadigmality in terms of the research procedures undertaken, and the use of complex linguo-cognitive modeling of the communicative and discursive space of professional communication as an innovative technique for parameterizing the professional activities of specialists in the field of development of the digital economy of the Russian Federation proving the novelty and validity of the results obtained in the proposed study.

Along with this, the applied potential of the research, expressed in the development of the author's methodology for mastering the conceptual and terminological apparatus of legal regulation of the digital economy, in the creation of a methodology for developing professional communicative competence of participants in professional communication, and in the generation of a number of databases, multimedia dictionaries, textbooks, and a series of online webinars, courses, and seminars also speaks about the significant practical value of the work carried out.

References

1. Akaeva, H. A., Alimuradov, O. A., & Latu, M. N. (2014). Applied and fundamental terminological systems as mutually-correlating verbal basics of professional communication: The statement of the problem. *Bulletin of Pyatigorsk State Linguistic University, 3*, 40–45.
2. Alimuradov, O. A., & Brudanina, A. V. (2009). An attempt at modeling the dynamics of concept formation and verbalization: The theoretical and methodological approaches and practical applications (the concept PARTING as represented in the English language). *Bulletin of Pyatigorsk State Linguistic University, 3*, 105–110.
3. Alimuradov, O. A., & Chursin, O. V. (2011). Frame modeling of specialized knowledge and its speech representation (on the material of terminological music vocabulary of modern English). *Theoretical and Applied Aspects of Speech Activity Studies, 6*, 6–27.
4. Alvesson, M., & Karreman, D. (2000). Varieties of discourse: On the study of organizations through discourse analysis. *Human Relations, 53*(9), 1125–1149. https://doi.org/10.1177/0018726700539002.
5. Bargiela-Chiappini, F., Nickerson, C., & Planken, B. (2013). *Business discourse*. Palgrave Macmillan. https://doi.org/10.1057/9781137024930.
6. Bargiella-Chiappini, F., & Nickerson, C. (1999). *Writing business: Genre, media, discourse*. Longman.
7. Belousov, K. I. (2010). Model linguistics and the problem of language reality modelling. *Bulletin of Orenburg State University, 11*(117), 94–97.
8. Chomsky, N. (1986). *Knowledge of language: Its nature, origin, and use*. Praeger. Retrieved February 18, 2022, from http://www.thatmarcusfamily.org/philosophy/Course_Websites/Readings/Chomsky%20-%20Knowledge%20of%20Language.pdf.
9. Drew, P., & Heritage, H. (1992). *Talk at work: Interaction in institutional settings*. Cambridge University Press. https://doi.org/10.5070/L451005175.
10. Gotti, M. (2008). *Investigating specialized discourse*. Peter Lang.
11. Irimiea, S. B. (2017). Professional discourse as social practice. *European Journal of Interdisciplinary Studies, 3*(4), 108–119. Retrieved February 18, 2022, from https://revistia.org/files/articles/ejis_v3_i4_17/Silvia.pdf.
12. Litvishko, O. M. (2021). To the issue of linguocognitive modeling of legal concept sphere. *Professional Communication: Top Issues of Linguistics and Teaching Methods, 14*, 87–96.
13. Litvishko, O. M., & Miletova, E. V. (2019). Socio-cognitive modeling of civil law discourse: Lexico-semantic aspect. *Cognitive Studies of Language, 37*, 770–775.
14. Litvishko, O. M., Shiryaeva, T. A., Tikhonova, E. V., & Kosycheva, M. A. (2022). Professional discourse: The verbal and visual semiosis interplay. *Research Result. Theoretical and Applied Linguistics, 8*(1), 19–40. https://doi.org/10.18413/2313-8912-2022-8-1-0-2.
15. Miletova, E. V. (2013). The experience of frame modeling of English-language arts discourse: The aspect of evaluation. *Bulletin of Higher Education Institutions: North Caucasus Region. Social Sciences, 4*(176), 110–115.
16. Shiryaeva, T. A. (2011). Metaphors as a factor of forming business awareness (on the material of modern English). *Bulletin of Pyatigorsk State Linguistic University, 3*, 128–132.
17. Shiryaeva, T. A. (2013). Modern business discourse: The experience of linguo-cognitive research. *Cognitive Studies of Language, 13*, 811–822.
18. Shiryaeva, T. A., & Litvishko, O. M. (2021). Cognitive-discourse mechanisms of generating new knowledge in English-language popular science editions. *Cognitive Studies of Language, 3*(46), 701–704.
19. Stankevich, G. V., Vilgonenko, I. M., Slepenok, Y. N., & Litvishko, O. M. (2021). Electronic document flow as evidence in civil proceeding: Evaluation features and application problems. *SHS Web of Conferences, 109*, 01039. https://doi.org/10.1051/shsconf/202110901039.
20. Tannen, D., Hamilton, H. E., & Schiffrin, D. (Eds.). (2001). *The handbook of discourse analysis* (2nd ed.). Blackwell Publishers.
21. Yokoyama, O. B. (2005). *Cognitive model of discourse and Russian word order*. LRC Publishing.

Social Consequences of the Information Economy

Ruslan S. Tolmasov, Elena V. Sibirskaya, Lyudmila V. Oveshnikova, Irina V. Cheremushkina, and Yulia I. Slepokurova

Abstract Strengthening the role of the information economy in global and inter-country processes produces the need to adopt a new paradigm for the development of society. The transformation of the paradigm has social consequences for the country, business structures, and the population. The study of the social consequences of the information economy produces the relevance of the research topic. The paper aims to shape the social consequences of the information economy. The authors considered the characteristics of the information economy and determined the forms of manifestation of the information economy. Moreover, the authors indicate the key subjects of the information economy, which are socially influenced; the social consequences of the information economy are formed. The research toolkit includes the method of grouping, the method of comparison, the method of classification by a single criterion, the method of the proposal based on the selected problematics, and the method of event modeling.

Keywords Social consequences · Information economy · Digital economy · Network economy · Electronic economy · Information product · Information flow

JEL Classfications M15 · L86 · O35

R. S. Tolmasov
MIREA –Russian Technological University, Moscow, Russia

I. V. Cheremushkina · Y. I. Slepokurova
Voronezh State University of Engineering Technologies, Voronezh, Russia

E. V. Sibirskaya (✉) · L. V. Oveshnikova
Plekhanov Russian University of Economics, Moscow, Russia
e-mail: Sibirskaya.EV@rea.ru

L. V. Oveshnikova
e-mail: Oveshnikova.LV@rea.ru

A. V. Bogoviz (ed.), *Big Data in Information Society and Digital Economy*,
Studies in Big Data 124, https://doi.org/10.1007/978-3-031-29489-1_7

1 Introduction

The change of conceptual foundations in the development of the economy has led to the deformation of social relations. The emergence of the information economy gave a new round in the development of civilization by studying the influence of information on economic decision-making. Despite this fact, two problems circulate in today's society related to understanding the essence of the information economy and the uncertainty of the process of social consequences from the introduction of this concept. The information economy is often understood as the patterns of using information factors in economic processes [1, 7–9]. This definition formats the system of human thinking towards the theoretical perception of the information economy, the postulates of which are reflected in the context of scientific publications, research, and foresights. The uncertainty of the process of social consequences of the information economy forms a negative message of changes, within which the role of a person is minimized. The criteria outlined above reduce the importance of the information economy for each person. In accordance with this statement, determining the essence of the information economy and highlighting the social consequences of its implementation are paramount tasks that allow us to form a system of understanding and development prospects in society through updating the chosen research topic.

The paper aims to study the social consequences of the information economy. Achieving this goal requires focusing on the following tasks:

- To consider the characteristics of the information economy;
- To determine the forms of manifestation of the information economy;
- To reflect the key subjects of the information economy that are impacted by the society;
- To form the social consequences of the information economy.

2 Materials and Method

The instrumental base of the research is regulated by the use of methods of data grouping, comparison, classification according to a single attribute, proposals based on the selected issues, and event modeling. The characteristics of the information economy are considered based on the analysis of existing theoretical aspects prescribed in scientific works by foreign and Russian scientists. The forms of manifestation of the information economy are determined by a clear specialization of norms, which are progressive transitions to modes of social relations in the presence of certain signs in the contemporary economy. The key subjects of the information economy are reflected in the form of economic agents of one of the key categories of microeconomics. Isolation of economic agents will form the social consequences of the information economy.

Thus, the information economy is understood as a transitional system of social relations due to changes in the nature of labor and the directions of economic processes [2, 12]. According to this definition, the information economy is not a concept or paradigm for the development of society. The information economy acts as a transitional link in the system of social relations. A distinguished transition is formed through changes in the nature of labor and the direction of economic processes. Changing the nature of labor involves obtaining new competencies that contribute to the transformation of the economic situation, the replacement of labor with capital, the formation of new social structures, the revision of demand for goods produced earlier, and the replacement of labor with another factor of production [19]. The direction of processes in the information economy is modified by the development of material forces based on the production relations of economic agents. Economic processes are influenced by the characteristics of the information economy (Fig. 1).

The information economy primarily reflects information as a value for economic agents [11]. The economic activity of an economic agent is studied based on the information, which provides for the use of information and communication technologies used at all stages of production. It is worth noting that the key elements of the information economy are the information product. Information is a resource that forms an information product [6]. An information product is declared information, formed to meet the needs of economic agents and intended for use by users in the process of distribution, redistribution, and consumption [17].

An additional link in the process of characterizing the information economy is the availability of tools that allow obtaining an information product. An information product is formed through the actions of economic agents interested in its creation. To this end, economic agents are trained in data collection and analysis. The

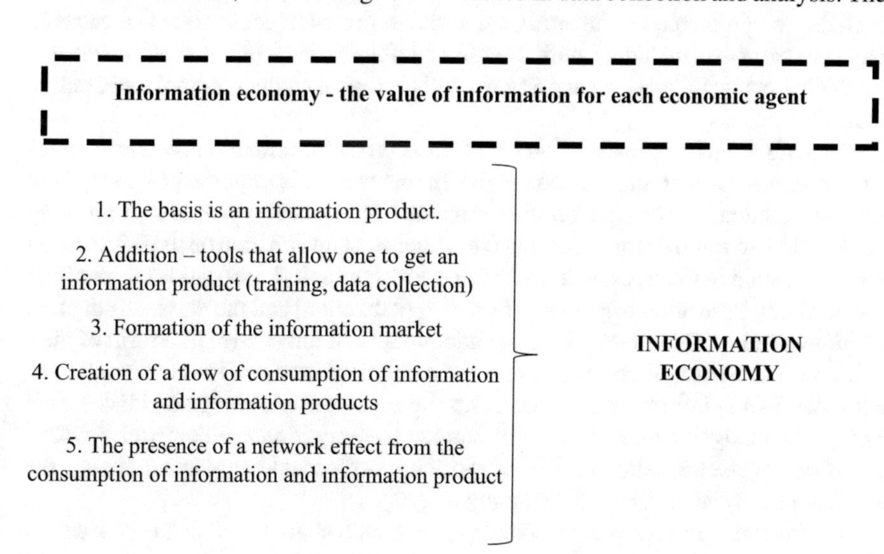

Fig. 1 Characteristics of the information economy. *Source* Compiled by the authors

selected criterion formulated the process of the importance of accumulating information within a single environment—an information product. To do this, a feature of the information economy is the mass character and global nature of the process of interaction between economic agents based on information that contributes to the distribution of the created benefits.

The information economy requires the formation of an information market. The information market presupposes the existence of a system of economic relations that regulates the products of intellectual labor on a commercial basis. The information market of the Russian Federation is heterogeneous. First, the centering of the processes of the information market does not allow forming and developing intellectual and information property in the regions [21, 22]. Second, the economic agents with access to information products are limited. This feature is manifested primarily in acquiring financial and commercial information, which is accessible to a limited number of people due to the high cost of information products in the Russian market [5].

3 Results

The information market of the Russian Federation is closely intertwined with the Internet segment. According to Mediascope, in 2020, 95 million people, or 78% of the country's population over 12 years old, used the Internet at least once a month in Russia [13]. Thus, the information economy determines not only the product specificity of this concept (information product) but also differentiates various areas and sectors within which the flow of information consumption is created. In accordance with this, e-commerce can be attributed to the information economy: (1) electronic payment services (including online retail or online travel), (2) marketing systems (including content marketing and SMM), and (3) digital content (e-books and online games).

Creating a flow of consumption of information and information products comes from consumers—external customers [3]. In this case, it is important to understand that information is a resource for the formation of information products. A properly constructed flow allows for increasing the influence of information on the information and production markets, as well as scaling the technological foundations throughout the economy (transition from innovation to digitalization) and the mass introduction of information technologies with the predominance of knowledge as a form of their achievement. These processes enhance the network effect from the consumption of information and information product. Despite sufficient knowledge and theoretical and practical significance of the information economy in the contemporary world, the forms of manifestation of this category are not considered when studying the foundations of the information economy (Fig. 2).

The information economy is a concept, the manifestation of which is carried out through forms that allow for the most structured differentiation of key processes and the identification of subjects—economic agents responsible for their implementation

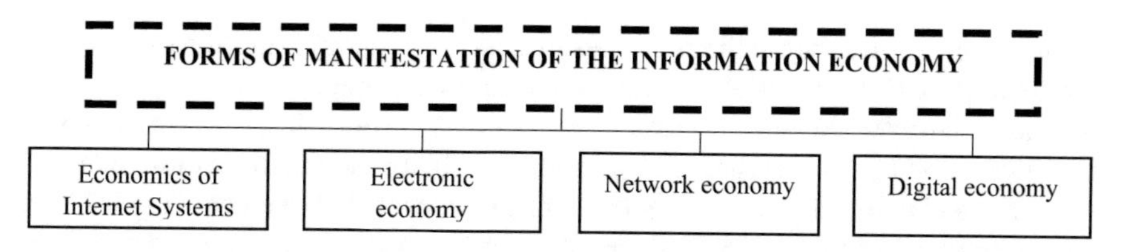

Fig. 2 Forms of manifestation of the information economy. *Source* Compiled by the authors

[14]. The first economic form of manifestation of the information economy is the economy of Internet systems. The economy of Internet systems is based on the use of a computer network of physical objects equipped with technologies for collecting and transmitting information, followed by intelligent data processing [15]. The economy of Internet systems combines various devices into a single computer network, which is aimed at collecting, analyzing, and evaluating data using the software. In general, the economy of Internet systems is aimed at working without human intervention. Simultaneously, users of the Internet systems economy are consumers who interact with each other based on access to Internet data. In the Russian Federation, the economy of Internet systems is tied to the implementation of industry projects to improve technology to change the economic structure. First of all, it includes the following:

1. Medical industry—conducting remote health monitoring or telemedicine through Internet systems. In this case, Internet systems allow patients to receive doctor's consultations through technology-smart devices combined with various applications. Additionally, Internet systems in medicine allow creating new forms of prescription drugs that contain microscopic sensors that can provide information about the state of the patient's internal organs. These developments have shaped the direction of Smart Health.
2. Housing and communal complex—the use of wireless sensors in garbage collection, housing management, and building maintenance. The introduction of Internet systems allows one to remotely manage energy resources and metering devices, overhaul, and current repairs. The main direction of Internet systems in this direction is Smart Home.
3. Transport sector—in this segment, Internet systems are used as communication technologies. An example is the presence of GSM modules, "smart" remote signaling, and the collection of statistics on driving style, determining the geolocation of the vehicle. Additionally, Internet systems for the transport sector allow reducing the intellectual connection of the information economy in this industry. The selected condition is associated with the development of a system consisting of integrated data collected in single information space and aimed at ensuring the management of these processes through advanced telematic technologies.

In general, it can be noted that Internet systems as a form of the information economy are produced by the presence of communication technologies that allow automating and structuring the processes of collecting and transmitting information for the development of various sectors of the national economy.

The electronic economy (e-economy) acts as a form of information economy based on network and intellectual specifics, involving the use of information and computer technologies to transform economic sectors [10]. The e-economy is based on the economy of things. In this case, the e-economy should include e-commerce, e-insurance, e-banking, e-money, etc. In addition to the highlighted structure, the e-economy affects all elements of macroeconomic regulation:

- The state manifests itself through e-government with the subsequent openness of data in all areas of the national economy of the country.
- Business is determined through remote interaction with the state of the small, medium, and large businesses and financial institutions.
- Households are determined through the use of social networks and e-commerce, including e-commerce, capital, marketing, and information exchange.

The network economy is a type of information economy driven by a focus on unlocking creative human capital by personalizing each person's basic needs through digital telecommunications [20]. The network economy is based on the use of the following elements:

1. Agile-net quick response—determines transformational changes, reactively adapting to them while building an algorithm to strengthen the role in new positions. Considering the Agile rapid response network at the level of business structures, it is worth noting their role in the interaction between manufacturers and customers on the issues of quickly reformatting the product range built with the wishes of consumers. Agile-net quick response includes a set of methods that build emerging problems in the form of a project. Agile-net rapid response is based on cascade models that implement the processes of the information economy in stages for each economic agent.
2. Networks of innovative systems allow one to create algorithms, the participants of which jointly solve the existing problems. Networks of innovative systems are primarily used to accumulate the business environment and open network assistants. Networks of innovative systems are formed within one enterprise. In this case, networks of innovation systems are internal network structures.
3. Risk hedging networks are personified with neural networks operating in parallel within the same process, focusing on tracking and minimizing the consequences of emerging risks.

As part of social responsibility to society, the network economy most fully reflects the system of meeting the population's needs with minimal stratification of citizens. The goal of the network economy is to maximize the satisfaction of the needs of each person in various activities that meet the capabilities of all members of society. The network economy acts as a system for the development of society based on technology.

A digital economy is a form of information economy based on the use of technologies in the field of redistribution of digital goods and services [18]. Despite the highlighted definition, the main focus of the digital economy in today's society is the development of digital infrastructure. Digital infrastructure is a prerequisite for the implementation of the digital economy. The digital infrastructure is formed based on ecosystems defined by the "digital competencies" of each member of society and the level of access to the Internet. The latter condition is due to overcoming the digital divide among the population of the Russian Federation.

The digitalization of the economy and its social component affects (1) the level of well-being and employment of the population (transformation of jobs with elements of remote work), (2) the establishment of social relationships (lack of attachment to the place of residence, interests, and leisure), (3) changes in the level of spending (reducing the cost of some services, stimulating the consumption of other services), and (4) human capital (continuous improvement of knowledge and skills). Despite the current priority, the transition to a digital economy began in line with the advent of the Third Industrial Revolution in 1969. Since the advent of the first programmable logistics controller and further application in the production of electronic and information systems that provided intensive automation and robotization of production processes, the digital economy has reorganized the causes and prerequisites for the development of society in accordance with the old concept of economic development. Digital technologies manifested themselves through means of assisting in data processing, performing complex calculations, or automating simple production tasks. An example would be sequentially turning on a conveyor, the supply of materials, and the processing of materials on machine tools. The use of digital technologies has made it possible to exclude human labor in routine monotonous operations. Further, digital solutions penetrated into the sector of entertainment—animation, visual effects, and video games. The digitization of the surrounding world has affected the daily life of every person. For example, the use of the Internet has given rise to many smart tools that provide transport and infrastructure systems for the city.

Inseparably with the digitalization of the economy, there is the use of IT technologies, which have turned from a simple tool into a strategic direction for the development of the country. The regulatory and legal consolidation of the process of forming a digital infrastructure for the use of IT technologies is associated with the declaration of the key foundations enshrined in the national project "Digital economy of the Russian Federation." The National program "Digital economy of the Russian Federation" includes the following federal projects: regulation of the digital environment, personnel for the digital economy, information infrastructure, information security, digital technologies, digital public administration, and artificial intelligence [16].

In general, the forms presented above within the framework of joint implementation in today's society allow implying the general system of the information economy. The transition to the information economy is manifested through the production of the main activity of economic agents, endowing them with key functions. Figure 3 shows the key subjects of the information economy, which are socially affected by the forms identified above. The economy of Internet systems differentiates the subjects of the information economy through (1) firms, (2) developers of ICT infrastructure, and (3) consumers—the public. A firm in the economy of Internet systems is an economic agent that combines online and offline activities. The presence of only one of the company's features as an economic agent (online or offline) does not allow us to single out this subject in the economy of Internet systems.

Economic agents developing ICT infrastructure are not limited by the specific characteristics of IT business entities. Economic agents that develop ICT infrastructure are state and municipal authorities that carry out information and electronic interaction between citizens. In the Russian Federation, the emergence of the economy of Internet systems through economic agents developing ICT infrastructure began in 2011, after the adoption of the federal target program "Electronic Russia" (validity

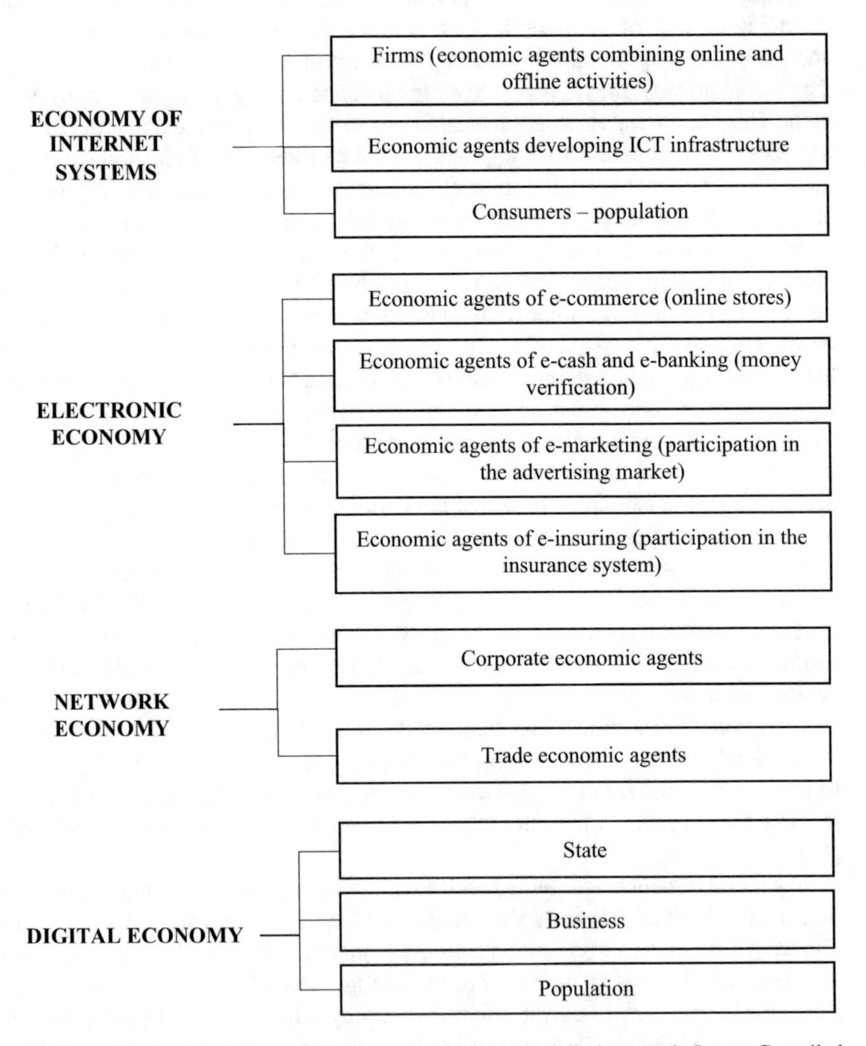

Fig. 3 Key subjects of the information economy, which are socially impacted. *Source* Compiled by the authors

period 2002–2011), the strategy for the development of the information society in the Russian Federation (validity period 2017–2030), and the state program "Information society" (validity period 2011–2020). The target audience of these regulatory documents is consumers.

The electronic economy subdivides the subjects according to the levels of e-directions. Economic agents of e-commerce are a set of online stores that implement an electronic payment system.

The economic agents of e-cash and e-banking are based on the verification of funds generated within the cryptographic protocol of electronic cash. Economic agents of e-marketing assume the presence of subjects participating in the advertising market. Economic agents of e-insure (participation in the insurance system) provide e-commerce insurance services. The presence in the structure of the electronic economy of only one type of agent does not allow us to fully structure and manifest the basic postulates of the information economy in society. The network economy is conditioned by the formation of corporate and trade economic agents. Corporate economic agents produce their economic activity through social networks. Within the network economy, corporate agents compete for limited informational advantages. Trading economic agents focus on the use of new tools for obtaining objective information as a specific advantage and scale their own activities (trade transactions and operations) through the use of the Internet environment.

As the final form of the information economy, the digital economy generalizes the groups of economic agents to the types of key subjects of macroeconomics. The main subjects of influence of the digital economy are as follows:

1. The population—through the national project "Digital economy of the Russian Federation";
2. Business—through the provision of financial support to IT companies and enterprises involved in information technology infrastructure;
3. The government—the developer of the strategic foundations of the country's digitalization and a member of the international group of informatization of the global economy.

The economic agents identified above act as implementers of the information economy and key actors in the influence of the social consequences of the information economy (Fig. 4).

The social consequences of the information economy are a multifactorial basis for changing socio-economic processes at the country level and at the levels of a particular region or municipality. The reduction of jobs (changes in the employment structure) is a primary social consequence of the information economy. First of all, this social consequence comes not only from the information economy but also from the processes of automation and the robotization of society. Thus, in 2020, about 48% of state-owned companies were fully involved in implementing a digital strategy using artificial intelligence in production and office activities. The information economy only produces the consequences of the changing employment system. Additionally, the employment system of the population is being transformed under

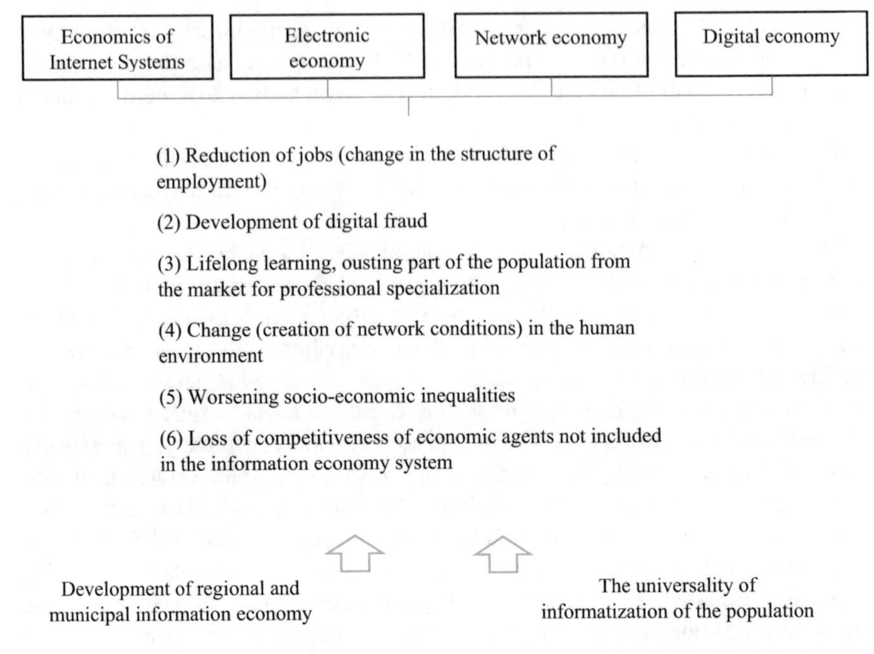

Fig. 4 Social consequences of the information economy. *Source* Compiled by the authors

the influence of the consequences of the coronavirus infection when a hybrid work system replaces the process of finding employees in the office.

Thus, the Future of Jobs report published at the World Economic Forum talks about replacing 85 million jobs with new digital places. According to the report's authors, new jobs will appear more slowly than old jobs will start to disappear. This feature will lead to an increase in unemployment and the need to retrain employees with a low level of qualification [24].

The development of digital fraud is due to the infliction of losses to individuals and legal entities based on the commission of illegal actions against them. According to RBC statistics, since the beginning of the COVID-19 pandemic in the Russian Federation, cases of digital fraud have seen an increase of more than 76% [12]. The growth was driven by phone and internet scams, which are the original types of digital scams. Nowadays, digital fraud affects such areas as lending (more than 3550 cases in 2020), receiving social benefits (more than 4000 cases in 2020), transfers from bank cards (more than 15,000 cases in 2020), insurance (more than 900 cases in 2020), and computer information (more than 245 cases in 2020). Simultaneously, the lack of a clear system undermines the well-being of the inhabitants of the Russian Federation and the information security of legal entities [4]. The social consequences of digital fraud destabilize public institutions and lead to the subsequent development of forms of theft through the client's personal account, mobile banking, the use of SIM cards, and issuance of credit cards without the client's knowledge.

Lifelong learning, which ousts part of the population from the market of professional specialization, is primarily due to the growth of the necessary skills and competencies of the information economy. The concept of lifelong learning is supported not by the principles of the information economy but by the environment formed in it. The factors of this environment are competition for solving creative (non-standard) business problems, replenishment of managerial experience, and the growing influence of the state on the labor market. Compliance with this factor requires constant replenishment of competencies, ranging from intrapreneurship, strategic vision, digital skills, and T-shaped specialization with subsequent goal setting and internal motivation. It should be noted that most people, especially those aged from 50 to 60, are not ready to rapidly develop their soft skills to model career specialization. This issue causes primary consequences due to rising unemployment (according to the report of the World Economic Forum, 75 million people worldwide will lose their jobs by 2022 due to the development of technology and innovation) [23].

Change (creation of network conditions) in the human environment produces the formation of network maps of the family, relatives, friends, and work. The network card will allow creating online communication between its participants, accumulating their ideas about life in the virtual world. Compared to this social impact, the change in the human environment will create a clear gradation between the younger generation—virtual users and "offline-visuals" who deny online conditions. The stratification of the population according to the level of entry into network conditions will exacerbate the consequences, not only formed in the labor market but also in intergenerational conflicts. The result of the problems highlighted above will be the aggravation of the socio-economic inequality of the population on a regional basis and individual consumption of information. The regional feature is the differentiation of subjects according to the level of socio-economic development. In this case, the irrational distribution of digital technologies accumulates a subjective division into developed and depressed regions. Individual consumption of information is due to the lack of data that allows structuring the professional activities of each person. The highlighted condition will exacerbate the problem of professional incompetence of the individual, which will enhance the transformation of the employment structure.

The loss of competitiveness of economic agents that are not included in the information economy system is produced for business structures. The loss of competitiveness focuses on denying the importance of entering the e-economy. First of all, the loss of competitiveness is due to the reduction of sales markets for these economic agents. Sales markets are being reformatted according to consumers to suit the conditions of the information economy. Social consequences for the economic agents of enterprises are concentrated on reducing the level of consumption of products and the priority of the enterprise as a tax subject of regional economies. Thus, the social consequences of the loss of competitiveness of economic agents that are not included in the information economy system are reduced to two components: (1) the bankruptcy of an enterprise and (2) a decrease in the importance of this agent for the regional economy.

Despite the social consequences identified above, the information economy is a necessary link for developing the country's national economy. The minimization of social consequences should be territorial in nature, consisting of the allocation of regions and municipalities as the primary link in the information economy. A countermeasure to weaken the social consequences of the information economy is the development of a regional and municipal information economy, as well as the introduction of universal informatization of the population.

4 Conclusion

Thus, the study on the social consequences of the information economy allowed us to draw some conclusions.

1. The category "information economy" does not have a clear definition in the scientific literature. In most cases, the information economy is understood as a transitional system of social relations due to changes in the nature of labor and the direction of economic processes. In this case, the information economy is not a concept or a paradigm for the development of society but a link in the transformation of social relations. The information economy includes a set of criteria reflected through (1) information product, (2) information market, (3) information flow, and (4) network effect from information consumption.
2. The information economy is produced by a number of forms that reflect the process basis for the manifestation of the conceptual norms of this category. The information economy is regulated by a combination of elements of the economy of Internet systems, electronic economy, network economy, and digital economy. The economy of Internet systems is based on the use of a computer network of physical objects equipped with technologies for collecting and transmitting information, followed by intelligent data processing. The electronic economy (e-economy)—acts as a form of information economy based on network and intellectual specifics, involving the use of information and computer technologies to transform economic sectors. The network economy is a type of information economy driven by a focus on unlocking the creative capital of the individual through the personalization of each person's basic needs through digital telecommunications. A digital economy is a form of information economy based on the use of technology in the field of redistribution of digital goods and services. In its totality, each of these forms allows highlighting the social consequences of the information economy.
3. The social consequences of the information economy are concentrated on a multifactorial basis of changes in socio-economic processes. Within the framework of this research, the following social consequences are highlighted: job cuts (changes in the structure of employment), the development of digital fraud, lifelong learning, changes (creation of network conditions) in the human environment, aggravation of the socio-economic inequality of the population on a regional basis and individual consumption of information, loss of competitiveness of economic agents that are not included in the system of the information

economy. To minimize the social consequences of the information economy, we propose creating a system for developing the regional and municipal information economy and introducing general informatization of the population.

References

1. Apatova, N. V., & Uzakov, T. K. (2018). Information economy and economic growth. Scientific Notes of the Crimean Federal University named after V.I. Vernadsky. *Series; Economics and Management, 4*(1), 13–22.
2. Babaev, B. D., Nikolaeva, E. E., Borovkova, N. V., & Babaev, D. B. (2020). Russia: Digital economy or industrial and information economy? In E. Popkova, & B. Sergi (Eds.), Digital economy: Complexity and variety vs. rationality (pp. 332–341). Cham, Switzerland: Springer. https://doi.org/10.1007/978-3-030-29586-8_39.
3. Bezzubov, D., & Stepankivska, N. (2017). Ensuring information security by entities in information economy. *International Scientific Journal Internauka. Series: Juridical Sciences, 1*(1), 20–23.
4. Central Bank of the Russian Federation. (2020, October 29). Overview of reporting on information security incidents when transferring funds. Retrieved May 18, 2022, from https://cbr.ru/analytics/ib/review_1q_2q_2020/.
5. Emirova, I. U., Bezuglova, Yu. V., & Igolkina, T. N. (2021). Information economy: Essence, genesis, main characteristics. *Bulletin of the Belgorod University of Cooperation, Economics and Law, 2*(87), 115–125. https://doi.org/10.21295/2223-5639-2021-2-115-125.
6. Giraev, V. K. (2019). Theoretical approaches and practice of assessing the investment attractiveness of regions. UEPS: Management. *Economics, Politics, Sociology, 1*, 17–29. https://doi.org/10.24411/2412-2025-2019-00003.
7. Grodskiy, V. S., & Khasaev, G. R. (2020). Digital economy – Information era: Retrospective analysis. In S. Ashmarina, A. Mesquita, & M. Vochozka (Eds.), Digital transformation of the economy: Challenges, trends and new opportunities (pp. 163–179). Cham, Switzerland: Springer. https://doi.org/10.1007/978-3-030-11367-4_16.
8. Kartashov, E. M. (2022). New operational relations for mathematical models of local nonequilibrium heat transfer. *Russian Technological Journal, 10*(1), 68–79. https://doi.org/10.32362/2500-316X-2022-10-1-68-79.
9. Kavanagh, M., & Rochford, O. (2015). Magic quadrant for security information and event management. Gartner. Retrieved January 29 2022, from https://scadahacker.com/library/Documents/White_Papers/Gartner%20-%20Magic%20Quadrant%20for%20SIEM%20-%202015.pdf.
10. Kirsanov, K. K. (2015). Theory of utility within the period of the conceptual proposition change. Naukovedenie, 7(2), 38. Retrieved January 29, 2022, from http://naukovedenie.ru/PDF/37EVN215.pdf18.
11. Kvilinskiy, A. S., Trushkina, N. V., & Rynkevich, N. S. (2019). Conceptual approaches to the definition of the term "information economy." *Problems of Economics, 3*(41), 147–155.
12. Lindell, D., Alekhina, M., Skrynnikova, A., & Balashova, A. (2020, August 31). The number of fraud cases has grown record-breaking amid the pandemic. What crimes did self-isolation contribute to RBC. Retrieved May 18, 2022, from https://www.rbc.ru/society/31/08/2020/5f48ea169a79477e21e25d9d.
13. Mediascope. (2021, January 12). Internet audience in Russia in 2020. Retrieved May 18, 2022, from https://mediascope.net/news/1250827/.
14. Molchanova, N. P. (2019). Information economics and innovations in finance. *Drukovsky Bulletin, 2*(28), 149–160.

15. Orlov, A. I., & Sazhin, Yu. B. (2020). Solidary information economy as the basis of a new paradigm of economic science. *Innovations in Management, 4*(26), 52–59.
16. Presidium of the Presidential Council for Strategic Development and National Projects. (2019). National program "Digital economy of the Russian Federation" (Approved by the minutes of the meeting of June 4, 2019 No. 7). Moscow, Russia. Retrieved May 18, 2022, from https://digital.gov.ru/ru/activity/directions/858/.
17. Protopopova, N. I., Grigoriev, V. D., & Perevozchikov, S. Y. (2019). Information and digital economy as an economic category. In E. Popkova, & V. Ostrovskaya (Eds.), Perspectives on the use of new information and communication technology (ICT) in the modern economy (pp. 300–307). Springer. https://doi.org/10.1007/978-3-319-90835-9_35.
18. Shishkin, V. V. (2019). Is the digital economy an information trap for entrepreneurs and another soap bubble? *Journal of Economy and Entrepreneurship, 5*(106), 1208–1212.
19. Sneps-Sneppe, M., Namiot, D., & Alberts, M. (2018). On digital economy issues looking from the information systems viewpoint. In *Proceeding of the 22nd Conference of FRUCT Association* (pp. 385–391). Retrieved January 29, 2022, from https://fruct.org/publications/acm22/files/Sne.pdf.
20. Sokolova, E. I. (2018). Information economy in public administration. *Innovative Management Technologies and Law, 1*(21), 24–27.
21. Ushakov, D. (2017). Information technologies within market economy: How communication tools became a field of activity. *EUrASEANs, 1*(2), 7–18.
22. Vasilyeva, E. V. (2020). Information economy: Innovation or innovation. Ratio et Natura, 1(1). Retrieved January 29, 2022, from https://ratio-natura.ru/sites/default/files/2021-06/informacionnaya-ekonomika-novovvedenie-ili-innovaciya.pdf.
23. World Economic Forum. (2020, January 15). The global risks report, 2020. Retrieved May 18, 2022, from https://www.weforum.org/reports/the-global-risks-report-2020.
24. World Economic Forum. (2020). The future of jobs report, 2020. Retrieved May 18, 2022, from https://www3.weforum.org/docs/WEF_Future_of_Jobs_2020.pdf.

Information and Instrumental Space of the Strategy of Reducing Technological Losses of Material Resources of Enterprises

Igor E. Mizikovsky, Ekaterina P. Garina, Natalia S. Andryashina, Elena V. Romanovskaya, and Victor P. Kuznetsov

Abstract The purpose of the study is to develop an information and instrumental space for the strategy of eliminating technological losses of material resources. The object of the research is resource consumption in the value stream, the subject is material costs. The article clarifies the concepts of «technological losses», «cost management», management accounting and internal reporting; the approach to the classification of technological losses is proposed, its use, as well as a set of techniques and methods for constructing the information and instrumental space of technological losses, will provide the possibility of localization and a significant reduction in the level of technological losses, ensuring economic stability and a subsequent increase in business profitability. The article presents and economically substantiates measures to eliminate technological losses of the enterprise. These measures are aimed at increasing the reliability and stability of technological processes, ensuring the financial stability of the enterprise, increasing the value of products and labor productivity, and reducing non-productive costs.

Keywords Reduction of technological losses · Information and instrumental space · Instrumentalization · Operationalization

JEL Classfications L23 · M11 · O21

1 Introduction

An increase in business profitability requires an in-depth and comprehensive restructuring of the information and instrumental space for managing an economic entity,

I. E. Mizikovsky
Lobachevsky State University of Nizhni Novgorod, Nizhny Novgorod, Russia

E. P. Garina · N. S. Andryashina · E. V. Romanovskaya · V. P. Kuznetsov (✉)
Minin Nizhny Novgorod State Pedagogical University, Nizhny Novgorod, Russia

N. S. Andryashina
e-mail: natali_andr@bk.ru

E. V. Romanovskaya
e-mail: alenarom@list.ru

A. V. Bogoviz (ed.), *Big Data in Information Society and Digital Economy*, Studies in Big Data 124, https://doi.org/10.1007/978-3-031-29489-1_8

including in terms of the development and implementation of break-even production methods. The transition to the construction and implementation of an effective cost management system based on the maximum possible exclusion of losses is a success factor of both micro- and macroeconomic nature.

The need to develop a policy for the systematic elimination of losses, the values of the parameters of which depend on the joint influence of many factors, such as internal (irrational organization of the production process, lack of coordination in interaction between departments, the use of outdated technologies and equipment used beyond the established service life, unauthorized replacement of materials, lack of employees with sufficient qualifications, etc.) and external (level of fulfillment of contractual obligations, loss of competitiveness, price discrimination, reduced access to credit resources, reduced investment attractiveness), is determined by the conditions of the functioning of an economic entity. The practical implementation of this policy should be based on the systematic incorporation into the complex fabric of the value stream of measures to improve operational efficiency and to optimize costs [4].

Research at several enterprises of the Nizhny Novgorod region on the topic of the issue showed that the implementation of «one-time» measures to reduce technological losses, implemented in the period 2016–2018, made it possible to increase the annual profitability of their technological costs on average from 6 to 17%, and more than 70% of the economic effect from these measures was obtained due to the localization and elimination of material losses. Thus, the elimination of losses of raw materials of the main production in volumes from 80 to 95%, made it possible to increase the annual profitability of production costs from 14 to 17%, to reduce the loss of electricity for technological needs by 30% (Fig. 1).

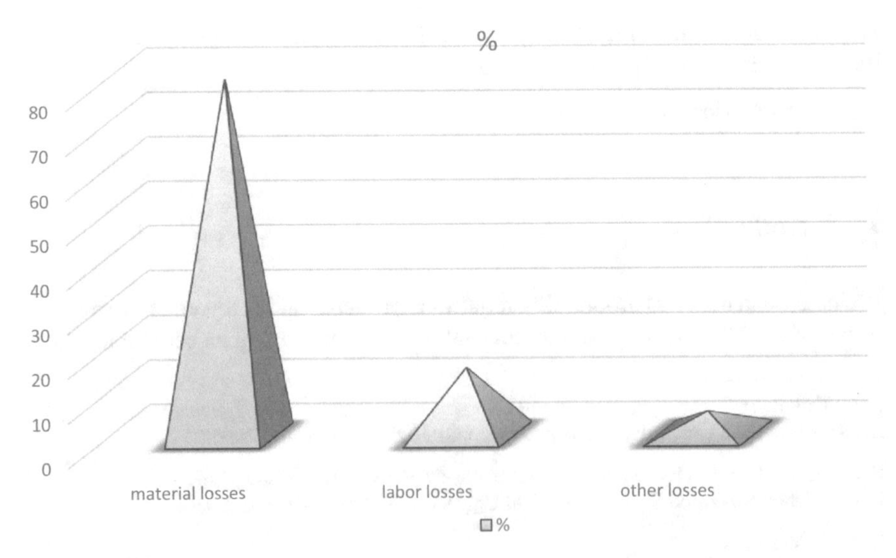

Fig. 1 Effects of the reduction of production losses on the increase in profitability of technological costs. *Source* Compiled by the authors based on [5]

At the same time, the use of various approaches to the formation of the composition of losses used at the studied enterprises; the ineffectiveness of the information potential of the management process; the absence of reference values and algorithms of accounting processes, which significantly reduces the results of the implementation of measures to ensure cost-efficiency, increases the likelihood of inaccuracy of the results obtained, should be noted. There are no special documented forms for presenting information about losses in the information field of enterprises, this of course implies the absence of structured and approved schemes for document flow and circulation of information about losses in the management system of an economic entity. There is also a «related» problem—the delay in the formation of information about technological losses, because there is a «secondary» processing of previously created documents, initially semantically not oriented towards information support for the development of managerial influences on the part of the management entity on the state of production activity losses.

The relevance of the application of algorithmically «floating» methods based on heuristics and empirical knowledge of individuals, as a rule, with a high level of professional intuition of workers who are not institutionalized in the space of the organization's competencies in the studied processes requires proof. This postulate serves as convincing evidence of the need for the formation and instrumentalization of permanent search and timely documented reflection of technological losses in order to build an information basis for making decisions to reduce them.

In the context of the formulated problem, it is advisable to determine *the purpose of the study*: improving the quality of information support of a complex-structured process of reducing the loss of material resources used in the value stream by an enterprise in the manufacturing industry. The objectives of the study are: clarification of the concept of «technological losses» for its operationalization in the practices of production and financial management, cost management, management accounting and internal reporting; development of an approach to the classification of technological losses, adapted to the conditions for making managerial decisions; construction of a model for instrumentalization of the formation of an array of data on losses, used as an information base for the development of managerial actions to reduce or completely eliminate them.

2 Methodology

The "boiler" model of the formation of production costs, which is often implemented in domestic accounting and management practices [1–3], does not imply a separate accounting for the cost of excess inventory of inventories that falls out of productive costs, correction of defects, many types of downtime of labor, equipment, machines and mechanisms, empty runs of transport, irrational internal movements of resources, etc., absorbing them in the general array of costs. Without this, by definition, it becomes impossible to carry out a set of measures to ensure cost-effectiveness.

This study proposes the use of the concept underlying the lean production system [6–11], the design of which assumes attributing to losses of all resources that did not provide the receipt of economic benefits in a given calendar period, where the concept of losses is fully defined and their classification was carried out on the basis of seven criteria: overproduction, waiting, moving, overworking, stocks, movement, correction. At the same time, both the definition of the studied economic category and the adopted classification need significant adaptation to the management process and the information requests of decision-makers formed in the context of its implementation. The approaches to the semantic identification of the concept of «loss» used in the concept do not imply a logical and semantic «tie» to the information support of the process of reducing losses, in particular, to the time (calendar) period of their occurrence, as well as to management functions, primarily to the estimated planning (budgeting) of technological costs and their results. Thus, filling the concept of «loss» with economic meanings does not affect the important parameters of structuring the information array for making managerial decisions to reduce and eliminate them. Proceeding from the task of qualitatively improving the information support of the process of developing management decisions within the framework of effectively planned, economically justified indicators of the strategy for saving resources; prompt formation of the most complete and reliable array of information relevant to this complex process, the authors think it appropriate to formulate the concept of «loss» as «a quantitative result of resource consumption that reduces the level of the planned financial result in the calendar period».

3 Results

The solution of tasks to reduce technological losses should be implemented within the information field of enterprises, where the tool for systematizing data on losses, their integration into the information space for managing costs and their results is to build a classification based on general criteria (Table 1).

The model of information support for reducing losses should be functionally oriented to the process parameters that identify their place of origin (an enterprise as a whole or a specific production unit); the type of resource and the process that caused them.

One of the important stages of instrumentalization of technological losses is their rationing, which is implemented in the context of each operation of the i value stream. The basis of this tool is the «rate of technological losses». Rationing of losses of material resources P_{ij} is carried out according to the formula (1) [12], of labor resources L_i (in enlarged form)—according to the formula (2) [13].

$$P_{ij} = N_{ij} - \left(M_{ij} + O_{ij}\right), \tag{1}$$

Table 1 Classification of technological losses in order to form an information base for making management decisions to reduce them

№	Classification criterion	Class names
1	By the effectiveness of the impact on the part of the subject of management	• Adjustable (reducible) • Non-adjustable (non-reducible)
2	By the possibility of planning (rationing)	• Planned (standardized) • Unplanned (unstandardized)
3	By likelihood of occurrence	• Expected (predicted) • Random (unpredictable)
4	By nature of occurrence	• System • Situational
5	By relation to the possibility of their application in decision-making	• Relevant • Irrelevant
6	By place of origin	• Workshop • Factory
7	By type of resource	• Material • Labor • Others
8	By type of production process	• Basic • Auxiliary • Service

Source Developed and compiled by the authors

where N_{ij}—j material consumption rate; M_{ij}—j useful material consumption; O_{ij}—technological waste due to the established production technology.

$$L_i = T_i - (P_i + S_i), \quad (2)$$

where T_i—working time consumption rate; P_i—time spent on changing the object of labor (its shape, size, appearance, physicochemical or mechanical properties, etc.), its state and position in space; S_i—time spent on auxiliary actions (installation and removal of a workpiece, machine control, tool approach and withdrawal, measurement of the workpiece, etc.).

In the context of the information space and decision-making technologies at the studied enterprises, other losses are advisable to structure on the basis of the value of illiquid assets in the broad sense of the meaning of this, unfortunately, not unambiguously defined, economic category. According to the authors, in this context, other losses should include all material and technological reserves that cannot be monetized. It is quite obvious that it is impossible to plan this kind losses, which usually arise as a result of management errors or other kinds of force majeure. At the same time, it is necessary to set a limit for other losses, for example, based on the parameters of breakeven production in order to optimize management and improve the quality of decision-making. The results of rationing and limiting material and labor technological losses make it possible to form an important information tool—a base of normalized technological losses, taking into account the following requirements: reflection of loss indicators based on calculated and analytical results; economic justification for subsequent changes; adaptive incorporation into information spaces of production and financial management, cost management, management and internal accounting, as well as making management decisions to reduce losses.

It should be noted that the base of normalized losses serves as an information basis for mapping them in the value stream, allowing monitoring losses (formula 3).

$$O_{ij} = F_{ij} - N_{ij}, \quad (3)$$

where F_{ij}—actual loss value; N_{ij}—value of standard losses.

An element of the information space of management accounting is the created reference book of the causes of deviations of losses, with their division into two categories: systemic and situational. In this guide, it is necessary to provide for a category of reasons affecting the occurrence of significant deviations that systematically exceed the materiality threshold. Along with this element of information support, there is a need to structure another equally important element—a directory of «culprits» for the occurrence of preferably significant deviations, which will make it possible to develop management influences on the activities of specific cost centers, to consider and form the values of key performance indicators of heads of structural technological units.

The proposed model of instrumentalization of information support for decision-making to reduce technological losses should be reflected in the accounting policy of the enterprise; in other internal standards (regulations and instructions): on planning, accounting and internal control of costs and calculating the cost of products (works, services); rationing of technological costs; schematization of document flow, creation and maintenance of an array of design and technological documentation, etc.

4 Conclusion

The research results correspond to the set goal of improving the quality of information support of a complex-structured process of reducing the loss of material resources used in the value stream by an enterprise in the manufacturing industry. On their basis, it became possible to clarify the definition of the concept of technological losses and its operationalization in the context of enterprise management in the real sector of the economy, to develop approaches to the classification and scaling of this economic category, to develop directions for its instrumentalization and legal regulation at the level of an economic entity.

Objectification of the indicator of technological losses, their operationalization and instrumentalization can significantly improve the quality of information support for making management decisions to reduce them, streamline economic calculations and significantly improve the accuracy of their results. The developed and economically justified measures to eliminate losses are aimed at increasing the reliability and stability of technological processes, ensuring the financial stability of the enterprise, increasing the value of products and labor productivity, reducing non-productive costs, levels of material consumption of production and illiquid assets, contributing to the introduction of a systematic approach to building and implementing a strategy for saving technological resources not only of enterprises of the real sector of the economy, but also of other industries.

References

1. Cheremisina S. V., & Lenchevskaya, N. V. (2007). Comparative analysis of the standard cost calculation method and the standard cost method. *Bulletin of Tomsk State University*, No. 300-2, 82–85.
2. Dubonosova, A. N. (2015). Cost management system: planned and actual costing. «Economist's Handbook». No. 10. Retrieved February 28, 2019, from https://www.profiz.ru/se/10_2015/.
3. Garin, A. P., et al. (2020). Research of the tools of influence on the behavior of market subjects. *Lecture Notes in Networks and Systems, 91*, 156–162.
4. Garina, E. P., Romanovskaya, E. V., Andryashina, N. S., Kuznetsov, V. P., & Shpilevskaya, E. V. (2020). *Organizational and economic foundations of the management of the investment programs at the stage of their implementation lecture notes in networks and systems, 91*, 163–169.
5. Hendriksen E. S., & Van Breda M. F. (1997). Accounting theory. Trans. from English; edited by Ya. V. Sokolov. Moscow: Finances and statistics, 576 p.: illustrated. (Series on accounting and audit).
6. Hobbs, D. (2008). *Introduction of lean manufacturing* (p. 320). Publishing House of Rowers Publisher.
7. Kamyshanov P. I., & Kamyshanov A. P. (2017). Financial and management accounting and analysis: textbook P.I. Kamyshanov, A. P. Kamyshanov. M.: INFRA-M, 592 p.
8. Levinson U., & Rerik R. (2007). Lean manufacturing: A synergetic approach to loss reduction. RIA «Standards and quality», 272 p.
9. Mann, D. (2009). Lean management of lean production of feathers. Trans. from Engl. Starshykov A. N. Bragin V. V. (Ed.). RIA «Standards and quality», 208 p. (Series «Lean Management»).
10. Rate setting expenditure for materials. Basic rules. (2021). Retrieved May 11, 2019, from http://docs.cntd.ru/document/1200012273.
11. Rother, M., & Shook, J. (2015). *Learn to see business processes* (p. 319). Alpina Business Books.
12. Savitsky, D. V. (2007). Advantages of the normative method of accounting for costs and calculating work at repair enterprises of railway transport. *Science and Technology of Transport, 1*, 42–46.
13. Vitkalova, A. P. (2007). «STANDARD-COST»—a system of operational management of production costs. *Bulletin of the Saratov State Technical University, 1*(28), 4, 178–185.

Big Data for the Digital Economy

Mathematical Model of Consumer Demand

Mikhail M. Ermilov, Sergey Y. Bityuckiy, Lyudmila G. Kudryavtseva, Liudmila E. Surkova, and Andrey A. Boltaevskiy

Abstract The paper aims to develop a mathematical model of consumer demand observed over a certain number of years. Compression (reduction) of a large amount of data in a mathematical model to compact mathematical objects—matrices and their eigenvectors—allows one to make certain conclusions about the general set of consumers in several markets at once. The basic mathematical apparatus used is Allen's approach, also known as the best average percentage. The research is based on one of the varieties of the ordinary least squares method (OLS). The paper considers the possibility of detecting the presence of periodicity in consumer demand. A mathematical criterion for accepting or rejecting assumptions about periodicity is also proposed. The proposed meta-mathematical model allows us to process large data sets on consumer demand of past periods and predict the periodicity of demand.

Keywords Demand · Homogeneity · Frequency · Random errors of observation · Matrix trace · Eigenvalues

JEL Classfications C020 · C3

M. M. Ermilov (✉) · S. Y. Bityuckiy · L. G. Kudryavtseva · A. A. Boltaevskiy
Russian University of Cooperation, Mytischi, Russia
e-mail: mermilov@ruc.su

S. Y. Bityuckiy
e-mail: s.ya.bitjuckij@ruc.su

L. G. Kudryavtseva
e-mail: lkudryabtseva@ruc.su

A. A. Boltaevskiy
e-mail: aboltaevskiy@ruc.su

L. E. Surkova
National Research Moscow State University of Civil Engineering (MGSU), Moscow, Russia
e-mail: surkovale@mgsu.ru

A. V. Bogoviz (ed.), *Big Data in Information Society and Digital Economy*,
Studies in Big Data 124, https://doi.org/10.1007/978-3-031-29489-1_9

1 Introduction

Currently, the digitalization of key processes is pursued to ensure the sustainable development of cooperation and the economy. The concept of the digital economy is described in various documents and publications, for example, in the work of Kleiner [8]. In a general sense, the digital economy proposes to carry out the transfer of all processes of production, distribution, and consumption, as well as their planning and forecasting, based on digital technology, using software and hardware. Mathematical models and processes of modeling economic activity form the basis of mathematical support for information systems. Only the use of mathematical apparatus [7], subsequently implemented in the software of the information system, allows solving a wide range of problems, including the economic sector and the cooperative sector in particular. The problems of interaction with consumers [12], the study of supply and demand, mathematical modeling of consumer behavior, taking into account random factors [4] are reflected in current research.

For several years, there have been systematic observations of seasonal fluctuations in demand in a certain fixed set of consumer markets. In this regard, there are several pressing issues related to the demand. In fact, these issues determine the relevance of research on market processes of supply and demand. This paper attempts to answer the following questions based on a mathematical model:

- Can we consider seasonal fluctuations in consumer demand to be similar in a certain sense at all points of sale?
- What is the profile of demand when considering different markets?
- Whether the processes of demand have a periodicity, that is, whether we can talk about their self-reproduction from year to year?

The purpose of this work is to develop a mathematical model of consumer demand observed over a certain number of years based on the compression (reduction) of a large amount of data to compact mathematical objects—matrices, their eigenvectors, and, accordingly, algorithms. The use of such a model will reduce the large number of mathematical transformations required to draw certain conclusions about the general set of consumers in several markets at once.

2 Materials and Method

The research method is based on mathematical modeling, in which the values of the society that allow for direct measurements are set as parameters. Under the conditions of economic and social realities, it is usually impossible to conduct experiments in the natural scientific sense. Consequently, in the social environment, only measurements of numerical data are acceptable, and this fundamental limitation should be made up for with sufficiently effective further mathematical processing of data sets.

The basic mathematical apparatus used is Allen's approach, also known as the best average percentage. The research is based on one of the varieties of the ordinary least squares method (OLS). One of the varieties of the methods known as the ordinary least squares (OLS) remains in the center. The OLS methods are described in many works [2, 5, 6, 9, 10, 13, 14]. Some books [1–3, 11, 13] give lengthy reviews of developed traditional methods of mathematical description, such as smoothing of temporal, as a rule, one-dimensional series, and various linearization methods, widely used for construction of forecasts with a rather short horizon. The world-renowned monograph by Rao [10] is now considered a classic. The specifics of the variation of the OLS method used in this research are that it uses a certain numerical characteristic of matrices, called the matrix product trace.

3 Results

We consider a situation where, over the course of a year or several years, there is demand from several consumers, who may be individual buyers or associations (e.g., consumer societies).

Suppose that the demand map for each such consumer can be represented by a matrix A_k, $(k = 1...n)$. Suppose we want to approximate the matrix A_k by the following dyad:

$$A_k = \mathbf{a}\mathbf{b}^T + \boldsymbol{\varepsilon}_k$$

where $\boldsymbol{\varepsilon}_k$—a random matrix. Consider the matrices A_k, ε_k as a system of rows:

$$A_k = \left(\mathbf{a}_{1,k}\ \mathbf{a}_{2,k}\ \dots\ \mathbf{a}_{n,k}\right)^T$$

$$\varepsilon_k = \left(\boldsymbol{\varepsilon}_{1,k}\ \boldsymbol{\varepsilon}_{2,k}\ \dots\ \boldsymbol{\varepsilon}_{n,k}\right)^T$$

It is assumed that random distortions of demand measurements for different categories of goods can be considered mutually independent. However, the measurements of the demand of the same category obtained during the same season (year) are, generally speaking, dependent. Mathematically, we can write down a general form of distortion covariance:

$$\left\langle \boldsymbol{\varepsilon}_{\alpha,k}\boldsymbol{\varepsilon}_{\beta,k}^T \right\rangle = C^{-1},$$

where C—is a symmetric positively defined matrix.

As a criterion, the parameters should be defined based on the minimum of the quadratic form:

$$\left\langle \begin{pmatrix} \boldsymbol{\varepsilon}_1^T \\ \boldsymbol{\varepsilon}_2^T \\ \dots \\ \boldsymbol{\varepsilon}_N^T \end{pmatrix} C\left(\boldsymbol{\varepsilon}_1 \ \boldsymbol{\varepsilon}_1 \ \dots \ \boldsymbol{\varepsilon}_N \right) \right\rangle = tr\langle \| \boldsymbol{\varepsilon}_i^T \| C \| \boldsymbol{\varepsilon}_j \| \rangle = tr \sum_k \langle \boldsymbol{\varepsilon}_k^T C \boldsymbol{\varepsilon}_k \rangle$$

Angle brackets usually indicate the result of statistical averaging. The above general type of quadratic criterion defines the general form of the corresponding Lagrange function:

$$L = S^2(\mathbf{a}, \mathbf{b}) + \mu \mathbf{a}^T C \mathbf{a} - \nu \sum_k \alpha_k,$$
$$S^2(\mathbf{a}, \mathbf{b}) = tr \sum_k \left(\alpha_k A_k^T - \mathbf{b}\mathbf{a}^T \right) C_k \left(\alpha_k A_k - \mathbf{a}\mathbf{b}^T \right) \quad (1)$$

It shows:

$$C = \sum_k C_k, \ C^T = C, \ \mathbf{a}^T C \mathbf{a} = 1 \quad (2)$$

The values $S^2(\mathbf{a}, \mathbf{b})$ is a measure of the deviation of the model from the real data. In some sense, the question of choosing non-negative numerical multipliers $\alpha_k \geq 0$ remains open. The very need for their introduction is of no doubt, given that the volume of total demand from different consumers can vary by tens, hundreds, or more times. Thus, varying only vectors $\mathbf{a}, \mathbf{b}$, it is impossible to make the values of all the differences $\alpha_k A_k - \mathbf{a}\mathbf{b}^T$ in the sum (1) simultaneously small. This problem has no analytical solution, but it can be solved by recurrence methods.

Let us describe the solution. First, we are not interested in the trivial case in which all multipliers are equal to zero. Therefore, we impose an additional requirement on them: the sum of the coefficients should be equal to one. Suppose a pair of vectors $\mathbf{a}, \mathbf{b}$ is found one way or another. By equating the derivative of the Lagrangian (1) with α_k, we get the equation:

$$tr\left[A_k^T C_k \left(\alpha_k A_k - \mathbf{a}\mathbf{b}^T \right) \right] = \nu$$

From this equality, we obtain an expression for the required multiplier with an arbitrary index:

$$\alpha_k = \frac{\nu + \mathbf{b}^T A_k^T C \mathbf{a}}{tr\left(A_k^T C A_k \right)} \quad (3)$$

The Lagrange multiplier ν is found based on the condition that the sum of all multipliers equals one:

$$\sum_k \frac{\nu + \mathbf{b}^T A_k^T C \mathbf{a}}{tr\left(A_k^T C A_k\right)} = 1$$

This is a linear equation that has a unique solution.

As a first approximation, when there are no estimates of the sought vectors $\mathbf{a}$, $\mathbf{b}$, the square root of the product can be taken as estimates of the coefficients:

$$\alpha_k = \left(tr A_k^T A_k\right)^{-1/2} = \frac{1}{\sqrt{tr A_k^T A_k}} \tag{4}$$

This is natural because the trace of the product of any matrix on its own transpose is, as can be proved, the sum of the squares of all its elements.

Now, at the research stage, we cannot exclude that the estimate (4) will prove to be the best in practical terms. Its undoubted advantage is that it does not need information about the unknown vectors $\mathbf{a}$, $\mathbf{b}$.

If we open the brackets in (1), we get:

$$L = tr \sum_k \left(\alpha_k^2 A_k^T C_k A_k\right) - \mathbf{b}^T A^T \mathbf{a} - \mathbf{a}^T A \mathbf{b} + tr \sum_k \left(\mathbf{b}\mathbf{a}^T C_k \mathbf{a}\mathbf{b}^T\right) + \mu \mathbf{a}^T C \mathbf{a}$$

Here, the following notation is introduced:

$$A = \sum_k \alpha_k C_k A_k$$

We need to find such vectors $\mathbf{a}$, $\mathbf{b}$, for which the function L takes a minimum value. According to the optimization theory, these vectors must satisfy the normalization condition (2) and the equations.

We write an expression for the differential dL for a variation of the vector $\mathbf{a}$ by some $d\mathbf{a}^T \neq 0$:

$$dL_{\mathbf{a}} = -2A\mathbf{b} + 2\left(\mu + \mathbf{b}^T \mathbf{b}\right) C \mathbf{a}$$

We do the same with the variation $d\mathbf{b}^T \neq 0$:

$$dL_{\mathbf{b}} = -2A^T \mathbf{a} + 2\mathbf{b}$$

Since at arbitrary $d\mathbf{a}$, $d\mathbf{b}$ both differentials should be zero at the optimum point, we need to perform a system of vector–matrix equations:

$$\begin{cases} A\mathbf{b} = \lambda C \mathbf{a} \\ A^T \mathbf{a} = \mathbf{b} \end{cases} \tag{5}$$
$$\left(\lambda = \mu + \mathbf{b}\mathbf{b}^T\right)$$

Let us check the consequences of the system (5).

1. Substitute the second equation of the system into the first one:

$$AA^T\mathbf{a} = \lambda C\mathbf{a} \tag{6}$$

Thus, the sought vector **a** is the eigenvector of the product of matrices AA^T in the measure of matrix C.

2. If the left and right parts of Eq. (5) are multiplied from the left by $\mathbf{a}^T$, we get:

$$\mathbf{a}^T AA^T\mathbf{a} = \lambda\mathbf{a}^T C\mathbf{a}$$

Let us note that, first, by the condition $\mathbf{a}^T C\mathbf{a} = 1$; second, the left side is the square of vector **b**, from which we get:

$$\lambda = \left(\mathbf{a}^T A\right)^2 = \mathbf{b}^2 > 0 \tag{7}$$

(by default, we assume that A is a matrix of full rank, so the inequality is strict).

3. Vice versa, if we express vector **a** through **b** from the first equation, and insert it into the second equation, then:

$$\lambda\mathbf{b} = A^T \cdot \lambda\mathbf{a} = A^T \cdot C^{-1}A\mathbf{b}$$

that is,

$$A^T C^{-1}A\mathbf{b} = \lambda\mathbf{b} \tag{8}$$

Consequently, the vector **b** also turns out to be an eigenvector, although of the other matrix $A^T C^{-1}A$. As can be seen from (6) and (7), the eigenvalue λ of both vectors is the same.

Now we find an estimate of the deviation value $S^2(\mathbf{a}, \mathbf{b})$ for the found vectors:

$$S^2(\mathbf{a}, \mathbf{b}) = tr\sum_k \left(\alpha_k^2 A_k^T C_k A_k\right) - \mathbf{b}^T A^T\mathbf{a} - \mathbf{a}^T A\mathbf{b} + \mathbf{b}^T\mathbf{b}$$

However, according to all the ratios obtained above:

$$\begin{aligned}
\mathbf{a}^T A\mathbf{b} &= \mathbf{a}^T \cdot \lambda C\mathbf{a} = \lambda \cdot \mathbf{a}^T C\mathbf{a} = \lambda \cdot 1 = \lambda;\\
\mathbf{b}^T A^T\mathbf{a} &= \left(\mathbf{a}^T A\mathbf{b}\right)^T = \lambda;\\
\mathbf{b}^T\mathbf{b} &= \mathbf{a}^T AA^T\mathbf{a} = \lambda.
\end{aligned}$$

Thus, the criterion deviation $S^2(\mathbf{a}, \mathbf{b})$ equals:

$$S^2(\mathbf{a}, \mathbf{b}) = \sum_k \mathrm{T}_k^2 - \lambda,$$
$$\mathrm{T}_k^2 = tr\left(\alpha_k^2 A_k^T C_k A_k\right) \tag{9}$$

Thus, according to (5) and (7), the optimal vectors $\mathbf{a}$, $\mathbf{b}$ are eigenvectors. We found that although the matrices in expressions (6) and (8) are different, their eigenvalues λ are the same. Since the minimum of the criterion corresponds to the optimum: $S^2(\mathbf{a}, \mathbf{b}) \to \min$, then, according to (9), it is necessary to choose those eigenvectors with the corresponding maximum eigenvalue λ.

It is advisable to find out how justified it is to combine the data of measurements of different years into one criterion in (1). Such aggregation is acceptable if the demand over a number of years conforms to the time homogeneity hypothesis. In itself, the time homogeneity hypothesis is not guaranteed because, generally speaking, consumer preferences can change over time; for a priori assumptions about them, it is necessary to test them experimentally in one way or another.

Let us describe one possible procedure for checking temporal homogeneity.

Throughout the observation interval, λ_k is calculated for each matrix A_k, which is the largest eigenvalue of the matrix $A_k^T C_k^{-1} A_k$, $k = 1, ..., N$. S_k^2 is also calculated.

For a perfectly homogeneous demand process, the value of the eigenvalue corresponding to the entire observation interval is equal to the sum of all λ_k. In reality, the value λ will be less. The deviation S^2 will be greater than the sum of the deviations $\sum_k S_k^2$. As a measure of homogeneity, we can take the ratio of the sum of deviations for all years of observation to the deviation for the entire period of observation:

$$\gamma = \frac{\sum_k S_k^2}{S^2} = 1 + \frac{\sum_k \lambda_k - \lambda}{S^2}$$
$$(0 \leq \gamma \leq 1)$$

The demand can be considered homogeneous if the value is sufficiently close to one.

4 Conclusion

It is shown that the known method can be extended to the case of periodic multidimensional economic processes. In the presence of periodicity, the matrices included in the model description are transformed according to certain rules. However, as in the classical work of Allen, the eigenvalues of these matrices remain the basis of the mathematical apparatus.

The application of the developed mathematical model of consumer demand, considering the seasonal fluctuations of demand, in a certain sense similar in all points of sales, will allow one to reduce a large number of mathematical transformations and extend the obtained results to the general set of consumers in several sales markets at once.

References

1. Allen, R. G. D. (1960). *Mathematical economics* (2nd ed.). Macmillan and Co LTD; St. Martin's Press.
2. Allen, R. G. D. (1980). *Index numbers in theory and practice* (L. S. Kuchaev Transl. from English). Moscow, USSR: Statistics. (Original work published 1975).
3. Dubrov, A. M., Mkhitaryan, V. S., & Troshin, L. I. (2003). *Multivariate statistical methods.* Finance and Statistics.
4. Ermilov, M. M., Surkova, L. E., & Samoletov, R. V. (2021). Mathematical modeling of consumer behavior, taking into account entropy. In A. V. Bogoviz, A. E. Suglobov, A. N. Maloletko, O. V. Kaurova, & S. V. Lobova (Eds.), *Frontier information technology and systems research in cooperative economics* (pp. 269–278). Springer. https://doi.org/10.1007/978-3-030-57831-2_28.
5. Green, W. H. (2002). *Econometrics analysis* (5th ed.). Prentice-Hall.
6. Green, W. H. (2016). *Econometric analysis* (Book 1) (A. V. Khodyrev, A. S. Stepanov, & B. N. Gafarov Transl. from English; S. S. Sinelnikov, & M. Yu. Turuntseva Eds.). Publishing House "Delo," RANEPA.
7. Intriligator, M. D. (2002). *Mathematical optimization and economic theory* (G. I. Zhukova, & F. Ya. Kelman Transl. from English). Iris-Press. (Original work published 1987).
8. Kleiner, G. B. (2017). System fundamentals of the digital economy. In *International Theoretical and Practical Conference "Institutional and Financial Mechanisms of Formation of Digital Economy."* Dubna, Russia: Dubna State University.
9. Magnum, Ya. R. Kamyshev, P. K., & Peresetsky, A. A. (2021). *Econometrics* (9th ed.) Publishing House "Delo," RANEPA.
10. Rao, C. R. (1968). *Linear statistical inference and its applications* (A. M. Kagan Transl. from English; Yu. V. Linnick Ed.). Moscow, USSR: Nauka. (Original work published 1965).
11. Romer, D. (2001). *Advanced macroeconomics* (2nd ed.). McGraw Hill.
12. Ryazanova, G. N. (2021). The system of interaction with the consumer in a modern organization. In G. B. Kleiner, & S. E. Shchepetova (Eds.), *Systems Analysis in Economics—2020: Collection of the 6th International Research & Practice Conference-Biennale SAE-2020* (pp. 188–191). https://doi.org/10.33278/SAE-2020.book1.188-191.
13. Taha, H. A. (2019). *Operations research: An Introduction* (10th ed.) (A. A. Minko, & A. V. Sleptsov Transl. from English). Dialectics. (Original work published 2016).
14. Tyurin, Y. N. (2011). *Multivariate statistics: Gaussian linear models.* Moscow University Press.

The Impact of Digital Employment Platforms on the Transformation of the Labor Market in the Digital Economy

Galina G. Goloventchik

Abstract Labor markets worldwide lag behind the rapid changes in the world economy. The ineffectiveness of these changes is shown through millions of people struggling with unemployment, underemployment, and insufficient earnings, not being able to find employment even with many job openings. The digital economy, which has an increasing impact on the world labor market, offers new opportunities to solve employment problems. Among these opportunities, it is important to point out digital labor platforms. The global community is entering the epoch of the digital platform economy. Platform employment is becoming more and more widespread; it can be termed a new institutional mechanism in the labor market. Platform employment has already significantly changed and continues to change the structure of the labor market in the developing digital economy, cardinally transforming the relationships between employers and employees. Even though hundreds of millions of people worldwide already use the services of digital platforms, their potential and capabilities are still developing. Currently, there is an asymmetry in the relations between the platform and the executors that are dependent on the platform and subordinate to it. However, most national labor legislations do not regulate platform employment. No one in the world knows how to build a relationship between the platform and contractors or what the legal framework for this relationship should be. The new challenges posed by the emergence of digital labor platforms must be addressed through a global social dialogue. Platform employees should be enabled to exercise their principal labor rights in full, regardless of their status.

Keywords Digital economy · Digital labor platforms · Platform employment · Self-employment · ILO · Regulation · Social protection

JEL Classfications J20 · J21 · J22 · J23 · J28

G. G. Goloventchik (✉)
Belarusian State University, Minsk, Belarus
e-mail: goloventchik@bsu.by

A. V. Bogoviz (ed.), *Big Data in Information Society and Digital Economy*,
Studies in Big Data 124, https://doi.org/10.1007/978-3-031-29489-1_10

1 Introduction

In recent years, platform employment has been the object of study by numerous foreign researchers and a number of major international organizations [1, 4, 8, 12, 14].

Belarusian and Russian research communities *have only recently begun to* study the problems of platform employment and its impact on the labor market. As of February 14, 2022, the search in the scientometric database of the Russian Science Citation Index gave 2197 publications on the keywords "digital platform," 1000 publications on the keywords "remote employment," and only 79 publications on the keywords "platform employment," of which 68 publications are dated to the years 2020–2022.

Zemlyanskaya and Semygina [19] were among the first to research the impact of international digital platforms on labor relations. The works of Chernykh [2, 3] and Shevchuk [15] cover the scale of platform employment and its socio-demographic characteristics. The development of platform employment on the example of foreign countries is studied by Glotova [6]. Goncharova [7] analyses the tendencies of the development of international digital platforms in the COVID-19 pandemic. Kursova [11] and Tomashevsky [17] devote their works to the questions of legal regulation of platform employment. The monograph by Savelieva [13] is the first large-scale attempt among Russian research works to reflect the key features of digital labor platforms in terms of labor organization. The report of HSE University's Institute of Social Policy [16] concerns platform employment, its characteristic features, its advantages, and the risks it causes in terms of population involvement and the economy as a whole.

The research aims to systematize the main trends in the development of platform employment. The research tasks are to study the characteristic features, advantages, and disadvantages of platform employment, as well as to suggest improvements to the regulation of labor through digital platforms.

The theoretical relevance of the research lies in forming a scientific approach to identifying specific features of the functioning of digital labor platforms. The practical relevance of the research lies in the possibility of using its results in further studies and in education.

2 Methodology

The research used general scientific methods: system and comparative analysis, generalization. One specific method—the analysis of the sources—was also applied. The sources encompass the reports of the International Labor Organization, HSE University's Institute of Social Policy, and data from the portal Crunchbase.

3 Results

According to the opinion of Doctor of Economics, Professor A. V. Babkin, "…digital platforms are hybrid structures aimed at forming mutually beneficial relationship of a significant number of independent economic agents in the common information zone and oriented at producing value by providing direct interaction and carrying out transactions among several groups of outside users" [5, p. 23].

Initially, the platform was a hybrid structure that provided direct interaction and carried out transactions among several groups of outside users. With the development of digital technologies, the term "platform" is also increasingly used to refer to:

- An ICT-enabled value-creating business model;
- Hardware, software, and network complexes;
- The actual enterprise with public infrastructure, which provides mutually beneficial interaction among outside manufacturers and consumers;
- The aggregate number of platform users.

Digital platforms promise the exponential growth of the private sector by eliminating the need for large start-up assets and scaling through network effects. The WEF research shows that new digital ecosystems may generate more than $60 trillion in revenue by 2025 (or over 30% of the global corporate income). With regard to the public sector, digital platforms and ecosystems are key strategic elements for creating smart cities and smart countries [10].

It is essential to mention digital employment platforms (or digital labor platforms [DLP])—these are actually digital markets that connect workers with those who will use their labor.

Platform employment is a non-standard form of employment; it uses online platforms and digital technologies for mediation between individual providers of services (performers registered on the platform) and customers [8].

In accordance with the approach of the International Labor Organization, digital labor platforms are divided into location-based and online web-based platforms. Within local digital labor platforms, the provision of work and services takes place within a certain physical area (in particular, this scheme is used to organize the work of Uber taxi drivers, online food delivery couriers, cleaners, and repairers). Within online digital labor web platforms, clients (customers) and contractors (employees) around the world interact via the Internet (physical contacts excluded), and the result is provided, as a rule, in digital form. The examples are the services of the online travel agencies (e.g., Tripadvisor or Booking), data analysis, software development, legal or financial services, audio and video processing, etc.

The development of the DLP has been one of the most noticeable social and economic changes in the labor market caused by COVID-19. The rapid growth in the number of digital labor platforms was seen even before the COVID-19 pandemic: from 107 in 2009 to 267 in 2019 (according to the Crunchbase database) (Fig. 1). However, ubiquitous lockdowns, combined with the mass transition of employees to remote work, have sharply increased the spread and the intensity of use of DLPs.

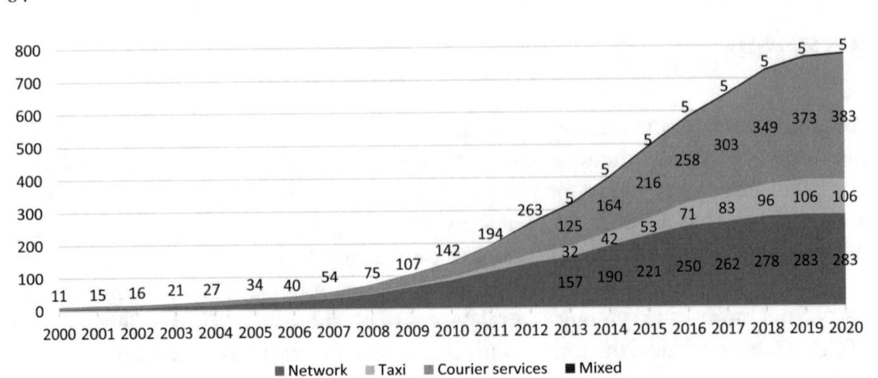

Fig. 1 The number of digital labor platforms. *Source* Compiled by the author based on ILO data [9]

Earlier large popular platforms like LinkedIn (774 million users in over 200 countries as of early 2022) were used mainly for recruiting high-skilled workers to well-paid workplaces. Now, they recruit semi-skilled laborers. There have appeared other types of digital platforms that connect workers with clients or companies to perform specific tasks or provide specific services. The role of such platforms is constantly growing in the market for "on-demand" workers. In the broad sense, they are temporary workers and, therefore, do not enter into standard part-time or full-time contracts with their employers. The digital platforms Lyft, TaskRabbit, Uber, Angie's List, Freelancer.com, and Upwork may serve as good examples of platforms that directly connect employees with clients and help companies find and hire qualified workers to perform a wide range of specialized tasks [18].

Until recently, there has been relatively little discussion of the impact of the rapid expansion of digital platforms of employment on the nature of labor and labor relations. The report of the International Labor Organization [9], released in February 2021, brought about significant changes in this matter. It states that DLPs open up new opportunities for millions of people worldwide and contribute to the reduction of structural unemployment. DLPs often provide jobs to those who have difficulty finding them in the traditional way, such as people of pre-retirement and retirement age, women with children, people with disabilities, youth, and representatives of national, religious, or other minorities.

However, there have recently arisen more and more questions regarding the prospects for the development of the DLP in terms of building up a conflict of interest among the main stakeholders—the platforms, employees, authorities, and society. The essence of the problem is that those who get a job through DLPs are self-employed, so the platform (which positions itself neutrally as an intermediary connecting the contractor with potential customers or a technological service simplifying the interaction between supply and demand) is not responsible for their pay level or rate of remuneration, working hours and working conditions (e.g., they sometimes have to work nights without any compensation for this). The issue of pensions for platform workers is equally important because the vast majority of employers do not make voluntary contributions to the pension fund, although they plan to live off pensions in the future. The work on the platform is paid regardless of qualifications and, as a rule, on a piecework basis; hence, the increased uncertainty and income instability for platform workers in faceless global competition, resulting in increased self-exploitation, lack of social protection, dangerous working conditions, health risks, and constant stress, exacerbated by a sense of powerlessness from the inability to influence the platform's operation. Additionally, there are suspicions about the "opacity" of the platforms' working algorithms, rating calculation rules, fines legitimacy, and the amount of commissions and fees levied on employees (the main components of algorithmic management are ratings, customer and consumer reviews, the skill level of employees, refusals of services and acceptance of work; algorithms control employees and track, monitor, and evaluate their work) [9].

During the COVID-19 pandemic, lots of digital platform employees cannot afford self-isolation because their income depends on the execution of orders, and they are not entitled to paid sick leave or health allowance. In 2020, the ILO conducted a survey of taxicab drivers and couriers (both freelancing or working for aggregator companies) in Chile, India, Kenya, and Mexico. The survey results showed that 32% of the respondents did not stop working throughout the crisis out of economic necessity despite the risk of contagion of COVID-19 [9]. In these countries, the incomes of nine out of ten taxicab drivers and seven out of ten couriers plummeted due to the crisis, aggravating the already difficult situation of digital platform employees during the COVID-19 pandemic.

However, in reality, DLPs possess significant power because they unilaterally determine the key labor parameters, employment conditions, and the salaries of formally independent workers. In fact, DLP workers are hired employees who answer to their "digital" employer, that is, to the platform algorithm, which structures the labor process in detail by determining the character, order, and limits of the possible human actions. For example, a courier may accept or reject a specific order being offered but cannot choose an order from the list available nor discuss the price and additional conditions. That is, an employee does not choose an assignment but is assigned to it, as practiced in traditional organizations, contrary to self-employment [15, p. 145]. The platform algorithm uses electronic monitoring systems and the readily accessible "big data" on the employees, which enables it to monitor them, assess their efficiency, and even manipulate their performance, encouraging the most efficient employees (e.g., they may get "beneficial" shifts). Simultaneously, less efficient workers get temporal or permanent "deactivation" (i.e., are fired) or some other disciplinary sanctions, which are, in fact, impossible to challenge. Therefore, as employers, the platforms should take care of their employees' health and safety, pay minimum wage, and comply with labor legislation.

European legislators are very much concerned about this problem. In this respect, some local platforms (in particular, delivery and taxi services) have been put under regulatory and legal control in the countries that tend to view these platforms as employers. Nowadays, several European countries have taken legal and judicial measures against the exploitation of employees by their platforms:

- The Supreme Court of France interpreted the relations between couriers and platforms as labor relations as early as 2018. As a result, these couriers are now considered as employees, and their contractors—whether it be Take Eat Easy, Uber, or Deliveroo—not just as technological companies but as employers, carrying social responsibility and obliged to protect their employees' health and safety on the job;
- In February 2021, the Supreme Court of the UK granted two Uber drivers the status of employees (rather than self-employed), which entitled them to minimum wage and holiday. This judicial decision affects tens of thousands of drivers and has an enormous influence on the Uber model;
- Since March 2021, delivery service drivers in Spain have been recognized as workers and are to be employed by the platforms.

These rulings have one common feature: they debunk the myth about platforms being "independent intermediaries" and admit the fact that the people they employ are, in fact, dependent workers entitled to all kinds of protection provided for by labor legislation.

4 Conclusion

The research has allowed us to point out the following tendencies in developing digital labor platforms.

- The last decade has seen rapid growth in the number and types of DLPs and the number of people performing their labor activity through platforms. DLPs are popular with workers due to the flexible working hours, free choice of orders, and a possibility for many contractors to work from anywhere in the world as they see fit. Employers see the main advantage of the platform business model in the reduction of transactional and other expenses. The transfer of many workers and employers to remote work and the active development of delivery services in 2020–2021 during the COVID-19 pandemic significantly strengthened the potential of DLPs, which provided new opportunities to earn a living to many people who had lost their jobs because of national lockdowns.
- Digital platforms determine the conditions and regulations of labor of their employees unilaterally because they are free from the bounds of the existing legislation on labor protection and social security for platform workers. Implementation of the business models adopted by labor platforms allows organizing the labor of many workers without the necessity of investing in fixed assets and taking on staff. This results in a whole range of negative socio-economic consequences, which affect the quality of labor resources and labor efficiency. Thus, the regulation of the activities of digital platforms is a considerable problem on the national and international levels because it affects many aspects of labor legislation concerning decent work.
- So far, the new platform labor relations seem clearly unfair in relation to DLP employees. This problem can be solved by further legal regulation of the platform

activity model as an independent form of labor relations, which, in turn, will allow effectively regulating platform employment. The use of platform algorithms should be made more transparent: people, and not digital applications, should monitor compliance with safe working conditions. Digital platforms, like all other companies, must disclose information about the people working on them, i.e., national authorities should have free access to the data on the employees of digital platforms.

- There is a clear need to rethink what a "platform" employer is and define its obligations and workers' rights in greater detail. Unlike the self-employed, platform workers should be entitled to the minimum wage, collective bargaining, regulated working hours, pensions, paid holidays, and sick leaves. There is a need for broader and more flexible requirements for labor standards that apply to piece work. Only part of these measures can be taken at the national level. Since digital labor platforms operate in many jurisdictions, it is vital to coordinate international policies in this area.

References

1. Berg, J., Furrer, M., Harmon, E., Rani, U., & Silberman, M. S. (2018). *Digital labour platforms and the future of work: Towards decent work in the online world*. ILO. Retrieved February 19, 2022, from https://www.ilo.org/wcmsp5/groups/public/---dgreports/---dcomm/---publ/documents/publication/wcms_645337.pdf
2. Bobkov, V. N., & Chernykh, E. A. (2020). Platform employment—The scale and evidence of instability. *The World of the New Economy, 14*(2), 6–15. https://doi.org/10.26794/2220-6469-2020-14-2-6-15
3. Chernykh, E. A. (2021). Socio-demographic characteristics and quality of employment of platform workers in Russia and the world. *Economic and Social Changes: Facts, Trends, Forecast, 14*(2), 172–187. https://doi.org/10.15838/esc.2021.2.74.11.
4. De Stefano, V., Durri, I., Stylogiannis, C., & Wouters, M. (2021). *Platform work and the employment relationship*. ILO Working Paper 27. ILO.
5. Geliskhanov, I. Z., Yudina, T. N., & Babkin, A. V. (2018). Digital platforms in economics: essence, models, development trends. Scientific and Technical Knowledge of SPbGPU. *Economic Sciences, 11*(6), 22–36. https://doi.org/10.18721/JE.11602.
6. Glotova, N. I., & Gerauf, Yu. V. (2021). Platform employment is the main trend of labor market development in modern conditions. *Economy. Profession. Business, 4*, 22–27. https://doi.org/10.14258/epb202151
7. Goncharova, D. M. (2021). Analysis of platform employment in Russia during the coronavirus. *My Professional Career, 1*(31), 133–137.
8. Hauben, H., Lenaerts, K., & Waeyaert, W. (Eds.). (2020). The platform economy and precarious work. Luxembourg: European Parliament. Retrieved February 10, 2022, from https://www.europarl.europa.eu/RegData/etudes/STUD/2020/652734/IPOL_STU(2020)652734_EN.pdf.
9. International Labour Office (ILO). (2021). World employment and social outlook: The role of digital labour platforms in transforming the world of work. ILO. Retrieved February 19, 2022, from https://www.ilo.org/wcmsp5/groups/public/---dgreports/---dcomm/---publ/documents/publication/wcms_771749.pdf.

10. Jacobides, M. G., Sundararajan, A., & Van Alstyne, M. (2019, March 25). *Platforms and ecosystems: Enabling the digital economy*. World Economic Forum. Retrieved February 18, 2022, from https://www.weforum.org/whitepapers/platforms-and-ecosystems-enabling-the-digital-economy.
11. Kursova, O. A. (2021). Legal regulation methods of non-traditional employment: ILO guidelines and experience of foreign countries. Tyumen State University Herald. *Social, Economic, and Law Research, 7*(2), 106–121. https://doi.org/10.21684/2411-7897-2021-7-2-106-121.
12. Larsson, A., & Teigland, R. (Eds.). (2020). *The digital transformation of labor: Automation, the gig economy and welfare*. Routledge. https://doi.org/10.4324/9780429317866.
13. Savelyeva, E. A. (2022). *Digital labor platforms: New forms of labor organization and regulation*. INFRA-M. https://doi.org/10.12737/1818511.
14. Schmidt, F. A. (2017). *Digital labour markets in the platform economy. Mapping the political challenges of crowd work and gig work*. Friedrich-Ebert-Stiftung.
15. Shevchuk, A. V. (2020). From factory to platform: Autonomy and control in the digital economy. *Sociology of Power, 32*(1), 30–54. https://doi.org/10.22394/2074-0492-2020-1-30-54
16. Sinyavskaya, O. V., Biryukova, S. S., Apothecary, A. P., Gorvat, E. S., Grishchenko, N. B., Gudkova, T. B., … Kareva, D. E. (2021). *Platform employment: Definition and regulation*. HSE University.
17. Tomashevsky, K. L. (2021). Platform employment: Between labor and civil tax law. *Justice of Belarus, 8*, 10–15.
18. Tyson, L., Mendonca, L. (2015, November 30). Digital job platforms changing labour market. *Gulf Times*. Retrieved February 20, 2022, from https://www.gulf-times.com/story/464745/Digital-job-platforms-changing-labour-market.
19. Zemlyanskaya, N. G., & Semygina, T. V. (2019). International online platforms and their impact on labor relations. In *Science, technology and innovations: Collection of scientific articles* (pp. 126–131). Pegasus Publishing.

Current Tools for Merchandising and Trading in the Digital Economy

Larisa V. Mikhailova, Sofia V. Ivanova, Elena N. Vlasova, Lyudmila N. Shmakova, and Evgeniya V. Vidyakina

Abstract The research is based on the problem of improving the customer centricity of retailers and the effectiveness of their activities in the digital economy. As a promising solution to this problem, the research offers retailers the use of advanced digital technologies and innovative application solutions in merchandising and trading. The paper examines the organization of trade and technological processes in the store selling a wide range of categories and subcategories of consumer goods. The objects of expert evaluation of the fundamental characteristics influencing the formation of the consumer value of the product are separate groups and types of non-food products, including household clothing and laundry detergents. The paper also analyzes the results of sociological studies and surveys of entrepreneurs and the population of Russian regions on retailing in the digital economy. As a result, the paper proposes a set of the author's recommendations for improving merchandising and trading practices in the activities of retailers in the digital economy. The contribution of the paper to the literature consists in the disclosure of features and prospects for improving the efficiency of retailers by increasing the flexibility of merchandising and trading based on the use of advanced features of the digital economy.

L. V. Mikhailova (✉) · S. V. Ivanova
Cheboksary Cooperative Institute (Branch), Russian University of Cooperation, Cheboksary, Russia
e-mail: larisacoop@yandex.ru

S. V. Ivanova
e-mail: golovanova_sofya@mail.ru

E. N. Vlasova
Ivanovo State Politechnical University, Ivanovo, Russia
e-mail: vlasovaen-ivanovo@mail.ru

L. N. Shmakova · E. V. Vidyakina
Kirov State Medical University of the Ministry of Healthcare of the Russian Federation, Kirov, Russia
e-mail: zontik-34@yandex.ru

E. V. Vidyakina
e-mail: vidjkina@mail.ru

A. V. Bogoviz (ed.), *Big Data in Information Society and Digital Economy*,
Studies in Big Data 124, https://doi.org/10.1007/978-3-031-29489-1_11

Keywords Merchandising · Trading · Competitiveness · Expertise · Assortment · Consumer demand

JEL Classification M310

1 Introduction

The key criteria of consumer preference at different times were price, assortment, and quality. These factors now come to the fore. This equally applies to large and small cities. A consumer is forced to navigate in the wide commercial range offered to purchase a quality product. The exact choice of assortment policy and the structure and qualitative characteristics of the assortment is among the most important factors of the profitable activity of the trade organization.

Timely response to innovations in merchandising and trading and rational application of commodity expertise tools play a critical role in helping retailers to solve important and major problems and bringing real competitive advantages in different areas: increase the volume of the product basket, justified by the calculated increase in revenue from consumers participating in various loyalty programs, and reduce the logistics costs incurred in providing and storing inventories. If the product or product category offered is personalized and the offer is targeted to customers, the response rate to marketing communications will increase.

2 Materials and Methods

The paper examines the organization of trade and technological processes in the store selling a wide range of categories and subcategories of consumer goods. The objects of expert evaluation of the fundamental characteristics influencing the formation of the consumer value of the product are separate groups and types of non-food products, including household clothing and laundry detergents.

During the research, the authors used general scientific methods, including systematization, comparative and descriptive analysis, and other methods, the preference in the choice of which was conducted based on the research algorithm. Particular attention is paid to heuristic and expert methods of commodity expertise [4], which are based on judgments supported by the generalized experience and intuition of the specialists, the demand for which is noticeably increasing in trade [7, 8].

The information base of the research is the scientific works of the authors of this work [6–10, 12], other Russian scholars [1–5], and statistical data [11].

3 Results

The success of a retail enterprise depends on the satisfaction of customers and the degree of this satisfaction. Knowing the competitors, their strengths and weaknesses, and their tactics and strategies allows one to create the prerequisites for success in competition [9].

For example, in the Chuvash Republic, a sociological study and surveys of entrepreneurs and the region's population were conducted in 2019 [11]. Among organizations and individual entrepreneurs, including those from the trade sector, 23.4% of respondents consider the level of competition to be high, and 5.9% consider it very high. The majority of respondents (43.3%) indicate a moderate nature of competition. Assessing the supply of goods and services in the target markets, more than half of the respondents noted a lack of offers or its complete absence.

Given the above, measures aimed at improving the competitiveness of the trade enterprise, first and foremost, should be aimed at the most important control points in the activities of the enterprise and the strengthening of promising sources of generation of key elements of the current business model and capabilities of the external environment.

If information and commodity resources are in orderly movement with fixed (control) points, it contributes to the effective use of all opportunities of the store to achieve high results of the organization [1, 5]. To solve the problems of optimization of trade activities, we proposed to automate trivial trade and technological processes [6].

The success of the trade enterprise largely depends on the range and totality of consumer properties of the products offered to customers [8, 10, 12]. The study of consumer demand and the dynamics of the sale of goods allows retailers to predict actions to achieve maximum customer satisfaction and the planned profit.

We considered the situation in the market of household apparel in the city of Cheboksary (Russia), taking into account the needs it meets and assortment diversity. Urban residents spend less money on clothing than on food because of the low average per capita cash income (e.g., the sale of outerwear is in second place). To characterize the segments of the Cheboksary market, the assortment of enterprises selling clothes and their price category was studied. In the city of Cheboksary, the specialized stores sell mainly factory and utilitarian brands of low and medium price categories, including "United Colors of Benetton," "Bugatti," "Truvor," "Luisa Cerano," "Giovane Gentile," "Calvin Klein Jeans," "ELIS," "Sela," "TVOE," "O'STIN," "Serginnetti," and others. There are no clothes in the higher price category.

Effective clothing trade is possible with the competent formation of quality and competitive range [8]. Many standards for homogeneous clothing groups impose additional requirements to its characteristics (including materials, packaging, and storage conditions), the implementation of which contributes to the mandatory requirements of safety and full satisfaction of the buyers' needs. Mandatory requirements for household apparel in terms of chemical, biological, and electrical safety, completeness and reliability of labeling, and confirmation of compliance in the form of declaration or certification are listed in the TR TS 017/2011 and TR TS 007/2011.

Experts quite often examine the degradation of the quality of research objects, considering their real condition and the presence of certain defects. Non-destructive organoleptic and measuring methods are predominantly applied during expert examinations of clothing quality. The primary means of quality assessment are normative documents and measuring instruments that have passed state tests. The questions asked in the examination and the identification characteristics of the objects determine the choice of theoretical and practical methods and means of their investigation. For example, for the categorical identification of clothing, it is necessary to make its characteristic consistent with the data of economic and statistical, and standard classification systems. Labeling clothing is one of the most critical information-analytical means of examination. The reliability of the labeling cannot be established without a chemical analysis of the fibrous composition of the material. The calculation method is used to find the overall level of quality reduction of objects, considering the identified defects and their impact on the quality of clothing.

Introducing new products into the commercial range is a risky step. The store must purchase and sell such goods, which will most fully meet the needs of consumers to successfully and actively develop, maintain, or increase the occupied market share.

Finding a balance between the interests of consumers and the capabilities of the store and manufacturers is also possible based on the results of commodity expertise. The intensity of innovation is a significant scientific and technological factor affecting the activities of the manufacturer and the store. According to experts, it is necessary to improve the formula of a product on average once every six months or a year so that this product does not bore the consumer. Such an innovative product may well be considered a laundry detergent, among which synthetic detergents have several advantages compared to other products (e.g., laundry soap). Synthetic detergents include special components providing a high level of consumer properties of the product, so there is a high supply level. It should be noted that the SMS formulas of the same manufacturer may differ dramatically from country to country. This does not mean that somewhere the product is of better quality. Textile materials and their coloring differ depending on the geographical location and national traditions of a particular country. The types of washing machines used also vary.

Thus, consideration of the features of the enterprise's external environment factors, combined with the results of targeted commodity expertise, will allow the store to strengthen its position in the consumer market. This will allow one to effectively manage the process itself, make adjustments in time, and, most importantly, achieve economic benefits in the activities of the retail store.

4 Discussion

The economic situation of the country and a particular region, the level of employment, and other factors impact the activities of the trade enterprise. Under the influence of economic sanctions, unfavorable sanitary and epidemiological conditions combined with scientific and technological progress, the current Russian consumer market poses new challenges to trade enterprises of different formats in the struggle for customers and their loyalty [3, 7].

The logistics approach in the organization of trade and technological process of the store can provide control over the material, information, and financial flows [2] and make optimal decisions within a single store and retail chain. The coordinated action of structural links, improving ergonomics, and reducing the time spent on the organization and maintenance of the trade and technological process in a modern trading enterprise contributes to applying the basics of the information concept of logistics. The marketing concept of logistics pays attention to the organization of the logistics process in sales to strengthen the competitive position of the retailer.

Many organizations in the retail of consumer goods face challenges that make consumer dynamics complex and slow. Due to objective and subjective limitations, stores often fail to use available data correctly and analyze it to provide a competitive advantage.

When decisions need to be made, regional retailers often tend to rely more not on expert judgment but on their own experience, sometimes on the verge of intuition. Characteristic features of such organizations are as follows:

- Disparate storage of data (e.g., operational information) and the resulting slow transition to analytics and mobile solutions;
- The movement toward expansion and, at the same time, inconsistent use of data;
- Understanding the need for action, but slow response.

It is not just a question of data analysis; the achievement of success requires the full integration of resources, including processes, organization, and systems.

The expertise of consumer goods needs to be used more extensively in trade, design, and industry. As a positive point, it can be noted that the results of the examination provide up-to-date information about consumer preferences, the specifics of using future products and their analogs, and the ways to achieve competitiveness of goods and the trade organization itself.

5 Conclusion

Today's retail chain is constantly expanding, using methods and techniques of innovative trade management practices. The management system in retail stores must also improve and use new tools and mechanisms for advertising, product sales while considering the role of planning at the stage of the organization of the trade and technological process.

The competitive place of trade enterprise on the regional market is determined by several parameters, including compliance of real consumer demand with the forecast, quality of planning, and maintaining the current range of goods with maximum usefulness for real and potential buyers. Automated collection and processing of information about customers and their value orientations, combined with an expert assessment of the consumer characteristics of the offered product categories, will allow one to optimize the architecture of sales and make personalized product recommendations, thus increasing the effectiveness of product sales and retaining key customers.

References

1. Alexandrova, L. Y., Mikhailova, L. V., Munshi, A. Y., Sorokina, N. D., & Timofeev, S. V. (2022). Using the logistics approach to assessing the competitiveness of an enterprise. In A. V. Bogoviz, A. E. Suglobov, A. N. Maloletko, & O. V. Kaurova (Eds.), *Cooperation and sustainable development* (pp. 357–364). Springer. https://doi.org/10.1007/978-3-030-77000-6_43
2. Antonov, G. D., Ivanova, O. P., Tumin, V. M., & Bodrenkov, A. V. (2021). *Supply and sales management of the organization: Textbook*. Moscow, Russia: INFRA-M.
3. Astrakhantseva, A. S., & Ignatyev, I. S. (2021). Approaches to managing assortment of non-food products. *Young Science of Siberia, 1*(11), 620–625. Retrieved 19 September 19, 2021, from https://mnv.irgups.ru/podhody-k-upravleniyu-assortimentom-neprodovolstvennyh-tovarov
4. Dolgushina, L. V. (2021). *Natural scientific methods of forensic investigations: Textbook*. Siberian Fire and Rescue Academy of EMERCOM of Russia. Retrieved October 9, 2021, from https://znanium.com/catalog/product/1354580
5. Gavrilova, M. V., Portnov, M. S., Rechnov, A. V., Philippov, V. P., & Egorova, G. N. (2021). The use of neural networks in predicting the economic performance of cooperative organizations. In A. V. Bogoviz, A. E. Suglobov, A. N. Maloletko, O. V. Kaurova, & S. V. Lobova (Eds.), *Frontier information technology and systems research in cooperative economics* (pp. 511–519). Springer. https://doi.org/10.1007/978-3-030-57831-2_54
6. Ivanova, S. V., & Mikhaylova, L. V. (2021). Trade and technological processes automation in retail trade in the context of innovative development of the economy. *Vestnik of the Russian University of Cooperation, 1*(43), 63–67.
7. Mikhailova, L. V. (2016). Diagnostics of detected defects as a result of the examination of leather shoes. We studied the characteristics of the degree of wear of shoes. The cause of the origin of defects. In S. N. Lebedeva, & A. P. Bobovich (Eds.), *Union of science and practice: Current problems and prospects for the development of commodity science* (pp. 59–63). Belarusian Trade and Economics University of Consumer Cooperatives.
8. Mikhailova, L. V. (2016). Formation of a competitive assortment using expert evaluation. In A. N. Avtonomov, G. V. Kalinina, N. D. Sorokina, & L. A. Taimasov (Eds.), *Management of assortment, quality, and competitiveness in the global economy* (pp. 130–134). Cheboksary Cooperative Institute (Branch) of the Russian University of Cooperation.
9. Mikhailova, L. V., & Gavrilova, M. V. (2020). Practical aspects of enterprise competitiveness assessment. *Vestnik of the Russian University of Cooperation, 3*(41), 61–66.

10. Mikhailova, L. V., & Vlasova, E. N. (2020). Trading assortment management based on consumer value orientations. *Vestnik of the Russian University of Cooperation, 3*(41), 55–60.
11. Ministry of Economic Development and Property Relations of the Chuvash Republic. (2020, January 24). A meeting of the expert group for monitoring the implementation of the standard of competition development in the subjects of the Russian Federation was held. Retrieved October 15, 2021, from http://minec.cap.ru/news/2020/01/24/proshlo-zasedanie-ekspertnoj-gruppi-po-monitoringu
12. Vlasova, E. N. (2012). Quantitative estimation of textile products competitiveness. *Textile Industry Technology, 2*(338), 20–23.

Empirical Testing of a Value-Based Management Model: The Historical Aspect of Foreign Experience

Olga A. Romanenko, Svetlana B. Efimova, Vlada E. Evdokimova, and Mariia V. Prianishnikova

Abstract As part of the study of the concept of value-based management (VBM), this article discusses research on the applicability of Discounted Cash Flow (DCF) models with the predictability of the current market value of common stock. The problem of reliability of forecasting the market price of shares based on DCF by analyzing the market value of highly leveraged transactions in comparison with the current value of the corresponding cash flow forecasts determines the difficulties that arise when trying to verify the validity of the model. The article also discusses studies aimed at determining the predictive power of various criteria of the company's value-based management model (VBM) in assessing the market price. The subject of the research was also studies in which attempts were made to confirm the influence of this model on the behavior of employees and the results of the company's activities. The scientific significance lies in the fact that the problems of empirical testing of these models presented in the study can contribute to the development of financial science and serve for further theoretical and practical developments in the field of company management based on the concept of value maximization. The practical significance of the research is that the presented results can be used by the heads of companies and business development departments, representatives of the business community who implement management programs based on the company's value.

Keywords Model · Company · Cash flow · Shareholder value · Efficiency · Stock

JEL Classifications F20 · F37 · F36 · F39 · G15 · G31 · G32 · G34 · G39

O. A. Romanenko (✉) · M. V. Prianishnikova
Yuri Gagarin State Technical University, Saratov, Russia
e-mail: olga_romanenko_@mail.ru

M. V. Prianishnikova
e-mail: marprya@mail.ru

S. B. Efimova
Saratov State Law Academy, Saratov, Russia
e-mail: efimovas@rambler.ru

V. E. Evdokimova
Moscow University for Industry and Finance "Synergy", Moscow, Russia
e-mail: vlada777evdokim@mail.ru

A. V. Bogoviz (ed.), *Big Data in Information Society and Digital Economy*,
Studies in Big Data 124, https://doi.org/10.1007/978-3-031-29489-1_12

1 Introduction

Value-based management is a set of management tools used to improve the company's operations in order to increase shareholder value. The financial journals are replete with articles praising the benefits of a value-based management system (VBM). The scientific literature in the field of finance is full of articles praising the advantages of a value-based management system (VBM). The question arises about the effectiveness of such a management system, whether VBM systems really help companies create shareholder value and whether they are really better than traditional methods of measuring performance. Ultimately, these are empirical questions in nature that require practical observations and confirmation. In this article, we consider issues related to these important aspects.

Does the discounted cash flow (DCF) valuation theory provide a reliable estimate of the common stock price? Many researchers note the fact that all VBM tools are based on the DCF theory. Thus, the DCF theory should provide reasonable guidelines for action so that VBM systems actually contribute to the creation of shareholder value.

Does the VBM criterion provide a reasonable forecast of the common stock price? This raises the question of how closely the VBM performance assessment tools are related to the stock price. This means that, despite the fact that the answer to the first question is positive, we still do not know whether our methods of measuring efficiency are sufficiently closely related to the cost of equity to use them as tools for managing shareholder value.

Does VBM have an impact on the activities of companies using this approach? Presumably, companies using the VBM system should achieve changes in the behavior of managers, i.e. changes that contribute to improving the efficiency of the company and are so highly valued by investors.

Is VBM a clearer criterion compared to such traditional accounting methods of measuring business efficiency as return on assets, profit and profitable growth? Supporters of VBM pay great attention to the disadvantages of traditional methods of calculating the profitability, while it becomes obvious that VBM tools are built on the company's accounting information system, although sometimes with significant modifications.

2 Methodology

The first of these issues has become the subject of research on the predictability of the common stock price based on the current value of future cash flows. These issues are reflected in the works of economists Bernard [1], Biddle et al. [2], Edwards and Bell [6], Gilson et al. [7], Kaplan and Ruback [10]. These researches are indirect testing of the relationship between VBM and stock prices, and the relationship between a specific VBM criterion and the stock price is not studied.

The second question is the subject of research, the direct purpose of which was to study the relationship between the criteria of VBM and the stock price. The works of Chen and Dodd [3], Kaplan and Rubak [10], Kim and Ritter [11], Kleiman [12] are devoted to these issues.

The third question concerned the impact exerted by VBM systems on various indicators of the company's operational efficiency and the market effectiveness. It was studied by Easton et al. [5], Hogan and Lewis [8].

The latter question is more complicated, since it requires comparing the effectiveness of VBM system with traditional systems based on accounting indicators. Nevertheless, some data concerning the latter issue were borrowed from the accounting literature by Dechow et al. [4], Jackson et al. [9], O'Byrne [14], Ohlson [15].

3 Results

Some researchers in the field of finance and economics, as well as many investors, argue that the equity valuation model using discounted cash flow is incompatible with the volatile nature of the stock market. It is often said that investors attach too much importance to fluctuations in profits over a short period. Such concern about short-term results is reflected in the stock market price, causing its excessive volatility. This phenomenon is not considered something new.

Many financial economists have a different opinion, and believe that it is possible to assess in the stock market the long-term perspective of a company's ability to create future cash flows. Moreover, they claim that when investors react to the profit announcement, they are actually reacting to information about the future, which, in their opinion, is embedded in the reports on current profits. Ultimately, the answers can only be found in the data. Does the valuation of shareholder value using discounted cash flow provide rational and reliable estimates of current market rates? This refers to estimates that are sufficiently accurate and enable the use of indicators of DFC values as a basis for the management of the company in order to create shareholder value.

In this regard, it seems important to consider the results of two studies on the effectiveness of DFC models in forecasting stock prices. Both studies use the Kaplan and Ruback model (1995), which is one of the variants of the DFC valuation model [10]. These studies were chosen as an example of this type of research, because they were conducted carefully and recognized by the scientific community. The authors conducted a study of the reliability aspect of forecasting the stock market price based on discounted cash flow by analyzing the market value of highly leveraged transactions (leveraged takeover and leveraged recapitalization) in comparison with the current value of their respective cash flow forecasts, which was a prototype of this kind of research and identifies the difficulties encountered when trying to verify the validity of the company's valuation model based on discounted cash flow.

The purpose of Kaplan and Ruback's study was to compare the forecasts of the discounted value of cash flow for a sample of 51 highly leveraged transactions with the actual value of the transaction. Here, the dependent variable that they are trying to predict is "the transaction value" (approximately corresponding to the market value of the company), and the independent variable that is used for forecasting is the DFC value of expected cash flows.

The DCF-based transaction valuation is calculated using the compressed adjusted present value (CAPV) model, which is a simplified version of the company valuation model with tax adjustments by Modigliani and Miller [13]. This model calculates the company's cost of capital as follows: the cost of the company's cash flows plus the current cost of interest payments due to tax savings resulting from the company's use of debt financing.

The sample included 51 highly leveraged transactions that took place between 1983 and 1989. The reason for choosing this particular group of companies was the availability of sufficient financial information to conduct a full assessment (the data of the Securities Commission and Stock Market Commission (SEC)). This information included estimated indicators for at least four years after the transaction, such as: (1) operating income before interest, depreciation and taxes; (2) depreciation; (3) planned capital expenditures; (4) estimated changes in net working capital.

Table 1 presents the results of the study, processed with three statistical indicators of forecast accuracy for three different DCF forecasts. These three ways to measure the accuracy of forecasts are in the first column. The first line contains the percentage of forecasts falling within the limits of plus or minus 15% of the market value of the transaction. The last two indicators of forecast accuracy are the mean squared error and the standard deviation. The reason why the average error of forecasts is not simply given is that positive and negative errors are mutually compensated, which will lead to an underestimation of the measured error. Therefore, it is necessary either to calculate the average from the absolute values of the forecast errors in order to estimate the standard deviation of the forecasts, or to square the forecast errors before calculating the average in order to determine the mean squared error of the forecasts. The three DCF forecasts differ in that the beta factor used in the capital asset pricing model (CAPM-based model) is identified. The beta forecast for the company uses the beta factor of the company itself; the beta forecast for the industry uses the average beta factor for the industry; and the beta forecast for the market uses a beta factor equal to 1.

Table 1 The accuracy of market price valuation forecasts by discounted cash flow

Methods of measuring forecast accuracy	Forecast using the company's beta factor (%)	Forecast using the average industry beta factor (%)	Forecast using the market beta factor (%)
Percentage within 15%	47.1	62.7	58.8
Standard deviation	21.1	18.1	16.7
Mean squared error	8.4	6.7	5.1

Source The table was compiled by the authors based on Kaplan and Ruback. The valuation of cash flow forecasts: an empirical analysis. Journal of Finance, 50, 4, pp. 1059–1093 [10]

The results indicate that the methods of determining DCF can be used to predict the actual value of the transaction, but with a significant error. For example, about 60% of DCF forecasts made using market or industry beta factors were within a range of plus or minus 15% of the transaction value. It should be noted that the mean absolute error of forecasts for each DCF model was 20%. In the same way, if we calculate the root of the mean squared error, we will see that the average error of forecasts is in the range from 22.5 to 29%. Consequently, it turns out that DCF estimates are acceptable, but DCF estimates differ from the transaction value with an average error of about 20%.

Gilson et al. [7] applied the CAPV model of Kaplan and Ruback to estimate the value of sixty-three open joint stock companies that arose as a result of the adoption of Chap. 11 (Bankruptcy Code, USA), and the value included in the forecasts with the help of cash flows in their reorganization programs. The research data indicate some limitations of the CAPV model in this case.

In any empirical study, the method used to select the companies included in the sample is important, since it is the method that determines the applicability of the research results. In this study, the authors started with a list of 1,342 applications on Chap. 11 submitted between 1973 and 1993. They determined that 377 companies by December 1993 had become open joint stock companies in accordance with Chap. 11. The shares of 134 companies from this group were listed on the NYSE, AMEX or Nasdag, and researchers were able to obtain official information about the final approved reorganization plans of 104 companies. The final sample of 63 companies includes all companies for which it was possible to obtain a cash flow forecast after reorganization for at least two years.

The value of the companies was estimated using the CAPV model of Kaplan and Rubak, as well as on the basis on the ratio of total capital and EBITDA (profit excluding taxes and depreciation) in the first year of the forecast for comparable companies. The analysis showed that the estimation errors that the authors calculate for both methods are considerable, and the errors are much higher than in the study of transactions of highly leveraged transactions by Kaplan and Ruback [10]. Nevertheless, the study still provides some data in support of the use DCF valuation method, which consists in the fact that CAPV estimates showed smaller error in forecasts than comparable valuation methods.

Unfortunately, the prediction errors for both methods were really large. For example, only 25.4% of forecasts with the use of DCF method fell into the range of plus or minus 15%, compared with 50–60% in the study of highly leveraged transactions (HLTs) of Kaplan and Ruback (1995). In addition, the mean deviation and the mean squared error in the study of the value of bankrupt companies were significantly higher than in the study of highly leveraged transactions.

In this regard, we should consider the study of Dechow et al. [4], recognizing that there are many other studies. This literature uses a specific variant of the DCF model called residual income. This model is based on three main assumptions. Firstly, the stock price at a particular time is equal to the current value of the expected future dividends. Moreover, the dividends should be adjusted for the risk, the level of profitability of which is required by the company's investors (it is assumed that it is unchanged for all future periods). The second assumption is called the "clean surplus accounting relation", according to which all items of income and expenses that affect the book value of the company's equity pass through the company's income and expense statement. Attention to this model in the accounting literature is associated with the inclusion of net profit and book value of equity in it.

Dechow et al. [4] evaluate the predictive ability of several versions of this model. They differ in how the time series of profit deviation develops over time. The study contains the relative average deviation of forecasts for two variants of the four models studied by the authors. These four models differ in indicators to assess the process of profit deviation (related to the existing trend), and the first set of models does not take into account external sources of information, which are different from those contained in the time series of profit deviations, while the second set includes other sources of information.

The results of the checks indicate that, on average, the absolute value of the forecast error is approximately 50% of the projected stock price. Moreover, the authors claim that their models more likely understated the cost of their own equity in relation to the securities market. Finally, they claim that the accuracy of their estimates is only slightly higher than the accuracy of past empirical studies using the results of analytical profit forecasts in conjunction with the traditional dividend discount model.

4 Conclusion

The constant growth in popularity of cost-based management systems testifies to its effectiveness. The recognition of VBM by manufacturing enterprises and numerous financial publications devoted to this issue are conclusive evidence to support it. After all, some industrial magnates are among the most ardent supporters of VBM. Nevertheless, it seems possible to single out three fundamental questions related to this controversy, the answers to which are not obvious at all. The following list provides a brief summary of the data concerning each of them.

The first question concerns whether the discounted cash flow valuation theory provides reliable estimates of stock prices. All VBM methodologies are based on the belief that the answer to this question is affirmative, and there is evidence to support such an opinion. However, the facts show that this theory provides very approximate predictions. This means that the discounted cash flow model provides estimates of the value of stocks that are fairly accurate, but have a significant forecast error. In this regard, the question arises: are the discounted cash flow models of equity sufficiently correct? Are VBM methods suitable for managing a company in order to create shareholder value?

The second question is: Do the VBM criteria provide acceptable forecasts of the current market value of common stock? When considering the available data, it is clear that most of the research conducted by consulting companies on this issue can be misleading. Too often, the criteria for measuring financial efficiency for a particular period are used as the only indicator of the stock price, even though it not consistent with the fundamental provisions of the discounted cash flow model. In fact, it would be surprising if the results of a company's activity for a particular period were enough to predict its value. The discounted cash flow theory, on which VBM methods are based, states that the value of a company is equal to the present value of all expected future cash flows.

The third question is: Does VBM have an impact on the activities of companies using this approach? There is evidence that companies implementing VBM system are really changing the way they manage their assets. In particular, companies using systems based on residual income or EVA often sell off or liquidate underutilized assets in an attempt to increase EVA. Such actions, of course, can have the positive impact on the stock price if the company has made excessive capital investments. Nevertheless, the question arises about what the company should do when these opportunities are exhausted.

And yet, recent studies of the performance of companies implementing VBM over a long period do not confirm significant differences in efficiency between companies that have introduced VBM and those that do not use it. No matter how disappointing this result may be for VBM supporters, one thing is certain: this comparison cannot assess what the efficiency of companies would be now if they had not decided to use VBM. In other words, if companies had not implemented VBM, things could be even worse now.

References

1. Bernard, V. L. (1994). Accounting-based valuation methods determinants of market-to-book ratios and implications for financial statement analysis. Working paper, University of Michigan, pp. 7–19.
2. Biddle, G. C., Bowen, R., & Wallace, J. S. (1997). Does EVA beat earning? Evidence on associations with stock returns and firm values. *Journal of Accounting and Economics, 24*(3), 275–300.
3. Chen, Sh., & Dodd, J. L. (1998) Usefulness of accounting earnings, residual income, and EVA: a value-relevance perspective. Working paper, Drake University, Iowa.
4. Dechow, P. M., Hutton, A. P., & Sloan, R. G. (1999). An empirical assessment of the residual income valuation model. *Journal of Accounting and Economics, 26*, 1–34.
5. Easton, P. D., Harris, T. S., & Ohlson, J. (1992). Aggregate accounting earnings can explain most security returns: The case of long return intervals. *Journal of Accounting and Economics, 15*(2–3), 119–142.
6. Edwards, E. O., & Bell, P. W. (1961). *The theory and measurement of business income*. Los Angeles, London, University of California press.
7. Gilson, S. C., Hotchkiss, E. S., & Ruback, R. S. (2000). Valuation of bankrupt firms. *Review of Financial Studies, 13*(1), 219–241.
8. Hogan, C. E., & Lewis, C. M. (2000). The long-run performance of firms adopting compensation plans based on economic profits. Unpublished manuscript. Owen Graduate School of Management. Vanderbilt University Nashville.
9. Jackson, A., Mauboussin, M. J., & Wolf, C. R. (1996). *EVA primer*. CS First Boston.
10. Kaplan, S. N., & Ruback, R. S. (1995). The valuation of cash flow forecasts: An empirical analysis. *Journal of Finance, 50*(4), 1059–1093.
11. Kim, M., & Ritter, J. R. (1999). Valuing IPOs. *Journal of Financial Economics, 53*, 409–446.

12. Kleiman, R. (1999). Some new evidence on EVA companies. *Journal of Applied Corporate Finance, 12*, 80–91.
13. Modigliani, F., & Miller, M. H. (1963). Corporate income taxes and the cost of capital: A correction. *The American Economic Review, 53*(3), 433–443.
14. O'Byrne, S. (1996). EVA and market value. *Journal of Applied Corporate Finance, 9*(1), 116–126.
15. Ohlson, J. A. (1990). A synthesis of security valuation theory and the role of dividends, cash flows and earnings. *Contemporary Accounting Research, John Wiley & Sons, 6*(2), 648–676.

Logistics Costs in Retailing Through the Prism of the Terms "Costs" "Expenditures," and "Expenses"

Olga S. Glinskaya, Raisa V. Kalinicheva, Ivan A. Chusov, Elena A. Ozornina, and Irina S. Djararah

Abstract Retail trade is one of the most developing sectors of the economy, the historical roots of which go back to the antiquity of humankind. Contemporary commerce, which is in dynamic development, constantly exists in a competitive struggle for a buyer. Retail chains use market research to manage sales. It should be noted that the purchase of goods is influenced by supply, demand, and pricing policy. The price of a retail product has several "added values" in the process of its formation. Reducing prices to increase purchasing power is one of the economic problems of chain retailers. The pricing policy is influenced by such a category as "logistics costs." On the one hand, it is quite clear what the logistics costs are—these are the costs associated with delivery, storage, pre-sale preparation, and so on. On the other hand, there is a controversy between the concepts of "costs," "expenses," and "expenditures." This paper focuses on the category "logistics costs" because managerial decisions in the field of logistics are of great importance, in which the correct interpretation of "logistics costs" to determine their place in the accounting trade organization is an important aspect. The research is based on the regulatory framework in the field of trade, accounting, and tax accounting, as well as approaches to the classification of logistics costs as applied to retail chains. Accounting for logistics costs and control over them is necessary for the successful functioning of retail chains, which should contribute to the sustainable development of the distribution of goods.

O. S. Glinskaya (✉) · R. V. Kalinicheva · I. A. Chusov · E. A. Ozornina · I. S. Djararah
Volgograd Cooperative Institute (Branch) of the Russian University of Cooperation, Volgograd, Russia
e-mail: oglinskaya@ruc.su

R. V. Kalinicheva
e-mail: rkalinihteva@ruc.su

I. A. Chusov
e-mail: chusov.ivan@mail.ru

E. A. Ozornina
e-mail: ozornina.70@mail.ru

I. S. Djararah
e-mail: idzhararah@ruc.su

A. V. Bogoviz (ed.), *Big Data in Information Society and Digital Economy*,
Studies in Big Data 124, https://doi.org/10.1007/978-3-031-29489-1_13

Keywords Retail · Logistics costs · Costs · Expenses · Sales management

JEL Classifications D80 · L81 · M49

1 Introduction

The trade industry significantly impacts the social and economic development of the country and its regions. Being a current format of trade, the retail chain has a favorable effect on the development of Russia's economy.

Logistics has a great influence on the sustainable development of retailers. Business processes in retail logistics aim to optimize the movement of goods to increase the competitiveness of retail chains and ensure their successful operation. Logistics structures of retailers solve many professional tasks, which are part of the logistics chain of the distribution of goods. The main functional tasks of the logistics structures of retail chains include the following:

- Arrangement of transportation;
- Storage mode according to technical requirements;
- Organization of quality storage;
- Market monitoring;
- Preparation of a product for sale;
- Other.

The performance of each logistics function requires a certain cost; it directly affects the profitability of the trading business.

The activities of retail chain organizations depend on the well-organized management of logistics costs [6]. Effective retail management in the form of retail chains becomes difficult without understanding and correct application of the terms "expenses," "costs," and "expenditures."

Publications of many authors who have devoted their research to the issue of interpreting the economic concepts of "expenses," "costs," and "expenditures" do not provide a unified answer. Some researchers consider "expenses" and "expenditures" as synonyms; others see a fundamental difference and distinguish the terms. Based on an array of information contained in the normative legal acts and educational and scientific literature, let us analyze the terms "expenses," "costs," and "expenditures."

2 Materials and Methods

The concepts of "expenses," "costs," and "expenditures" are reflected in the normative legal acts. Article 320 of the Tax Code of the Russian Federation specifies the procedure for determining the costs of trade operations: "taxpayers engaged in … retail trade, form the expenses of sales (hereinafter the costs of circulation), considering the following features." Thus, the regulatory documents use the terms "expenditures," "expenses," and "costs of circulation" as synonyms.

According to the Regulations on Accounting "Organization expenses" (PBU 10/99), the organization's expenses are a decrease in economic benefits resulting from the disposal of assets (money or other property) or the creation of obligations, leading to a decrease in the capital of the organization, except for the reduction of contributions by decision of the participants (property owners) [1].

According to various normative legal acts, these concepts are defined as: "expenses—… expenditures (losses) …," "expenses …—(costs of circulation) …," "circulation costs—… cost assessment …," "… expenses (circulation costs) …," "labor expenses—labor costs …"; in this regard, we can conclude that the legislator does not see a significant difference between these concepts, and, from the legal point of view, their identity is traced.

According to a large accounting dictionary, costs are the amount spent on something, expenses; elapsed expenses, which include the production costs of using products and services in the production of final products; expenses relating to products already sold and the corresponding sales revenue, etc. [3, p. 377].

Raizberg et al. determine expenses as the monetary expenditures of enterprises, entrepreneurs, and private producers for the production, circulation, and sale of products [14].

Based on the above, we can note that the concepts of "expenses," "costs," and "expenditures" are mostly characterized as synonyms. The definitions given in the dictionaries basically contain such words and phrases as "payments," "costs incurred in the process of…," "monetary expenses," "costs," "expenditures," "monetary form of expenses," and "amount spent on something."

Let us explore the similarities and differences of these concepts presented in the works of Russian and foreign authors.

In the book "Management accounting," Horngren, Foster, and Datar consider costs as the amount of money paid for purchased goods or services [8, p. 61].

An American researcher R. Anthony relates expenses to the cost of production—"a decrease in assets or an increase in liabilities." According to Anthony and Reece, expenses are resources used to earn income in the current period [2, p. 46].

According to Drury, the cost is an oft-used word that describes the monetary measurement of all resources used in the process of acquiring goods or services [4, p. 33].

According to Hendricksen and Van Breda, expenses represent the use of goods and services in generating income [7, p. 243].

Interpretations given by foreign researchers have certain points of contact. Thus, the costs and expenses are a cost or set of resources used to generate income. According to Anthony, costs are a "reduction of assets." In Russian sources, the decrease in assets is associated mainly with expenses.

Kuter, a famous scholar in the field of accounting, believes that "costs and expenses are related by the fact that both are accompanied by a disposal of assets," but the causes of liabilities are different. However, costs are not associated with a decrease in economic benefits of the economic entity because, when the cost occurs, there is the withdrawal of an asset. Instead, the organization receives another asset, without which it is impossible to carry out business activities, equal in amount to the withdrawn one.

Professor Sokolov, the Honored Scientist of the Russian Federation, believed that "expenses" are characterized by uncertainty; synonyms of "expenses" are "costs," "expenditures," and "losses." "Expenses"—"are something that reduces an asset or increases a liability"; "costs"—"are a portion of the expenses that will become such in the next reporting periods"; "expenditures"—"are part of the costs associated with the processes of selling works and services (costs of circulation)" [15, p. 446, 447]. In his book "Fundamentals of accounting theory," Sokolov noted that "in the Russian language, at least three words have exactly the same meaning: expenses, costs, and expenditures; attempts to distinguish their content are meaningless, these terms reflect actions" [15, p. 171].

In our opinion, the definition of costs given by Sokolov accurately and capaciously characterizes the essence of costs as an economic category.

Professor Manyaeva believes that "the concept of 'organization costs' in financial accounting should be identified with expenses on ordinary activities" [12, pp. 61–62].

Consequently, all expenses for the consumption of various resources required to implement trade activities can be considered costs, which are transformed into expenses and are written off in the reporting period after the sale of goods.

Koske and Mishuchkova conducted a comparative analysis of the use of the terms "costs" and "expenses" in Russian accounting standards and revealed that there are no unified approaches to the definition of costs. The term "costs" is used in 12 accounting regulations; in some regulations, the use of the term "costs" exceeds the use of the term "expenses." The authors note that the concept of "costs" should be fixed at the level of legislation because the interchangeability of these terms can lead to a distortion of their meaning [11, p. 55].

Considering the judgments of Russian researchers, we note that costs can be classified into incoming, which is a monetary expression of acquired resources, and outgoing (expired)—spent resources that are written off as expenses of the reporting period. Costs are identified with the resources necessary for a business entity to carry out its activities.

For example, Vrublevsky considers "expenses" as a generalizing term that integrates the costs and the expenditures of production of a certain (reporting) period [17, p. 51].

The analysis shows that some scientists understand the concept of "costs" broader than the concept of "expenses," but these categories may be equal in total terms in some moments.

Professor Kerimov understans expenses as financial flows and their movement in the organization. Thus, expenses are all payments for resources that are made by business entities during their activities; made payments are recognized as expenses only at the time of the sale of products [9, p. 58].

Khoruzhiy and Yusupova differentiate the concepts of "costs" and "expenses." Under costs, they understand the cost estimate of all resources consumed by the company. Expenses are considered as the reduction of economic benefits. The authors emphasize that "costs are the use of resources, and expenses are when resources leave the company" [18, pp. 25–31].

According to professor Druzhilovskaya, costs are "a decrease in economic benefits resulting from a decrease in assets or an increase in liabilities, expressed as a decrease in equity not related to transactions with owners." In a later work, the author indicates that the cost is "the monetary evaluation of the use of material, labor, and other resources, leading to an increase in assets in the case of acquisition for payment or production of these assets" [5, p. 30; 37].

Professor Kondrakov believes that the concepts of "costs" and "expenses" should be distinguished. He considered costs in terms of the cost of resources and the expenses according to the concept reflected in the PBU 10/99 [10, pp. 142–143].

Costs are payments for the use of resources; they become expenses only in the structure of the cost of goods sold. In turn, expenses are associated with the moment of sale [16, pp. 241–245].

In "Basic approaches to the interpretation of cost terms," Professor Mizikovsky and Ulyasheva classify costs as an assessment of resources used by an organization to acquire an asset, while expenses are the withdrawal of assets from the organization's balance sheet [13, p. 211].

The idea of distinguishing between the concepts of "costs" and "expenses" is held by many Russian scientists who believe that the boundary between these economic terms is fundamental. The following summarizes the opinions of the authors presented above: costs are interpreted from the perspective of the cost of various resources used during the entrepreneurial activity of the business entity; expenses are regarded as an outflow of assets, a decrease in economic benefits associated with the sale of assets.

3 Results

Thus, costs are characterized by the use of various resources, the amount of which is expressed in monetary terms; costs relate to certain goals and objectives of the economic entity. In turn, the costs in the process of their consumption (sale of goods) are expired and charged to expenses of the reporting period; the same costs that have not been consumed remain in the form of an asset (inventory) of the organization.

The terminology that is given in the regulations shows that expenses can mean costs, and costs can mean expenses. However, these definitions can be distinguished in terms of management accounting and the time factor (expense recognition). Within the framework of management accounting, each business entity has the opportunity to develop its own classification of costs, methods of control, and ways to minimize costs. The basis for distinguishing between costs and expenses is the time factor. Thus, all resources used by the retail chain to carry out its activities are costs (assets); at the moment of sale of goods, costs are transformed into expenses, and part of the unsold goods remain as stocks (assets).

In our opinion, it is possible to consider expenses as part of the cost, the part that is completely consumed within a certain period to generate income. Costs in the retail network are the monetary value of purchased goods for subsequent resale to the retail customer; part of the costs incurred at the time of the sale will be written off as an expense.

Currently, there is no consensus on the identification or fundamental distinction between the studied concepts. Some prominent authors consider these terms to be synonymous and interchangeable, while others see them as essentially different.

The controversy over the economic concepts of "expenses" and "costs" does not cease even nowadays. This fact indicates the relevance and importance of the problem of normative consolidation of the concept of "costs" as opposed to the current concept of "expenses."

4 Conclusion

Based on the research, we can state that it is advisable to conduct analytical accounting of logistics costs according to the nomenclature of logistics costs, which can be formed by each retail network according to their needs. This will allow making prompt management decisions to reduce costs and increase revenues of a retail chain, which, in turn, will enhance competitiveness and sustainable development.

References

1. Ministry of Finance of the Russian Federation. (1999). Regulation on accounting "Organization expenses" PBU 10/99 (adopted May 6, 1999 No. 33n). Moscow, Russia. Retrieved February 18, 2022, from http://www.consultant.ru/document/cons_doc_LAW_12508/0463b359311dddb34a4b799a3a5c57ed0e8098ec/
2. Anthony, R. N., & Reece, G. S. (1993). *Accounting: text and cases* (A. M. Petrachkov Transl. from English; Ya. V. Sokolov Ed.). Finance and Statistics. (Original work published 1989)
3. Azrilian, A. N. (Ed.). (1999). *The great accounting dictionary*. Institute of New Economics.
4. Drury, K. (2003). Management accounting for business decision: A textbook (Transl. from English). UNITI-Dana. (Original work published 2004).
5. Druzhilovskaya, T. Y. (2007). On the question of the relationship between the concepts of "costs" and "costs." *Vestnik KG IPPE, 2*(7), 28–30.
6. Glinskaya, O. S., & Skorikova, I. S. (2011). Classification of logistics costs in retail chain organizations. *Audit and Financial Analysis, 2*, 48–51.
7. Hendriksen, E. S., & Van Breda, M. F. (1997). *Accounting theory* (I. L. Smirnova Transl. from English; Ya. V. Sokolov Ed.) Finance and statistics. (Original work published 1992).
8. Horngren, Ch., Foster, J., & Datar, Sh. (2005). *Cost accounting: A managerial emphasis* (Transl. from English) (10th ed.). St. Petersburg, Russia: Peter. (Original work published 2000)
9. Kerimov, V. E. (2001). *Theory and practice of organizing management accounting at industrial enterprises*. ITC "Marketing."
10. Kondrakov, N. P. (2013). *Management accounting: A textbook*. INFRA-M.
11. Koske, M. S., & Mishuchkova, Yu. G. (2015). Expenses in the Russian accounting: Economic substance and statutorization. *International Accounting, 32*, 51–63.

12. Manyaeva, V. A. (2010). Classification of the organization's expenses in strategic management accounting. *Problems of the Modern Economy, 2–3*, 61–66.
13. Mizikovsky, I. E., & Ulyasheva, L. G. (2016). Basic approaches to the interpretation of cost terms. State and municipal management. *State and Municipal Management. Scholar Notes, 3*, 206–212.
14. Raizberg, B. A., Lozovsky, L. Sh., & Starodubtseva, E. B. (Eds.). (2010). *Modern economic dictionary*. INFRA-M.
15. Sokolov, Ya. V. (2003). *Fundamentals of accounting theory*. Finance and statistics.
16. Sytnik, O. E., & Ledneva, Yu. A. (2009). The economic essence of the categories "costs", "expenses", and "expenditures" and their industry characteristics. *Bulletin of the North Caucasus State Technical University, 4*, 241–245.
17. Vrublevsky, N. D. (2002). *Management accounting of production costs: Theory and practice*. Finance and statistics.
18. Yusupova, A. V., & Khoruzhy, L. I. (2012). Differences between the concepts of "expenditures", "expenses", and "costs", their reflection in Russian and international accounting standards. *Accountant's time, 11*, 25–31.

Internal Control in the Integrated System of Accounting and Analytical Information Formation

Vera V. Darinskaya, Irina V. Bratko, Natalia S. Demidova, Gulzira U. Bekniyazova, and Elena V. Ponomarenko

Abstract Significant changes in information support of economic entity management system are characterized by a new quality of collection, recording, analysis, and interpretation of obtained data. Nowadays, it is appropriate to speak of integrated data formation and utilization system, which went beyond the limits of accounting (financial, management) records and in full includes the data to provide all management functions. Accounting and analytical information in any required classification should not only have a retrospective orientation but be able to respond to the needs of operational and strategic planning. The major challenge in the successful functioning of the system of formation of accounting and analytical information is the effective connection of management technologies with information systems used by the economic entity. Therefore, integrated systems become essential, i.e., those that meet multiple management needs and consequently combine (fully or partially) the functions of different information programs. The availability of advanced, perfect systems for information collection and processing that fully meet the management needs of the economic entity is not a guarantee of their successful operation, improvement of the organization efficiency, prevention of negative situations, etc. Managing the work of the structural units of the organization in the new conditions of information support, their adaptation to the changing principles of interaction within the company, the continuous improvement of the organizational structure of the economic entity are the tasks, the solution of which determines the effectiveness of

V. V. Darinskaya (✉) · G. U. Bekniyazova
Russian University of Cooperation, Mytishchi, Russia
e-mail: v.v.darinskaya@ruc.su

G. U. Bekniyazova
e-mail: tillabek@bk.ru

I. V. Bratko · N. S. Demidova · E. V. Ponomarenko
The Military University of the Ministry of Defense of the Russian Federation, Moscow, Russia
e-mail: tatarnova@yandex.ru

N. S. Demidova
e-mail: tasik19852008@rambler.ru

E. V. Ponomarenko
e-mail: hell40@rambler.ru

A. V. Bogoviz (ed.), *Big Data in Information Society and Digital Economy*,
Studies in Big Data 124, https://doi.org/10.1007/978-3-031-29489-1_14

the economic entity at the present time. The implementation of the control function of management in large organizations with an established and well-functioning structure is of interest. For several decades, many large and medium-sized companies have created services responsible for internal audits. The position of such a structure in the integrated system of formation of accounting and analytical information will be illustrated by a Russian aviation company, which produces alighting gear, hydraulic aggregates, and systems of advanced airplanes and other aircrafts of all types.

Keywords Integrated system of formation of accounting and analytical information · System of internal audit · Risk management and monitoring · Accounting monitoring

JEL Classification M2

1 Introduction

Among the existing accounting models, the most complete, rigidly structured, and systematized is the continental one, which is the basis of Russian accounting. The main factor influencing the development of system accounting is its purpose for making management decisions both at the level of the economic entity and external interested users. Nowadays, within the situation of instability and uncertainty, the integrity, unity, completeness, and versatility, established in the Russian accounting system, which is able to provide information to government institutions and internal users with the necessary depth of detail, acquire particular relevance. Efficient use of resources is highly demanded in the present economic conditions in the country and the world. Nevertheless, we should remember the value-based management, which is not only an increase in business worth but an increase in its value to the owners and consumers. Necessary parameters can be presented in the accounting system in the form of key performance indicators and constantly monitored, assisting in making timely and informed management decisions [1].

The study showed that integrated accounting is not the provision of information to financial management but the communication function of management, aimed at managing financial flows, the dynamics of assets, business transactions, etc.

We should consider the peculiarity of the present day, when planning, as a function of management, is increasingly focused on financial indicators rather than on general economic indicators. Technical and economic indicators are of secondary importance; the first and foremost is the financial result. When making management decisions, the focus is on the prospective analysis and economic situation forecasting at the same time. Apparently, the general instability of the situation does not suggest that the performance targets will be achieved; their main purpose is the possibility of rapid reorientation and adaptation of the economic entity to the changing economic environment [2].

All available resources of the organization, including the established and successfully operating system of internal audit, should be used to the maximum extent possible. Internal audit as part of the integrated system of accounting and analytical information will allow timely tracking and identifying causes of deviations in key performance indicators, focusing on its main task of confirming the data in the financial statements.

2 Methodology

During the study, the theoretical basis was the scientific papers of Russian scientists devoted to the internal control system as an element of strategic management accounting. The study focuses on the connection between the information architecture of the economic entity and its strategy, the use of information technology in manufacturing facilities.

The works of foreign scholars concentrate on the transformation of traditional employee behavior into a strategic advantage to the company [3].

The empirical research method consisted of observing the organization of work and operation of the accounting and analytical systems of enterprises belonging to the aviation concern. The empirical method also consisted of the study of an integrated system of formation of accounting and analytical information, carried out to determine its unknown features, undeveloped qualities, and missing elements [4]. Overall, this should help to confirm the correctness of the assumption that the existing internal control in the integrated system of formation of accounting and analytical information is important and can be quickly adapted to the requirements of the present day.

The results of the experimental situation modeling of the actualization of the internal control services within the integrated system of forming accounting and analytical information were confirmed by practical actions at the enterprises belonging to the aviation concern.

The analysis and synthesis of the results of the theoretical and practical part of the study, comparison and generalization of the data obtained allowed us to prove the correctness of the assumptions on the possibility of effective use of existing structural units of internal control made at the beginning of the paper.

3 Results

Large companies may have several structural units that perform audit functions within the economic entity. The implementation of the management audit function in the studied organization is distributed among three services. A comparative characterization of the units that carry out the audit functions of the production enterprise, which supports the conclusions of the study, is available at https://figshare.com/ with the identifier https://figshare.com/s/702cbf7b1e1ea1d9b2c1.

The orientation of planning on financial indicators and financial results in general leads to an increasing role of the internal audit service (internal financial control). In the studied organization, this structural subdivision mainly satisfies information requests of top managers and business owners [5].

The effectiveness of this structural unit can be assessed by analyzing the operational effectiveness of internal financial controls implemented through:

- A sample of operations in the course of which operational efficiency testing is performed;
- Development and approval of a testing plan, including identification of specific testing methods;
- Comparison of the results of the effectiveness of internal financial control testing.

The sample of operations is selected in accordance with primary accounting documents and accounting registers for accounts.

The employee testing the control procedure independently sets the sample size and the source of its formation. Nevertheless, there is a certain dependence of the sample size on the frequency of the control procedure, which is established in the local normative acts of the organization [6]. The dependence of the sample size on the frequency of the internal financial control procedure, which confirms the conclusions of the study, is available at https://figshare.com/ with the identifier https://figshare.com/s/a7aa747a74059b93128a.

The Internal Audit Service assesses the probability of the risk of misstatement of the data presented in the financial statements (accounts) (FS). The analyzed organization uses a number of levels when misstatement is nearly impossible, possible, or exists as an exception.

To determine the probability value of each of the levels listed above, an appropriate table is used (Fig. 1).

It should be noted that a specific testing method is chosen according to the set goals and objectives of the internal financial control procedure.

The analyzed corporation uses the methods shown in Fig. 2.

As part of the assessment of the probability of formation of the risk of misstatement of the financial statements of the analyzed organization, the following scale of assessment is used:

- High risk;
- Low risk.

The most frequent factors leading to misstatements in the financial statements of the analyzed organization are as follows:

- Manipulations resulting from the method of sales revenue recognition;
- Possible disregard by the management of certain established standard procedures;
- Processing by non-automated information systems, a part of operations subject to automatic processing;
- Incorrect reflection of complex non-standard events and transactions;
- Distortion of the scale of operations;

Significance of misstatement risks in FS				
	Critical	Average risk of misstatement in FS		
	Medium	Low risk of misstatement in FS	Average risk of misstatement in FS	
	Minimal	Low risk of misstatement in FS	Low risk of misstatement in FS	Average risk of misstatement in FS
		Nearly impossible	Possible	Most definitely
		Probability of the risk of misstatement		

Fig. 1 Table for assessing the overall level of risk of misstatement in financial statements. *Source* Compiled by the authors

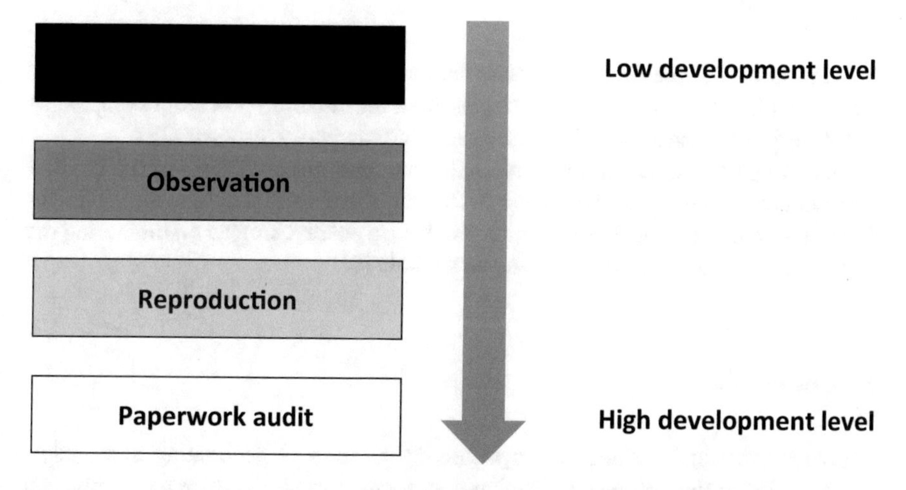

Fig. 2 Testing methods within the internal control system. *Source* Compiled by the authors

- Accounting entries that are based on the judgements and estimates of individual officers, particularly in relation to provisioning and write-off of receivables;
- Failure to take timely account of changes in domestic and foreign legislation;
- Staff turnover in the positions of highly qualified specialists, including managerial personnel.

Once the risk assessment is completed, the results are coordinated with the Accounting Process Control Department (APCD). Then, in accordance with the results obtained, a plan for the formation and subsequent implementation of new financial control procedures is agreed upon, which should compensate for the risks of misstatement of the corporation's financial statements that have been identified. The Control and Risk Management Department is also notified, and joint work is carried out in terms of business process improvement [7].

It is debatable whether it is appropriate to have three structures in a company performing control functions. Their functionality and orientation of received information are different. Nevertheless, from the point of view of optimizing the organizational structure of an economic entity, it is possible to redistribute tasks of internal control service between other subdivisions [8].

An additional argument for the importance of the internal financial control service is the presence of state defense orders and the production of products that fall into the category of weapons and military objects. Therefore, internal financial control and audit are carried out, taking into account the system of confidentiality.

In particular, internal financial control over most intangible assets (objects) is protected by national security information, such as:

- Research and development costs;
- Patents;
- Industrial prototypes;
- Flight models.

In its turn, this creates significant difficulties in the formation of the staff of the internal control service (admission to classified information) and strict compliance with a certain sequence of actions, accuracy, and increased responsibility.

Analysis of the effectiveness of internal financial control at the enterprise enables more accurate planning of the system of measures that will maximize the effectiveness of internal financial control, which will have a positive impact on improving the financial performance of the company as a whole [9].

4 Conclusion

Internal control in the integrated system of formation of accounting and analytical information remains a significant area of focus, especially given the company's specifics. It allows supplementing the integrated accounting system with necessary information focused on a wide range of issues related to the validity of financial reporting data, the appropriateness of internal processes, and calculations of the effectiveness of business processes.

Nevertheless, the constant improvement of the used management technologies and information systems requires regular updating of the organizational structure, leading to changes in the performance targets not only as a result of the external environment.

Integrated accounting and analytical systems that meet several requirements of management and, as a consequence, fully or partially combine the functions of different information programs and structural units of the company can qualitatively and fully meet the information needs in terms of financial performance indicators, from those presented in the accounts to those characterizing the increase in business worth, increasing its value to owners (shareholders) and consumers [10]. The flexibility of the existing accounting system in the country allows performing this in the most optimal way with minimal cost.

References

1. Basovsky, L. E. (2013). *Modern strategic analysis*. INFRA-M.
2. Borovitskaya, M. V. (2018). Internal control system as an element of strategic management accounting. *Azimuth of Scientific Research: Economics and Management, 7*(1), 61–65.
3. Danilin A., & Slusarenko, A. (2009). Architecture and strategy. In *The "Yin" and "Yang" of enterprise information technology*. Internet University of Information Technology.
4. Karpova, T. P., & Karpova, V. V. (2017). Internal control in the economics of corporations. *Accounting, Analysis and Audit, 4*, 56–66. https://doi.org/10.26794/2408-9303-2017-4-56-66
5. Kryatova, L. A. (2010). Internal audit in the system of internal control of public catering organizations. In *Socio-economic problems of the cooperative sector of the economy: Materials of the III International Scientific and Practical Conference of Young Scientists, Teachers, Employees, Postgraduate Students and Applicants*. Mytishchi, Russia: Russian University of Cooperation.
6. Laro, W. (2009). *Office-Kaizen: Transforming office operations into strategic advantage*. Grevzov Publisher. (Original work published 2002).
7. Meyer, K., & Davis, S. (2007). *Living organization* [It's alive: The coming convergence of information, biology, and business] (A. Stativka Transl. from English). Dobraya kniga. (Original work published 2003).
8. Mintzberg, H. (2011). *Act effectively! Best management practices* [Managing] (V. Kuzin, & A. Stativka Transl. from English). Piter. (Original work published 2009).
9. Pisarenko, A. S. (2016). The role of accounting and analytical information in the strategic management of cash flows of the organization. In *Accounting: Achievements and Scientific Prospects of the XXI century: Proceedings of the International Scientific-Practical Conference of the Accounting Department* (pp. 487–490). Kantsler.
10. Williamson, O. E., & Winter, S. G. (2001). *The nature of the firm* (M. Ya. Kazhdan Transl. from English; V. G. Grebennikov Ed.). Delo.

Application of Big Data in Digital Business by Sectors

Methodological Tools for the Generation and Revenue Accounting in a Construction Company in Accordance with International Standards

Tatyana V. Bulycheva, Irina V. Volgina, and Tatiana V. Zavyalova

Abstract The order of formation and recognition of revenue in a modern company plays an important role. Revenue expresses the level of economic performance of the company. A radical change in the system of standards regulating the procedure for accounting for revenue forces the specialists of the accounting apparatus to change the current approaches to revenue generation. The main purpose of this article is to consider various situations of revenue recognition by construction companies under the current version of IFRS 15. Particular attention is paid in the article to the harmonization of approaches of domestic practice to the generation of revenue in accordance with the norms of IFRS. In the article, two methods are given for calculating the share of revenue as obligations are fulfilled—according to the resources spent and the results obtained. In the course of the research, the following methods were used: synthesis, abstraction, classification, analysis, modeling. Within the framework of IFRS 15, a settlement scheme is proposed between developers and shareholders-buyers of the dwelling through escrow accounts. The article suggests the use of several methods of generating revenue for a construction company. The results method or revenue approach, in which revenue is recognized based on estimates of the value of goods, scope of work or services delivered to the buyer or customer on the current date. A formula for calculating the degree of completion of work by the results method in a construction company is proposed.

Keywords IFRS · Revenue · Construction production · Developer · Contractor · Results method · Resource method · Assets · Commitment · Innovative approach

JEL Classification M41

T. V. Bulycheva (✉) · I. V. Volgina · T. V. Zavyalova
Saransk Cooperative Institute (branch) of the Russian University of Cooperation, Saransk, Russia
e-mail: m-tatyana@list.ru

I. V. Volgina
e-mail: ivolgina@ruc.su

T. V. Zavyalova
e-mail: tprnina@rambler.ru

A. V. Bogoviz (ed.), *Big Data in Information Society and Digital Economy*, Studies in Big Data 124, https://doi.org/10.1007/978-3-031-29489-1_15

1 Introduction

"The globalization of international financial markets and cash flows has created strong incentives for regulators to harmonize accounting standards worldwide" [6]. Before the entry into force of the new international standard for revenue accounting, construction organizations had a separate standard—IAS 11. Now they must account for revenue differently: as the obligations of the developer or contractor to the property buyer or the customer of construction work are fulfilled.

The purpose of this article is to consider various situations of revenue recognition by construction organizations under IFRS 15 [1].

To achieve this goal, the following tasks are set:

- to analyze the revenue recognition procedure in accordance with IFRS 15;
- determine the main difference between IFRS 15 from the previous IFRS (IAS) 18 and IFRS (IAS) 11 when calculating the degree of completion of work;
- consider the procedure for recognizing revenue as obligations are fulfilled;
- to propose ways to reflect these operations on accounts.

The contribution to the study of accounting features in construction organizations was made by foreign scientists-economists: King, Varkulevich, Bubnovskaya and others.

At the same time, the issues of revenue generation and accounting and the use of new approaches are relevant to this day.

2 Methodology

In the issue of revenue recognition, IFRS 15 "Revenue from contracts with buyers" pays special attention to the fulfillment of all promises made by the developer to the buyer. For example, it is impossible to formally recognize revenue only on the basis of:

- signed acts of KS-2;
- certificates of KS-3;
- the act of acceptance of the completed works by the acceptance commission represented by Gosstroynadzor (Inspection of state construction supervision) of the city, the developer, the general contractor and the designer, as well as the State Fire Service of the Ministry of Emergency Situations.

In the context of IFRS 15, the scheme of settlements between developers and equity buyers of housing through escrow accounts fits well. It should be reminded since July 01, 2019, the procedure for settlements between property buyers and developers has changed in Russia [2].

Now all the money of the shareholders is stored in escrow accounts in authorized banks. The developer will get access to the buyers' funds only when he fulfills all obligations to future homeowners under the terms of the contract. For example, when the developer will hand over a turnkey object with interior decoration and at least one apartment from the newly constructed apartment block will be registered as the property of the buyer.

Construction is characterized, as a rule, by the long-term nature of the work. As a result of such work, a product in the form of real estate is created. The construction can last more than a year, so it will not be possible to recognize all the revenue strictly on a certain date of the act of acceptance of the completed works until control over the goods is fully transferred to the buyer. This means that it is necessary to recognize the share of revenue as all the obligations of the developer to the buyer are actually fulfilled. This meets criterion "b" or criterion "c" of revenue recognition from paragraph 35 of IFRS 15, namely: in the process of fulfilling an organization's obligation to perform, an asset is created, control over which the buyer receives as it is created; or the fulfillment of its obligation by the organization does not lead to the creation of an asset that the organization can use for alternative purposes, and at the same time the organization has a legally protected right to receive payment for the part of contractual work performed to date.

Even with the phased handover of work, all controls are transferred to the customer not immediately, but partially. Therefore, he recognizes revenue according to the degree of completion of these works and fulfillment of obligations [3]. For example, the contractor handed over to the developer the basement and the first floor of an apartment building. This does not mean that his obligations are fulfilled, because under the contract there may be obligations for installing water supply, sewerage, electricity, etc. In addition, there are transitional preparatory works for the next stage of construction, without which the house cannot be completed. Therefore, in practice, there is a combination of separate obligations to be fulfilled under a construction contract and those obligations that are not separately identified, but without their fulfillment, the object cannot be commissioned (paragraphs 29, 30 of IFRS 15).

In IFRS 15, there is a concept of variable consideration: discounts and other expenses that relate to this concept should be estimated at the expected cost or at the most probable value. Regardless of whether royalties are documented, revenue must be recognized minus the expected or most probable amount of variable consideration that relates to it. (paragraphs 53–57 of IFRS 15).

We will analyze each group of reverse payments under IFRS 15 from the supplier. The main task of such an analysis is to determine whether the group meets the concept of variable consideration and the three criteria for cost analysis: separability and benefit, the relationship between the price of the service and the price of delivery of the goods, the fair value of the service.

Information services are recognized as variable consideration. Usually the cost of these services depends on the volume of supplies, so they cannot be considered separable. In addition, it is difficult to determine the market price of the service, since the service is exclusive. They are taken into account as a decrease in the transaction price.

Logistics services. They are recognized as variable consideration, since they are usually within the framework of a single contract and depend on the volume of deliveries. It is difficult to determine the market price of the service, since the service is exclusive. They are taken into account as a decrease in the transaction price.

Marketing promotion activities. Are recognized as variable consideration. They can be provided both within the framework of a supply contract and as a result of separate contracts. In this case, the service is inseparable from the main delivery contract due to the fact that there is an indirect connection with the volume of delivery, as well as direct inseparability due to the fact that without these measures, the sale of goods on the network is very difficult. They are taken into account as a decrease in the transaction price. However, in this group, it is possible to distinguish, for example, such expenses as participation in exhibitions, which are usually separable and have no relationship with the main supplies, except for increasing the value and recognition of the supplier's goods in this network.

3 Results

To calculate the share of revenue as obligations are fulfilled, two methods are used: according to the resources spent and the results obtained (paragraph 41, B14 19 of IFRS 15).

A method of results or a profitable approach. Revenue is recognized based on estimates of the value of goods, scope of work or services delivered to the buyer or customer on the current date. In Russia, the share of revenue by the results method is often considered by contractors who perform work in stages. If three floors of a 20-storey residential building are built and the cost of such works is calculated according to the acts of KS-2 and KS-3, then the contractor recognizes the revenue not at the time of signing the acts, but at the end of the reporting period in proportion to the amount of these works in total revenue.

The formula for calculating the degree of completion of work by the results method:

$$D = \frac{C}{R \times 100\%} \tag{1}$$

D—% degree of completion;

C—cost of work performed;

R—contract revenue.

Resource method or cost approach. This method of calculating the share of revenue can be found in the accounting of a developer who sells real estate to final customers—individuals [4].

After all, it is impossible to gradually hand over an apartment under construction to a client in an apartment building and sign acts with him at the end of each stage. For them, construction is a continuous process during which resources are consumed.

However, it is necessary to exclude expensive equipment and materials from the calculation of the share of revenue if they were delivered to the construction site, but not installed. This method is also good for renovation of buildings, reconstructions, restoration.

The formula for calculating the degree of completion of work by the resource method:

$$D = \frac{\sum A}{T \times 100\%} \tag{2}$$

D—% degree of completion;

A—actual costs incurred;

T—Total expected costs for the object.

Companies have the right to use any method of calculations. The main thing is that it corresponds to the nature of the contract. In addition, the method should be applied consistently to similar contracts. As a rule, the method of calculating the degree of completion (fulfillment) of obligations to the buyer is fixed by construction organizations in their accounting policy for certain groups of contracts. It is important to remember that the estimate of the revenue share will not correspond to the amounts of invoices issued to the customer during the period.

The main difference between IFRS 15 and the previous IFRS (IAS) 18 and IFRS (IAS) 11 is the exclusion from the calculation of the share of revenue of the cost of services, goods or works over which the supplier has not yet transferred control to the client [5].

For example, with the resource method, it is necessary to exclude from the calculation the costs of purchasing materials that have not actually been used for the construction of a building or the performance of work. IFRS 15 provides detailed recommendations on the calculation of resources by the method when purchasing so-called "uninstalled" materials. The margin from the sale of equipment that has been delivered but has not yet been installed at the customer's facility is excluded from the calculation of the revenue share.

According to IAS 11, such purchases were recognized in the contractor's profit immediately at the time of delivery to the buyer. When using the resource method, such purchases were included in the calculation of the percentage of the degree of completion of the contractor's obligations to the customer, which overestimated the share of revenue in the current period.

In practice, builders use several standard contracts for the construction of the same object or several objects. For example, the construction of a residential building and an underground parking lot for the same customer is executed by different contracts.

IFRS 15 sometimes requires combining such contracts. For example, if contracts for the construction of objects with one customer are made out almost simultaneously, and the amount of compensation to be paid for the construction of one object depends on the amount of payment for another, then such contracts need to be combined.

Contracts are also combined if there is one obligation to fulfill for the construction of two objects. This is the main difference from IAS 11, where contracts were accounted for separately and revenue was also recognized separately.

Example. Accounting of revenue under combined contracts.

The customer engaged a contractor to build a house and an underground parking lot. Two contracts were signed simultaneously. The price of the contract for the construction of a house is 40,000 thousand rubles, the price of the parking lot contract is 4,000 thousand rubles. Parking lot is built immediately with the construction of the house, so the contractor provided a discount for this kind of work. According to experts, in the real estate market, the price of a detached house sale is 40,000 thousand rubles, house detached parking—8000 thousand rubles. Estimated costs under contracts: house—30,000 thousand rubles, parking—2000 thousand rubles. As of 30.09.2021, the cost of the house amounted to 20,000 thousand rubles, for parking—1,000 thousand rubles.

The contractor's specialists have combined the contracts into one and recognize the revenue for two obligations to be fulfilled during the construction of both facilities. To calculate revenue, it is necessary to determine the relative price of a separate sale of each object. By house: 36,667 thousand rubles (44,000 thousand rubles × (40,000 thousand rubles/48,000 thousand rubles)). By parking: 7333 thousand rubles (44,000 thousand rubles × (8000 thousand rubles/48,000 thousand rubles)).

On 30.09.2021, the following transactions will be made in the construction company's accounting (Table 1).

As a rule, within the framework of construction, the same contractors provide an additional set of works and services. For example, cleaning the territory from construction debris, restoring the environment after the construction of a building, landscaping work, etc. All these works and services can be considered as one obligation to perform or divided into several.

IFRS 15 requires construction companies to divide services and work into several obligations to be performed if they are recognized as "distinct". In other words, if

Table 1 Revenue recognition methodology for two liabilities

Accounting	Accounts	Debit	Comment
Revenue, thousand rubles • for the construction of a house • for the construction of a parking lot		(24 445) (3667)	Revenue recognized: House: 24,445 thousand rubles (36,667 thousand rubles × (20,000 thousand rubles/30,000 thousand rubles) Parking lot: 3,667 thousand rubles (7333 thousand rubles × (1,000 thousand rubles/2,000 thousand rubles))
Asset under the contract, thousand rubles	(28 112)		Asset recognized under the combined agreement

Source compiled by the authors

such work is not closely related to the main type of work, such as the construction of an apartment building. The customer may benefit if he hires another contractor for such additional services at lower prices than his contractor. If the result is two or more obligations to be fulfilled, then the proceeds should be divided in proportion to the individual prices of the components of the contract. Each component is a distinct service or job. Each component is a distinct service or job.

If the main construction works are inextricably linked with other types of work, coordinated by a single package of documents, the specifics of their implementation is related to the quality and nature of the construction of the building, then these are indistinguishable services and works that should be recognized as one obligation to perform. For example, the development of a foundation pit, groundwater drainage, concreting and construction of the foundation walls of the building—in this case, it is necessary to take into account the revenue, as in the example of combining contracts.

Example. Revenue accounting for the distinct services and works.

Within the framework of one contract, a construction company performs works on the construction of a residential complex and interior decoration of apartments. The total price of the contract is 200,000 thousand rubles. The price of a separate sale (the market value of services) is 20,000 thousand rubles, and the construction of a house is 120,000 thousand rubles. According to the contract, there are two obligations to be fulfilled. The revenue from the construction of the house and interior decoration services will be recognized as the work is completed. The degree of completion of works on 30.09.2021 for construction is 70%, for interior decorating—40%. In accounting, the contractor will make the following transactions (Table 2).

Usually, the modification of construction contract is associated with a change in the scope of work and prices for types of work. In IFRS (IAS) 11 there was little information about the accounting procedure for the changed terms of contracts. And IFRS 15 provides detailed guidance on how to regard a contract modification as a change to the current contract or as the conclusion of a new one. In addition, most often in IFRS (IAS) 11, the modification of the contract led to the emergence of two contracts with separate recognition of revenue according to the degree of completion of work.

Now, according to IFRS 15, if new volumes of works and services are added to the contract and they are not recognized as distinct, it is possible to continue the previous contract and adjust the amount of revenue on a certain date (paragraph 21b of IFRS 15).

Similar actions are carried out in the case of a change in prices for the previous amount of work. If the prices are agreed with the customer and he is ready to continue cooperation, such modification does not lead to termination of the contract and you can continue the same.

When a new distinct service or work has been added to the contract, and its price is lower than the market price (the price of a separate sale), then you will have to terminate the previous contract and conclude a new one. Accordingly, it is necessary to stop recognizing revenue under the old contract and charge it under the new one (paragraph 21a of IFRS 15).

Table 2 Revenue accounting methodology for the distinct services and works

Accounting accounts	Debit	Credit	Comment
Revenue, thousand rubles The total amount of revenue is: • for the construction of a house; • for the interior decorating works		(120 000) (11 428)	Distribution of the transaction price between the obligations under the contract: by construction of a house: 171 429 thousand rubles. (200 000 = (120 000/140 000)); for interior decorating works: 28,571 thousand rubles. (200 000 × (20 000/140 000)) At the reporting date, revenue from work is recognized as obligations are fulfilled based on the degree of completion of work: • construction of 120,000 thousand rubles (171,429 thousand rubles × 70%); • interior decorating 11,428 thousand rubles (28,571 thousand rubles × 40%)
Asset under the contract, thousand rubles	131 428		The asset under the contract is recognized as the amount of revenue for the construction of a house and interior decorating works at the reporting date

Source compiled by the authors

If the price of a distinct service corresponds to the market price, you should make a separate contract for this type of service. Then, in future periods, revenue will be accounted for a new type of services, works (paragraph 20 of IFRS 15).

Example. Revenue accounting for contract modification.

On December 1, 2020, the developer signed a contract with the contractor for the construction of a suburban cottage settlement. The performance obligations of the builder included the construction of 20 turnkey houses. 01.09.2021 the developer revised the contract and included the construction of a fitness center and the replacement of 50-L boilers in houses with 100-L ones.

The construction of a fitness center is a new distinct work, and the replacement of boilers of another volume in houses is a modernization of the previous work, which is considered indistinguishable. For an indistinguishable boiler replacement service, the contract cannot be terminated, but continue the previous one. The only thing is to adjust the revenue at the reporting date.

The cost of work on the construction of houses is 20,000 thousand rubles. The price of the construction of the fitness center corresponds to the market price—3000 thousand rubles. The cost of work on the replacement of boilers—2000 thousand rubles.

Prior to the modernization of the contract, the degree of fulfillment of obligations under the initial scope of work was 40%. The revenue that was recognized before the modernization, taking into account the degree of completion of the work, is 8000 thousand rubles. (40% × 20,000 thousand rubles). The new revised percentage of completion of work after the modernization of the contract by the method of spent resources was 38%. The revenue to be recognized under the upgraded contract is 8360 thousand rubles ((20,000 thousand rubles + 2,000 thousand rubles) = 38%). It turns out that at the time of modernization of the contract, it is necessary to adjust the previously accrued revenue by making a cumulative adjustment in the amount of 360 thousand rubles (8360–8000).

From the point of view of prices, the cost of constructing a fitness center corresponds to the market one, which means that a new contract must be concluded for its construction and revenue for the new construction object must be recognized separately from 01.09.2021.

4 Conclusion

Although the standard is dedicated to revenue, it also refers to cost accounting, in particular for tenders. Modern construction companies often pay for participation in tenders for government orders. The standard allows you to capitalize the preliminary additional costs associated with the conclusion of the contract and the start of construction, if without such costs the contract would not have been concluded.

IFRS 15 allows only additional costs for concluding a contract to be considered for capitalization. Therefore, only those costs that would not have been incurred if the contract had not been concluded are subject to consideration. An example is the commission for the sale, which is payable only if the contract is won. Or payment for participation on electronic platforms for a successful tender. Conversely, the running costs of doing business, such as general expenses for lawyers, are not subject to consideration, since these costs would have been incurred regardless of whether the contract was received (paragraph 93 of IFRS 15).

IFRS 15 provides users with clear and transparent information about revenue and accounts receivable, both in terms of the amounts actually recorded and in terms of disclosures. However, along with this, IFRS 15 requires the use of a larger number of estimates based on actual business models.

In our study, we propose an integrated approach to revenue accounting, which allows us to accumulate various methods contained in IFRS 15. This takes into account the specifics of the activity, the scale of the company, the legal relationship between the customer, contractor, subcontractor. A very important tool is the application of a settlement scheme between developers and shareholders-buyers of the dwelling through escrow accounts. This innovative revenue generation mechanism should become an important tool in the accounting practice of financial results in construction companies.

References

1. IFRS 15 Revenue from Contracts with Customers. (2021). IFRS. Retrieved September 20, 2021, from https://msfo.ru/directory/msfo/standards/ifrs/ifrs-15/
2. International Financial Reporting Standards (IFRSs) par. 6, 2020. – IASCF, 2020. Contributions to Economics. (9783319552569), c. 207–210. https://doi.org/10.1117/968-3-230-55254-6_51.
3. King, T. A. (1992). *More than a numbers game: A brief history of accounting* (p. 21). Wiley.
4. Raihi-Belkaoui, A. (2004). *Accounting theory.* Cengage Learning Business Press.
5. Statement of principles for financial reporting. The Accounting Standards Board Limited (1999). Retrieved September 23, 2021, from https://www.frc.org.uk/getattachment/31cb1973-82a6-439b-bf44-8fffad5b20da/Statement-of-Principles-for-Financial-Reporting-1999.pdf.
6. Varkulevich, T. V., & Bubnovskaya, T. V. (2020). Mechanism of transformation of the financial reporting of companies-participants of international markets in the format of IFRS: Practical aspects. *Azimuth of Scientific Research: Economics and Administration, 9*(3), 97–101.

Russian Practice of Ensuring the Economic Security of Production Enterprise on the Basis of Corporate Accounting System

Oksana V. Kadesnikova, Sergey A. Kadesnikov, Rezeda R. Safina, Olga A. Kireyeva, and Rasul O. Kholbekov

Abstract In the conditions of turbulence and dynamism of financial and economic processes, increasing the economic security of manufacturing enterprises is an urgent issue, which is ensured by the introduction of corporate information systems into accounting processes to improve the operational efficiency of the business. The paper reveals the practical aspects of implementing a corporate information system using the software products of the leading Russian company "1C" in the activities of a large manufacturing enterprise, including benchmarking of software products for the digitalization of accounting processes in companies and analysis of business processes, which are also included in the corporate information environment and business analysis risks. The author's approach to assessing the economic effect expected as a result of the introduction of software products to automate the basic business processes of an industrial enterprise is disclosed. The obtained results prove the importance of using corporate information systems in management practice to prevent risks associated with human factors and, in general, ensure the information security of an enterprise.

Keywords Economic security · Corporate information system · Manufacturing enterprise · Software product · Implementation · Economic effect · Risk reduction

JEL Classifications M40 · M41 · M49 · G39

1 Introduction

One of the areas of economic security of enterprises in the context of the global digitalization of society is information security. As correctly noted in scientific periodicals [1–4], digitalization, on the one hand, is an advanced engine of the scientific and technological progress of society; on the other hand, it possesses threats

O. V. Kadesnikova (✉) · S. A. Kadesnikov · R. R. Safina · O. A. Kireyeva
Ufa State Petroleum Technological University, Ufa, Russia
e-mail: kadesok@mail.ru

R. O. Kholbekov
Tashkent State University of Economics, Tashkent, Uzbekistan

A. V. Bogoviz (ed.), *Big Data in Information Society and Digital Economy*, Studies in Big Data 124, https://doi.org/10.1007/978-3-031-29489-1_16

from cyber criminals caused by the insecurity of corporate information systems. The object of economic security of industrial enterprises is its information support, which is formed in the accounting and reporting system.

Until recently, information support of enterprises fit into the concept of accounting and reporting information systems. Indeed, the area of accounting and reporting has become the first object of automation and informatization [5].

The evolution of information technologies, the introduction of digital competencies, the economic development of enterprises, and the emergence of new business processes have led to the need for complex automation of activities and, as a result, to the creation of corporate information systems (CIS) and reporting in the XBRL format [6].

The main goal of the corporate information system is to increase the company's profits through the most efficient use of all its resources, including resources that form the economic security system. No one doubts that the implementation of the task of complex automation of business processes by creating CIS is an expensive project. Simultaneously, the corporate information system is not only a tool for automating the activities of an enterprise but also a valuable resource for strategic development and an intangible asset that can generate income for the company in the future, thereby increasing the value of the company, which is a key factor in increasing competitiveness and business efficiency. Among the advantages provided by the development and implementation of CIS, in the author's opinion, the following ones can be distinguished:

1. Standardization and unification of business processes;
2. The information component of the management function, which involves planning, analysis, and forecasting, is being improved;
3. Accounting systems are improved through the use of integrated information space (CRM);
4. More informed operational and strategic decisions are made;
5. Potential errors of employees are reduced due to the uniformity of the procedure for entering information and the availability of current control procedures;
6. The productivity of personnel is increased by automating the functions they perform, increasing the productivity of operations;
7. The quality of customer service (CRM) and relationships with supplier counterparts (SRM) are improving;
8. The number of completed orders on time is increasing;
9. The time to market for innovative products is shortening;
10. Management costs are reduced;
11. The efficiency of working capital management is increasing by reducing inventory.

To assess the effectiveness of CIS implementation, it is necessary to determine the relationship between the cost of a project to develop and support the information environment and the increase in the efficiency of an economic entity. At the conceptual design stage, it is vital to determine the expected economic effect from the implementation of the system, which is directly related to the depth of coverage of all business processes, because CIS increases business efficiency only when it is fully implemented in all company processes.

2 Materials and Methods

It is advisable to start the assessment of the effectiveness of creating a corporate information system of the studied economic entity, which provides for the transition to the digitalization of business processes using 1C software products, from analysis of the current situation and problems that have developed at the enterprise in connection with the use of heterogeneous software. The most pressing issues that should be addressed include the high costs of maintaining systems and collecting consolidated information, the significant number of disparate contractors, a limited list of fully automated business processes, the presence of hidden business processes, and the absence of conditions for improving efficiency.

Currently, the studied economic entity uses an information system that ceases to meet current requirements for running an efficient business due to the following circumstances: the lack of a developed partner network and representative offices of a maintenance service company in the city of Ufa; closed program code for third-party developers; the cost of maintenance is set exclusively; many operations are performed manually; there are no conditions to reduce the cost of ownership of the program. While using this information system, there are risks of increasing costs for ownership and maintenance, the impossibility of scaling the software product, and the inefficiency of the enterprise. The main risk is the lack of integration of the enterprise's economic security management subsystem with the overall business process management system, which significantly reduces the overall level of economic security of the manufacturing enterprise, bringing it closer to a critical value.

Additionally, the company uses programs of its design, which also have several disadvantages: for example, to maintain them permanently, it is required to maintain a staff of unique employees; there is no guarantee of compliance of software products with regulatory requirements and a single platform for automation; the cost of their maintenance is constantly increasing; product development is haphazard; there are no conditions for reducing the costs of program ownership.

An important factor affecting the level of economic security in terms of the information support function also includes the potential threat of risks of rising costs for owning and maintaining the software products used in-house, the departure of key developers, the presence of incorrect calculations, and, as a result, fines, penalties, and forfeits.

Analyzing all information mentioned above, we can conclude that there is a high degree of risk and the need to introduce a new corporate information system to increase the level of economic security of the studied economic entity. The authors analyzed and compared the functionality of the software "AS IS"–"TO BE" for each business process in fourteen structural divisions of the studied enterprise (Table 1).

Table 1 Comparison of the functionality of the software "AS IS"–"TO BE" at the studied enterprise

Department	Number of business processes in a department		Share of business processes with complete or partial absence of the necessary functionality of the current software (AS_IS) (%)
	Total (pcs.)	With the complete or partial absence of the necessary functionality of the current software (AS_IS) (pcs.)	
Customer service	65	23	35.38
Financial department	22	12	54.55
Accounting department	82	20	24.39
Planning department	34	24	70.59
Contract service	13	11	84.62
Mechanics department	38	24	63.16
Department for technical supervision and additional services	11	7	63.64
Chief technologist and environmental protection service	24	11	45.83
Logistics and economic support service	31	25	80.65
Occupational safety and human resources service	67	22	32.84
Services of water supply and sewerage facilities	92	54	58.70
Technical development service for water supply and sanitation systems	20	9	45.00
Service for the operation of water supply and sanitation systems	47	24	51.06
Enterprise security and protection service	31	11	35.48
Total for the enterprise			48.01

Source Compiled by the authors

The analysis demonstrated that 48% (almost half) of the necessary functionality of the software used for business processes participants is missing. However, these problems can be solved by implementing the proposed architecture with the implementation of the functionality of the 1C configuration. Automation on the 1C platform will allow taking advantage of the following:

1. The use of the 8.3 technology platform;
2. The presence of a developed market of the system supports specialists;
3. The absence of dependence on the unique developer.

These factors, taken together, allow the studied company to receive a guarantee of information technology support (consultations on the software product, implementation of legislative changes, development of new functionality, acceptance of development requests, and technical support) and the choice of accompanying organization strictly controlled by the vendor—Firm 1C, which fully complies with the law and the conditions for the provision of quality services to increase the level of information security.

3 Discussion

One of the initial key factors determining the effectiveness of the implementation of a corporate information system to ensure the economic security of the enterprise, considered by the authors, is the determination of the optimal set of functionalities of the implemented configuration. The authors and the group of involved experts, which included representatives of the studied economic entity, developer company, and teachers of related disciplines from one of the leading universities in the region, ranked software products on the 1C platform using one of the methods of comprehensive comparative assessment—the method of the sum of points by a discrete scale on a segment from 1 to 3. One point was assigned to the software product that least satisfies the requirements of the customer enterprise, two points—the requirements are met to average degree, and three points—the requirements are met to the maximum extent. The criterion for choosing the best configuration is the maximum score for all compared indicators. A comparison of the functionality of the software considered for implementation was carried out in terms of the key integrated business processes of the studied enterprise.

The ranking results indicate that it is advisable for the studied enterprise to choose the software product "1C: Integrated Automation" as the target corporate information system.

As noted in periodic sources, compared with other corporate information systems, this product is most suitable for organizing internal control, tax monitoring by the state tax service [7, 8], and for organizing business process planning system within a corporate group [9, 10].

The plan for creating the target system at the studied enterprise provides for three options, including the same set of ten stages of implementation, differing in terms of execution, the number of services, and the cost of owning a software product. Job titles include the following stages of automation:

- Stage 1—warehouse logistics;
- Stage 2—accounting and tax accounting;
- Stage 3—payroll and personnel policy;
- Stage 4—water supply and sanitation services. Conducting mutual settlements;
- Stage 5—provision of motor transport services to the divisions of the enterprise;
- Stage 6—treasury department;
- Stage 7—needs for inventory and procurement;
- Stage 8—claim work for legal entities and individuals;
- Stage 9—budgeting;
- Stage 10—implementation of procurement procedures by the contract service.

While implementing option 1, the work on stages 2 and 3 can be performed in parallel. Simultaneously, stage 3 can begin only after the end of stage 1 and must be completed before the start of the commercial operation of stage 2. The commercial operation of stage 3 must take place at the beginning of the calendar year. Steps 4 and 5 must be carried out together. Stage 5 can only start after the end of stage 2.

Works on stages 7 and 8 can be performed simultaneously. Stage 9 is desirable to perform one of the last, after the automation of all enterprise departments.

In the process of implementing option 2, work on stages 2 and 3 are performed simultaneously, while stage 2 can only begin after the end of stage 1 and must be completed before the start of the commercial operation of stage 4. Work on stages 4 and 5 can be performed simultaneously. Work on stage 6 must start after the end of stage 2; work on stage 8 must start after stage 4.

While implementing option 3, work on stages 3, 4, 5, 6, 7, 8, 9, and 10 must be carried out in parallel, while stage 3 can only begin after the end of stage 1. Work on stage 5 must start after the end of stage 2; work on stage 7 must start after stage 3.

In variants 2 and 3, the cost of maintaining all products during the implementation period is approximately the same.

In variant 1, the cost of maintenance is more expensive due to the fact that the current ERP system will have to be supported for one year longer because the payroll block will only begin to be implemented in 2021.

Variant 3 is the shortest in terms of time: work on stages 1, 2, 3, and 4 are carried out almost in parallel. However, it is necessary to attract more resources for simultaneous work. There is a big risk that the contractor will not be able to provide all projects with the required volume of resources.

Variant 2 involves the introduction of key modules (stages 1, 2, and 3), which are currently being conducted in the current ERP system. This approach will ensure a quick transition to the 1C platform, which will remove the risk of termination of maintenance of the current ERP system by the developer and reduce the annual cost of product maintenance. It must be considered that the introduction of the billing system will be carried out only in 2021.

Option 1 provides benefits in terms of the priority implementation of the billing system (during 2020) and partially removes the risks in terms of maintaining the ERP system (implementation of Stages 1 and 3 models).

4 Conclusion

As a result of the project, it is planned to create conditions for increasing the enterprise's income by reducing the indicators by 5–10%. Together, this will provide an economic effect in the amount of 250.34 million rubles.

The economic effect is the sum of indicators "reduction of costs for material resources," "reduction of production costs," "reduction of accounts receivable," "reduction of operating and administrative costs," and "reduction of labor costs." The economic effect amounts to 250.34 million when the first variant is implemented.

The payback period of the project is nine months. The payback period of the project is calculated according to the formula (1): 38.5/(250.34/60) = 9 months.

$$\text{Payback period of the project} = \frac{\text{Project cost}}{\text{Economic effect/Average useful life}} \quad (1)$$

where:

- The cost of the project—38.5 million rubles.
- Economic effect—250.34 million rubles.
- Average useful life of the implemented software—60 months (5 years).

In addition to the economic effect expected from the implementation of the system, it is necessary to analyze the depth of coverage of all business processes because CIS only increases the level of economic security when it is fully implemented in all company processes to minimize risks. The expected (planned) risk reduction is observed for each module (business process).

Considering the expected economic effect from the implementation of the modules (Table 2) and risk minimization, it would be more profitable to apply the implementation plan for Variant 1. The expected effect from the automation of the provision of water supply and sanitation services will be 37,307,000 rubles, while the effect from the automation of payroll personnel and personnel policy will amount to 21,662,400 rubles.

5 Results

The research results indicate that it is not enough just to buy a standard configuration of the software product on the 1C platform to increase the level of information security of a manufacturing enterprise. Additionally, it is necessary to install and configure programs properly, adapt to the specifics of doing business, carry out the procedure for transferring data from previously used accounting systems, train specialists, and provide support in putting programs into operation.

The transition to using a corporate information system on the 1C platform at the studied enterprise will reduce the annual costs of maintaining the software complex

Table 2 Evaluation of the effectiveness of the implementation of a corporate information system in terms of reducing risks for each business process

Name of works	Project start	Implementation cost (rub.)	Economic effect (rub.)	Payback period	Planned (expected) risk reduction
Stage 1—Automation of warehouse logistics	March 2020	2,955,500	20,360,000	9 months	1. Risk of termination of maintenance of the existing ERP
Stage 2—Automation of the provision of water supply and sanitation services, as well as maintaining accounts	May 2020	8,288,250	37,307,000	14 months	1. The risk of termination of maintenance of self-written programs; 2. The risk of the need to independently refine programs in connection with changes in the legislation of the Russian Federation
Stage 3—Automation of accounting, tax accounting	July 2020	6,527,800	41,844,200	10 months	1. Risk of termination of maintenance of the existing ERP; 2. The risk of termination of maintenance of self-written programs for unloading and sending regulated reporting; 3. Human factor in manual data collection for tax reporting

(continued)

Table 2 (continued)

Name of works	Project start	Implementation cost (rub.)	Economic effect (rub.)	Payback period	Planned (expected) risk reduction
Stage 4—Automation of personnel payroll and personnel policy	February 2021	3,598,000	21,662,400	10 months	1. Risk of termination of maintenance of the existing ERP; 2. The risk of termination of maintenance of self-written programs
Stage 5—Automation of claim work for legal entities and individuals	April 2021	2,749,900	36,287,000	5 months	1. The risk of termination of maintenance of self-written programs
Stage 6—Automating the provision of motor transport services to the divisions of the enterprise	August 2021	3,469,500	23,927,000	9 months	1. Risk of termination of maintenance of the existing ERP; 2. The risk of termination of maintenance of self-written programs
Stage 7—Treasury automation	February 2022	1,863,250	10,831,200	11 months	1. Human factor in manual data processing; 2. Risk of losing Excel files (human factor, computer viruses, and technical failures)
Stage 8—Automation of the formation of requirements for inventory and purchases	February 2022	2,955,500	25,634,200	7 months	1. Human factor in manual data processing; 2. Risk of losing Excel files (human factor, computer viruses, technical failures)

(continued)

Table 2 (continued)

Name of works	Project start	Implementation cost (rub.)	Economic effect (rub.)	Payback period	Planned (expected) risk reduction
Stage 9—Budgeting automation	August 2022	3,212,500	21,662,400	9 months	1. Human factor in manual data processing; 2. Risk of losing Excel files (human factor, computer viruses, technical failures)
Stage 10—Automation of the execution of procurement procedures by the contract service	January 2023	2,827,000	10,831,200	16 months	1. Human factor in manual data processing; 2. Risk of losing Excel files (human factor, computer viruses, technical failures)

Source Compiled by the authors

by more than 20 times (136,472 rubles against the existing 2,900,000 rubles per year). The enterprise will have the opportunity to reduce the cost of maintaining software through the introduction of circulation programs, reduce debt through the use of mechanisms for claim work, reduce fines by ensuring compliance of calculations with legal requirements, and increase efficiency by maximizing the automation of manual labor. In addition to reducing the cost of ownership and maintenance, the key benefits of the transition will be an increase in the level of automation of the studied enterprise and optimization of current business processes in terms of reducing information risks, which contributes to an increase in the level of economic security of the enterprise as a whole.

References

1. Ergasheva, S. (2020). Development of digital systems of enterprises of agriculture and water management. *Asian Journal of Technology & Management Research (AJTMR), 10*(1), 41–44. Retrieved February 18, 2022, from http://www.ajtmr.com/papers/Vol10Issue1/Vol10Iss1_P5.pdf.
2. Leybert, T. B., & Khalikova, E. A. (2020). Current tendencies of transformation of the Russian practice of decision making in business systems. In E. G. Popkova & A. V. Chesnokova (Eds.), *Leading practice of decision making in modern business systems: Innovative technologies and*

perspectives of optimization (pp. 13–26). Emerald Publishing Limited. https://doi.org/10.1108/978-1-83867-475-520191003.
3. Popkova, E. G., Bogoviz, A. V., & Sergi, B. S. (2021). Towards digital society management and "capitalism 4.0" in contemporary Russia. *Humanities and Social Sciences Communications, 8*(1), 77. https://doi.org/10.1057/s41599-021-00743-8.
4. Popkova, E. G., De Bernardi, P., Tyurina, Y. G., & Sergi, B. S. (2022). A theory of digital technology advancement to address the grand challenges of sustainable development. *Technology in Society, 68*, 101831. https://doi.org/10.1016/j.techsoc.2021.101831.
5. Ergasheva, S. T., & Mannapova, R. A. (2022). Accounting policy of the enterprise: Development and amendments. *Asian Journal of Technology & Management Research (AJTMR), 11*(2), 21–27. Retrieved February 18, 2022, from http://www.ajtmr.com/papers/Vol11Issue2/Vol11Iss2_P5.pdf.
6. Astafeva, O. V., Astafyev, E. V., Khalikova, E. A., Leybert, T. B., & Osipova, I. A. (2020). XBRL Reporting in the conditions of digital business transformation. In S. Ashmarina, M. Vochozka, & V. Mantulenko (Eds.), *Digital age: Chances, challenges and future* (pp. 373–381). Springer. https://doi.org/10.1007/978-3-030-27015-5_45.
7. Vanchukhina, L. I., Leybert, T. B., Khalikova, E. A., & Shamonin, E. A. (2018). The main directions of development of information production of Russia's tax system. In A. P. Sukhodolov, E. G. Popkova, & T. N. Litvinova (Eds.), *Models of modern information economy: Conceptual contradictions and practical examples* (pp. 265–274). Emerald Publishing Limited. https://doi.org/10.1108/978-1-78756-287-520181027.
8. Vanchukhina, L. I., Galeeva, N. N., Rudneva, Y. R., Rogacheva, A. M., & Shamonina, T. P. (2020). Development of internal corporate control in the conditions of tax monitoring. In E. G. Popkova & A. V. Bogoviz (Eds.), *Circular economy in developed and developing countries: Perspective, methods and examples* (pp. 215–229). Emerald Publishing Limited. https://doi.org/10.1108/978-1-78973-981-720201031.
9. Arif, N., & Tauseef, S. M. (2015). *Integrating SAP ERP financials: Configuration and design* (A. Zyabrikov, Trans. from English). Publishing House Expert-RP. (Original work published 2011).
10. Vanchukhina, L. I., Leybert, T. B., Khalikova, E. A., & Luneva, N. N. (2018). Modern approaches to operational planning in oil refinery using the pins software product. *Quality – Access to Success, 19*, 123–129.

Features of the Development of the Digital Segment of the Agro-Industrial Complex During the Pandemic

Alexey S. Molchan, Olga Yu. Frantsisko, Olga Yu. Smyslova, Yulia A. Kleymenova, and Alexey N. Shevchenko

Abstract The paper aims to study the conceptual foundations of the formation of the digital segment of agriculture and its advantages for the national economy and determine the vector of state policy to create a basis for implementing the Agro-Industrial Complex 4.0 (AIC 4.0) model in Russia. The authors identified the main elements of the AIC 4.0 model: a digital base for decision support systems in the AIC, digitalization of production, analytics and big data, and digitalization of sales. The subjects of economic relations in the AIC 4.0 model are defined (agro-industrial enterprises, hi-tech companies promoting new technologies, the government, consumers of agricultural products, scientific and educational organizations, and financial intermediaries). The authors highlighted the conflicts of interests between subjects in developing the digital segment of agriculture in Russia. It is proved that the existing conflicts of interests form three key problem areas that hinder the implementation of the AIC 4.0 model: undeveloped information and communication infrastructure of the agricultural sector; lack of qualified personnel to ensure the introduction of IT technologies in agriculture; the imperfection of state financing and support of agriculture. The measures aimed at creating the foundations of the AIC 4.0 model in Russia are defined: the development of new standards and rules governing the process of

A. S. Molchan (✉)
Financial University under the Government of the Russian Federation (Krasnodar branch), Krasnodar, Russia
e-mail: economresearch@mail.ru

O. Yu. Frantsisko
Kuban State Agrarian University named after I.T. Trubilin, Krasnodar, Russia

O. Yu. Smyslova
Financial University under the Government of the Russian Federation (Lipetsk branch), Lipetsk, Russia

Y. A. Kleymenova
Kuban State Technological University, Krasnodar, Russia

A. N. Shevchenko
Voronezh State University of Forestry and Technologies named after G.F. Morozov, Voronezh, Russia

A. V. Bogoviz (ed.), *Big Data in Information Society and Digital Economy*, Studies in Big Data 124, https://doi.org/10.1007/978-3-031-29489-1_17

digital transformation of the industry; the creation of new institutions for the development of innovations in the agro-industrial complex; the development of an innovation transfer system; financing projects in the field of digital infrastructure development; support for industry-specific educational institutions and stimulation of specialized research activities; expansion of forms of state support for the agro-industrial sector.

Keywords Agro-industrial complex 4.0 · Digital technologies · Agriculture · Cost reduction · Problem areas · Digital infrastructure

JEL Classifications Q13 · Q16

1 Introduction

During the period of digitalization, new types of business activity are intensively developing; they ensure the formation of digital segments of various sectors of the economy. For example, in the financial services market, digital technologies have provoked the development of the segment of innovative finance (Fintech). In the field of education, the use of the Internet, big data, cloud technologies, virtual and augmented reality technologies, and artificial intelligence has led to the emergence of a segment of high-tech education (EdTech) [1, 2]. The agro-industrial complex did not stand aside either. Here, business processes are automated according to industry specifics and rely on breakthrough digital technologies. With the digitalization of the industry, the digital segment of the agro-industrial complex is being formed [3]. The volume of the global information technology market in the agro-industrial complex reached $17.44 billion by the end of 2019; a large share of sales (39%) fell on North America, followed by the Asia-Pacific region and Europe [4]. Digitalization of the industry carries multiple advantages [5]. The main advantages for enterprises implementing digital technologies are cost reduction and increased productivity [6]. Simultaneously, the use of digital and geoinformation solutions, robotics, genetics and breeding, integrated solutions for agriculture, and biologization of intensive forms of agriculture allows for reducing dependence on the effects of climatic and biological factors. However, the introduction of digital technologies in agriculture occurs unevenly in countries and regions of the world, contributing to the strengthening of socio-economic and technological inequality and becoming a risk factor for national economies, which increases during the pandemic [7, 8].

2 Literature Review

Digitalization creates conditions for improving agricultural production in terms of the transition to the Agro-Industrial Complex 4.0 (AIC 4.0) model [9]. Researchers focus on studying the factors contributing to the formation of the digital segment of the agro-industrial complex. In particular, Orlova et al. [10] believe that such factors are as follows:

- The transition to new technological order (when food production depends not on natural and climatic factors but on technologies to increase yields and productivity and prevent losses);
- Changes in value chains (value-added is concentrated in knowledge-intensive sectors such as genetics and breeding, the IT sector, industrial design, and engineering);
- The growing influence of large integrator companies taking control of larger areas of food systems;
- The shift in demand from traditional food raw materials to products that correspond to the value orientations of new generations;
- The strengthening of the role of factors of "sustainability" and product safety.

Goncharova [11], Anischenko and Shutkov [12], and Altukhov et al. [13] focused on the problems of applying new digital technologies in the field of agriculture in their works.

3 Methodology

The paper aims to reveal the conceptual foundations of the formation of the digital segment of agriculture, highlight its advantages for the national economy, and determine the vector of state policy to create the basis of the AIC 4.0 model of Russia.

Research objectives are as follows:

1. To form a comprehensive understanding of the nature of economic relations that ensure the development of the AIC 4.0 model;
2. To identify problem areas of the development of the digital segment of the agro-industrial complex in Russia, including during the pandemic;
3. To determine the directions of state policy focused on implementing the AIC 4.0 model in Russia.

Research methods are a method of theoretical analysis, method of systematization, method of economic analysis, graphical method, analysis, and synthesis.

4 Results

The concept of Industry 4.0, successfully extended to the area of industrial production, has long gone beyond its limits. In particular, in the field of agriculture, the potential of Industry 4.0 is manifested through the implementation of new approaches to production and consumption based on the collection of big data, its processing, and its use for performing actions and operations independently of a person [1, 14]. The main elements of the AIC 4.0 model are as follows [15]:

- Digital monitoring of agribusiness performance and food security based on IoT;
- Automated production in agriculture: smart agriculture and robotic food production;
- AI-enabled intelligent management decision support in agribusiness for increased productivity and climate resilience;
- Integration of links and increasing transparency of agribusiness value chains based on e-commerce, cloud technologies, and blockchain.

Let us highlight the features of economic relations in the AIC 4.0 model and identify their problem areas that hinder the development of the digital segment of agriculture. The subjects of economic relations in the AIC 4.0 model are as follows:

- Agricultural enterprises;
- Hi-tech companies promoting new technologies;
- Government;
- Consumers of agricultural products;
- Scientific and educational organizations;
- Financial intermediaries.

The tools mediating economic relations in the AIC 4.0 model are the Internet, online sales services, distance learning technologies, big data, cloud services, new IT technologies, smart sensors, etc.

In the process of implementing economic functions, multilevel conflicts of interest arise between the listed entities (Table 1).

The designated conflicts of interest highlight three key problem areas that hinder the implementation of the AIC 4.0 model:

1. Undeveloped information and communication infrastructure of the agricultural sector;
2. Lack of qualified personnel to ensure the introduction of IT technologies in agriculture;
3. The imperfection of state financing and support of agriculture.

Free and open access to information resources, in other words, the availability of a developed ICT infrastructure ensures the optimization of production processes, allows enterprises to reduce costs significantly, and contributes to an increase in production indicators in terms of the volume of raw materials, products, and indicators of financial and economic activity [16]. The Ministry of Agriculture of the Russian Federation has developed a concept and approved a departmental project "Digital Agriculture" within the framework of the segment development program until 2024 to solve this problem. One of its goals is to create an information and telecommunications infrastructure for the functioning of the digital platform of digital agriculture [16]. Indeed, the basic condition for the digitalization of agriculture is the availability of an ICT infrastructure that determines the access of agricultural producers to the Internet and the possibility of using digital technologies and online services to promote products. Meanwhile, according to the latest available data from the All-Russian Agricultural Census of 2016, the situation with the access of agricultural enterprises to digital infrastructure remains very tense. For example, only 69.5% of agricultural enterprises are covered by telephone, and only 46.9% of economic entities are connected to the Internet. Among peasant (farm) enterprises, telephone coverage has increased from 24% (in 2006) to 48.6% (in 2016). Nevertheless, 50% of farms do not have a telephone connection, and only 15.9% of farms and individual entrepreneurs are connected to the Internet [17, 18].

Meanwhile, the departmental project "Digital Agriculture" assumes the allocation of funding for the formation of digital infrastructure in 2019–2024 in the amount of 6100 million rubles (with total project financing in the amount of 300,000 million rubles), which is 2% of the total project budget. These funds will be used, in particular, to provide server capacities and form requirements for the development of regional information and telecommunications infrastructure to deploy digital sub-platforms for agricultural management at the regional and municipal levels. However, the allocated project budget for achieving this task is clearly insufficient, given the extent of Russian territories and the number of farmlands. Probably, the presence of an undeveloped digital infrastructure of the agro-industrial complex explains the choice of an extensive development path followed by Russian agriculture. It is easier for companies to work extensively and strengthen mechanization rather than improve technologies and invest in digital products. Moreover, in recent years, the industry has significantly re-equipped and received new equipment by attracting investment (Table 2).

The priority of the extensive path of agricultural development is evidenced by the dynamics of the acquisition of new technologies by agricultural enterprises. From 2017 to 2020, the number of new technologies and specialized software purchased by Russian agricultural producers decreased by more than five times in animal husbandry and crop production [20].

The second problem area hindering the implementation of the AIC 4.0 model is the lack of qualified personnel to ensure the introduction of IT technologies in agriculture. The technological and digital lag of the Russian agro-industrial complex from foreign markets is only increasing. For comparison, according to AB InBev Efes, in the USA, Germany, and the UK, the share of IT specialists from the total number of employees in the agro-industrial complex exceeds 4% (4.3%, 4.5%, and 4.1%, respectively), while in Russia this figure is only 2.4% [21]. There are regional

Table 2 Investments in fixed assets aimed at the development of agriculture, 2017–2019

Indicator	2017	2018	2019
Investments in fixed assets for the development of agriculture (in actual prices), million rubles	400,509	431,720	445,378
As a percentage of the total investment in fixed assets	3.3	3.2	3.1
As a percentage of the previous year (in constant prices)	102.0	103.4	97.1

Source Compiled by the authors based on [19]

initiatives in terms of training specialists for digital agriculture. Thus, in 2018, the ANO "Competence Center for Digitalization of the Agro-Industrial Complex" was established in the Tambov Region [22]. However, the tasks of the Center are more limited to the creation of a digital platform for agriculture and the provision of services to small and medium-sized businesses, rather than training and retraining agricultural specialists [23]. Tambov universities have also joined the digitalization of the economy through science. Based on two regional universities, a center for collective use "Digital Engineering" has been created, which allows studying advanced technologies in the field of mechanical engineering, metalworking, and digital design, as well as conducting research and development work. The Center for GIS Technologies and Precision Agriculture was also launched. It is a place of educational and project activities for students and an interregional platform for training and retraining specialists in the field of precision agriculture and geoinformation systems.

However, in general, the existing problem in Russia is aggravated by two circumstances: low prestige of agricultural professions and insufficient attention to vocational guidance of young people, explained by the gap in the quality of life between the city and the countryside, as well as differentiation in income and access to infrastructure, career, and professional growth opportunities; isolation of educational programs and formed knowledge and skills of young professionals from the qualification requirements of business and from the tasks of agricultural science [10]. The discrepancy between the current system of agricultural education and the place of the agro-industrial complex in the country's economy is also evidenced by its critical lag behind world leaders. For example, in the global ranking of QS universities in the subject area "Agriculture" for 2018, there is only one Russian specialized university—Russian State Agrarian University—Moscow Timiryazev Agricultural Academy (RGAU-MSHA), which is part of the group of universities occupying 201st–250th places [10]. In solving this problem, it is also not necessary to rely on the implemented departmental project. From 2019 to 2024, it is planned to allocate 5368 million rubles (i.e., 1.8% of the total project budget) to create a system of continuous training of specialists of agricultural enterprises to form their competence in the field of the digital economy. This amount of funds is extremely small, considering that the funds are planned to be spent within six years. Among the tasks for which budget funds will be allocated, it is necessary to highlight the following:

- Systemic modernization of digital training for agro-industrial complex and enterprises, the synergistic effect of which is to support knowledge-intensive employment and reduce unemployment, as well as the most complete and effective staffing of digital agro-industrial complex (650 million rubles);
- Support for lifelong learning among workers and entrepreneurs in the agro-industrial complex based on distance higher education (3300 million rubles);
- Advanced training and professional retraining of personnel for the agro-industrial complex based on mass mastering of digital competencies (1355.5 million rubles);
- Stimulating demand for educational programs to train digital workforce for the agribusiness at various levels of the post-secondary education system (62.5 million rubles) [16].

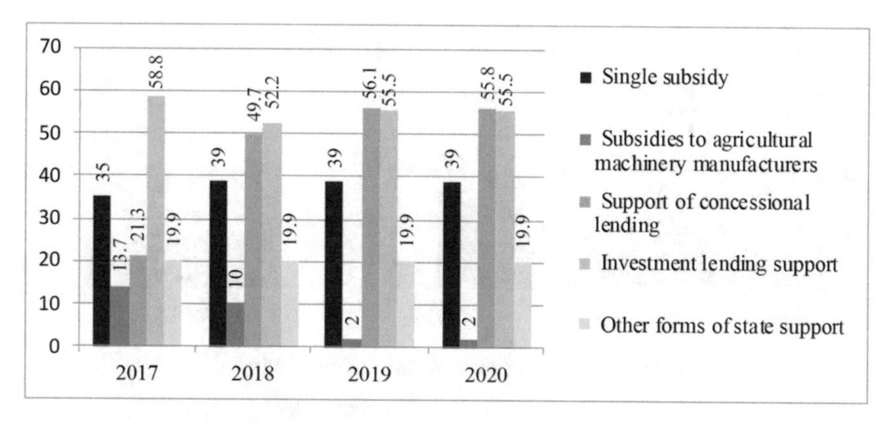

Fig. 1 State subsidies for the development of agriculture, billion rubles, 2017–2020. *Source* Compiled by the authors based on [25]

In 2020–2021, the COVID-19 pandemic contributed to the "personnel shortage." Agribusiness enterprises have a number of responsibilities for the organization of work and monitoring employees' health [24]. In practice, the conflict of interests between agricultural enterprises and the state results in underfunding of the agricultural sector and a lack of state support tools for agricultural producers implementing digital technologies. The contribution of agriculture to the formation of Russia's GDP is at the level of 5%, while the annual amount of state support does not exceed 1–1.2% from the consolidated state budget, and state financial support for farmers' incomes in Russia is no more than 3.5%. For example, in 2018, 7640 billion rubles were allocated from the federal budget for the implementation of programs in the field of healthcare, education, and the development of an accessible environment in rural areas, while state subsidies for the development of agriculture amounted to only 170.8 billion rubles (Fig. 1).

Meanwhile, in countries with a developed agricultural sector, these figures are ten times higher than in Russia: from 15% in Australia to 80% in Switzerland [26, 27]. The existing volume of state subsidies to agriculture cannot contribute to an increase in technological innovations of agricultural enterprises. The largest volume growth of investments is typical only for crop production against the background of stagnant values in animal husbandry. In terms of the intensity of costs, the agricultural sector is gradually approaching the values of the food industry and is reducing the gap from other industries (1.2% as of 2018). The dominant share is occupied by capital investments: in the purchase of machinery and equipment, as well as engineering. Simultaneously, the share of investments in research and development, and, accordingly, their importance, remains at a fairly low level (12% in agriculture), which is another argument in favor of the weak demand of agricultural enterprises for domestic developments. Investments in other types of "intellectual" innovations are even less significant; the share of positions "training and training of personnel,"

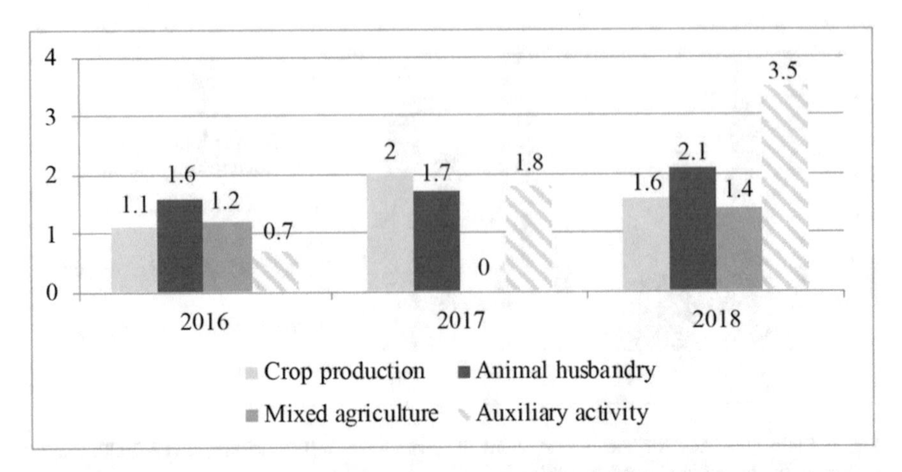

Fig. 2 Share of innovative products in total agricultural production, %, 2014–2018. *Source* Compiled by the authors based on [19]

"acquisition of new technologies," and "marketing research" in total does not exceed 1% [10].

The lack of proper state funding is reinforced by the imperfection of informal institutions, adherence to old forms and management methods at the micro-level of rural areas, and the imperfection of the informal institution of trust at the macro level. The latter is partially manifested in an intangible internal "resistance" to the process of implementing IT technologies and alertness to the ongoing transformation processes [28]. The consequence of insufficient investment in new technologies and R&D is a low share of innovative products in the overall production structure and low growth rates. In agriculture, this indicator increased from 1.4 to 1.9% over the period 2016–2018, while the livestock sector achieved greater "innovation" (2.1%) than crop production (1.6%) (Fig. 2).

Comprehensive multidimensional assessment and analysis of the degree and pace of digital modernization of agro-industrial complex at the level of the regional economy involves the consideration of indicators such as the pilot implementation of innovative models of digital organization of agribusiness enterprises, the speed of institutionalization of the most successful practices of this implementation, the flexibility of legal support for digital innovation in agriculture, and the availability of applied solutions to support digitalization of agriculture in regional systems of electronic government [25]. It currently remains low. According to experts, there are only individual elements of the implementation of the AIC 4.0 model in Russia. We are currently not talking about the integrated implementation of digital systems.

The COVID-19 pandemic has become a considerable problem for agricultural producers worldwide. In North America, it has jeopardized access to agricultural labor and complicated global supply chains. In 2020, due to the difficult epidemiological situation, enterprises in the USA, Canada, and Mexico were suspended, negatively affecting sales of such agrotechnological products as sensors, robots, etc. COVID-19 worsened logistics, transport accessibility, and trading activities in Russia, which became a big problem for the market [4]. However, in this challenging period, according to analysts, one of the most promising areas for venture investments will be the agro-industrial sector. Startups that introduce innovations in agriculture are still not enough; the demand for technology in the sector is growing, and investments in agricultural technology can bring investors multiple profits. In general, among the measures to create the foundations of the AIC 4.0 model in Russia, it is necessary to highlight the following.

First, to switch to the AIC 4.0 model, new standards and rules are needed that would regulate the entire process of digital transformation of the industry, the planned large-scale digitization of agricultural lands, the creation of a training system for the agro-industrial complex, and the launch of a single national digital platform in the agro-industrial complex. It is necessary to create new institutions to develop innovations in the agro-industrial complex, whose activities could compensate for the "bottlenecks" and limitations of existing support tools and expand the set of these tools. Bridging the technological gap between industry leaders and small producers is possible by developing an innovation transfer system.

Second, there is a need for large-scale financing of projects in the field of digital infrastructure development, including by attracting business structures and investment funds to equity participation in these projects.

Third, support of branch educational institutions and stimulation of specialized research activities are required. The first condition becomes feasible by increasing the involvement of students in practical work on farms and agricultural enterprises with long-term applied practice, strengthening the career guidance work of universities and colleges in schools to increase interest in the agricultural sector, and conducting explanatory work for the able-bodied population about the demand for specialists in the agricultural sector. The creation of social infrastructure near enterprises or the offer of additional conditions to the standard social package (e.g., the transfer of employees from the nearest territories or the provision of housing, food, and insurance) will help increase the attractiveness of professions in the agricultural sector. The second condition is associated with the creation of research centers on the basis of agricultural universities for the introduction and testing of new innovative technologies for the agro-industrial complex.

Fourth, it is necessary to expand forms of state support for the agricultural sector. In particular, state support for leasing and agricultural insurance can become mechanisms for the development of the industry. The agricultural insurance market with state support is the fastest-growing segment of voluntary property insurance in Russia, while positive dynamics are observed in all directions. According to the results of eight months of 2021, it increased by 45% in livestock insurance and by 37% in crop production.

5 Conclusion

The main elements and the subjects of economic relations in the AIC 4.0 model are highlighted. The specifics of conflicts of interest between subjects of economic relations in the AIC 4.0 model that hinder the development of the digital segment of agriculture in Russia are revealed.

It shows how conflicts of interest emerging between subjects form problem areas of digital agriculture. The problems hindering the implementation of the AIC 4.0 model, including the undeveloped information and communication infrastructure of the agricultural sector, are highlighted, including the lack of qualified personnel to ensure the introduction of IT technologies in agriculture and the imperfection of state financing and support of agriculture.

A set of measures for developing the basics of the AIC 4.0 model in Russian conditions is proposed. The authors substantiate the need for the formation of institutions for the digital transformation of agriculture, the expansion of financial and other forms of state support for the industry, the intensification of efforts to develop the digital infrastructure of the agricultural sector, the support of industry-specific educational institutions and stimulation of specialized research activities to solve the problem of training qualified personnel to ensure the introduction of IT technologies in agriculture.

Data Availability Data on conflicts of interest between subjects of economic relations in the Agro-Industrial Complex 4.0 model (Table 1) is available at https://figshare.com/ with DOI https://doi.org/10.6084/m9.figshare.19228281.v1.

References

1. Gusev, V., Smyslova, O., Ioda, J., Karpunina, E., & Shevtsov, N. (2020). Industry 4.0 technologies: A tool for overcoming contradictions of socio-economic development or a source of new threats? In *Proceeding of the 35th IBIMA Conference* (pp. 7654–7671). Seville, Spain.
2. Karpunina, E., Gubernatorova, N., Daudova, A., Stash, Z., & Kargina, L. (2020). The spillover effects of the digital economy. In *Proceedings of the 36th IBIMA Conference* (pp. 942–954). Granada, Spain.
3. Dung, L., & Hiep, N. (2017). The revolution of Agriculture 4.0 and sustainable agriculture development in Vietnam. In *Proceedings of the International Conference Emerging Issues in Economics and Business in the Context of International Integration* (pp. 317–328). Hanoi, Vietnam: National Economic University Press.
4. Research and Markets. (2021, April 14). *Global $41.17 Billion Agritech Market Forecast to 2027—Proliferating Use of Precision Farming Gaining Momentum*. Retrieved November 20, 2021, from https://www.businesswire.com/news/home/20210414005576/en/Global-41.17-Billion-Agritech-Market-Forecast-to-2027---Proliferating-Use-of-Precision-Farming-Gaining-Momentum---ResearchAndMarkets.com.
5. Astakhova, T., Kolbanev, M., Romanova, A., & Shamin, A. (2019). Model of digital agriculture. *International Journal of Open Information Technologies, 7*(12), 63–69.
6. Panova, A. (2020). Agriculture 4.0: Problems and prospects. *International Research Journal, 7–3*(97), 160–164. https://doi.org/10.23670/IRJ.2020.97.7.100.

7. Klerks, L., & Rose, D. (2020). Dealing with the game-changing technologies of Agriculture 4.0: How do we manage diversity and responsibility in food system transition pathways. *Global Food Security, 24*, 100347. https://doi.org/10.1016/j.gfs.2019.100347.
8. Sen, A. (1999). *Development as freedom*. Oxford University Press.
9. Fedotova, G., Gorlov, I., Glushchenko, A., Slozhenkina, M., Mosolova, N., & Mosolova, D. (2019). *Agriculture 4.0: Digital trends in the development of agriculture*. SPHERE.
10. Orlova, N., Serova, E., & Nikolaev, D. (2020). Innovative development of the agro-industrial complex in Russia. Agriculture 4.0. In *Proceedings of the XXI International Scientific Conference on Problems of Economic and Social Development*. Moscow, Russia: HSE University.
11. Goncharova, A. (2019). Development and distribution of scientific innovations as a growth factor for the economy of the AIC. *World Science, 7*(28), 7–11.
12. Anischenko, A., & Shutkov, A. (2019). Agriculture 4.0 as a promising model for scientific and technological development of agrarian sector in modern Russia. *Food Policy and Security, 6*(3), 129–140. https://doi.org/10.18334/ppib.6.3.41393.
13. Altukhov, A., Dudin, M., & Anishchenko, A. (2019). Global digitalization as an organizational and economic basis for the innovative development of the agro-industrial complex of the Russian Federation. *Market Economy Problems, 2*, 17–27. https://doi.org/10.33051/2500-2325-2019-2-17-27.
14. Dubovitski, A., Kleimentova, E., Karpunina, E., & Cheremisina, N. (2019). Ecological and economic foundations of effective land use in agriculture: The implementation prospects of food security. In *Proceedings of the 33rd IBIMA Conference "Education Excellence and Innovation Management Through Vision 2020"* (pp. 2687–2693). Granada, Spain.
15. Espolov, T. (2020). *Digitalization is a key factor in the development of agriculture*. Retrieved November 20, 2021, from http://www.eurasiancommission.org/ru/act/prom_i_agroprom/dep_agroprom/actions/Documents/4%20Есполов.pdf.
16. Ministry of Agriculture of the Russian Federation. (2019). *Departmental project "Digital Agriculture."* Moscow, Russia: Rosinformagrotech. Retrieved November 20, 2021, from https://mcx.gov.ru/upload/iblock/900/900863fae06c026826a9ee43e124d058.pdf.
17. Federal State Statistics Service of Russian Federation. (2017). *Preliminary results of the All-Russian Agricultural Census of 2016 for the subjects of the Russian Federation*. Moscow, Russia. Retrieved November 20, 2021, from http://www.gks.ru/free_doc/new_site/business/sx/vsxp2016/VSHP2016_tom2.pdf.
18. Maksimova, T., & Zhdanova, O. (2018). The digitalization strategy implemented in the Russian agro-industrial complex: Possibilities and constraints. *Theory and Practice of Social Development, 9*(127), 63–67. https://doi.org/10.24158/tipor.2018.9.9.
19. Federal State Statistics Service of Russian Federation. (2021). *Agriculture, hunting and forestry*. Retrieved November 20, 2021, from https://rosstat.gov.ru/enterprise_economy.
20. EMISS. (2021). *The number of new technologies (technical achievements) acquired by the organization, software tools since 2017*. Retrieved November 20, 2021, from https://www.fedstat.ru/indicator/58773.
21. Gavrilov, A. (2019, November 1). *Agro-industrial complex with big numbers*. Expert. Retrieved November 20, 2021, from https://expert.ru/south/2019/11/apk-s-bolshoj-tsifryi/.
22. Agrocenter. (2018). *Competence Center for Digitalization of the Agro-Industrial Complex*. Retrieved November 20, 2021, from http://agrocenter.pro/.
23. Kommersant. (2018). *The economy goes into "digit."* Retrieved November 20, 2021, from https://www.kommersant.ru/doc/3834926.
24. RBC. (2021, November 10). *Agro-industrial complex 2021: New opportunities and global challenges*. Retrieved November 20, 2021, from https://chr.plus.rbc.ru/partners/618905817a8aa963aed6ebbf.
25. Federal Ministry of Food and Agriculture of Germany. (2020). *An extended overview of the development of digitalization of agriculture in the Russian Federation. Status and prospects. As of April–May 2020*. Retrieved November 20, 2021, from https://agrardialog.ru/files/prints/rasshirenniy_obzor_razvitiya_tsifrovizatsii_selskogo_hozyaystva_v_rf_aprel_may_2020.pdf.

26. Klimova, N. (2013). Features of regulational actions of state government for foreign countries in agribusiness. *Polythematic Online Electronic Journal of the Kuban Agrarian University, 90*. Retrieved November 20, 2021, from http://ej.kubagro.ru/2013/06/pdf/45.pdf.
27. Nally, D. (2016). Against food security: On forms of care and fields of violence. *Global Society, 30*(4), 558–582. https://doi.org/10.1080/13600826.2016.1158700.
28. Bondarenko, T., & Isaeva, E. (2017). Conceptual approaches to improving the mechanism of lending to the agricultural sector. In M. P. Bobkova (Ed.), *Financial strategies and models of economic growth in Russia: Problems and solutions* (pp. 48–56). Auditor.

Model of Formation of the Production Program of Enterprises

Valentina N. Sidyakova, Anna I. Orlova, Zhanna V. Smirnova, Natalia S. Andryashina, and Ekaterina P. Garina

Abstract In this article, the authors analyze the model of the production program of the enterprise. The authors consider the problems of modern business in enterprises that are engaged in the production of products. At present, one of the most urgent problems for the lack of funds of enterprises is the efficiency of allocation of production resources. In the process of theoretical substantiation of the production program, which reflects the main directions and tasks of the enterprise development in the planned period, production and economic relations with other enterprises, the profile and degree of specialization and combination of production are considered. An analysis of the proposed methods for organizing the production program of an enterprise in the study showed that an integrated approach is needed. The authors considered a mathematical model for calculating the profit of an enterprise, which will depend on the volume of output. The introduction of the recommendations developed by the authors into the activities of enterprises will bring a significant economic effect.

Keywords Mathematical model · Production program · Profit · Enterprise · Economic efficiency

JEL Classifications R11 · R12 · R58 · Q13 · Q18

V. N. Sidyakova (✉) · A. I. Orlova
Institute of Food Technologies and Design—A Branch of the State Budgetary Educational Institution of Higher Education "Nizhny Novgorod State Engineering and Economic University", Nizhny Novgorod, Russia
e-mail: valy-0573@mail.ru

Z. V. Smirnova · N. S. Andryashina · E. P. Garina
Minin Nizhny Novgorod State Pedagogical University, Nizhny Novgorod, Russia
e-mail: z.v.smirnova@mininuniver.ru

N. S. Andryashina
e-mail: natali_andr@bk.ru

A. V. Bogoviz (ed.), *Big Data in Information Society and Digital Economy*, Studies in Big Data 124, https://doi.org/10.1007/978-3-031-29489-1_18

1 Introduction

Today, Russian enterprises have to solve problems that in recent years, marked by stable and fairly rapid economic growth, have not been on the agenda.

In crisis situations, the main problem for enterprises in the main industries is the financial and economic situation. The main task of the management of enterprises is the formation of an anti-crisis strategy for managing production resources, which is aimed at optimizing the production program in a crisis economic situation. The work of enterprises largely consists of the formation of effective production programs. The development of economic and mathematical models for production programs will help enterprises overcome the production crisis.

2 Methodology

An analysis of modern economic literature has shown that the works of Russian and foreign economists reveal the concepts of economic justification for the content of production programs. The issues of managing the formation and optimization of the production program were studied in the works of Afanasiev and Suvorov [1], Sklyarenko and Prudnikov [2], and Volkov and Sklyarenko [3].

Based on the foregoing, it becomes relevant in terms of theoretical substantiation of the calculation model of the production program of enterprises at the economic level of development.

3 Results

The goal of organizations in market conditions is not only profit, but also the financial stability of the enterprise as a whole. Today, the profit of the enterprise is far from in the first place of economic relations.

Each head of the enterprise considers the work of his production in different ways: some consider the production of goods to be the main thing in their enterprise, others the payment of wages.

According to William J. Stevenson [4] «Leaders should not lose sight of the important role that planning plays in the overall success of the organization. A well-designed plan provides a competitive advantage; unsuccessful—reduces the competitiveness of the company». «Planning is the development and adjustment of a plan, which includes foresight, justification, concretization and description of the activities of an economic object in the short and long term. Planning is central to the management of the economy to achieve the goal, based on the balance and sequence of production operations and the solution of social problems» [5].

At present, one of the most urgent problems for the lack of funds of enterprises is the efficiency of allocation of production resources. At the same time, enterprises should pay great attention to the selection of products that meet the requirements for self-sufficiency and contribute to the expansion of sales through revenue [6]. These requirements are possible only with the development of enterprise resource management strategies.

The main plan for the development of the enterprise is the production of certain types of products, while the commodity producer promptly offers goods that correspond to production activities that meet the requirements of certain buyers.

Each developing enterprise fights for its place in the economic services market, while the client is a key link that makes a profit.

The production program of organizations should include a pre-thought-out nomenclature of goods that are in demand.

In the literature, there are various synonyms for the term production program planning, such as: the formation of an assortment of an enterprise, the formation of an economic portfolio, the development of a master production schedule (Master Production Schedule-MPS), the development of a plan for production and sale of products, etc. [1, 7].

Planning the range of products manufactured by the organization is considered as an independent production program. The main objective of the production program of organizations is the integrated nature of the management of all internal activities of the enterprise: marketing, financing and management.

«The production program spelled out the directions and objectives of the development of the enterprise in the planned period, production and economic relations with other counterparties, the profile and degree of specialization and combination of production» [8, 9].

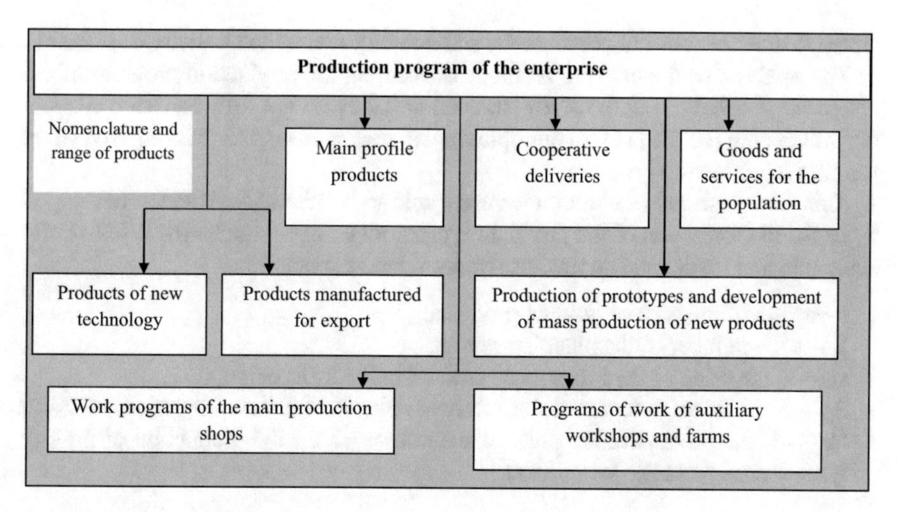

Fig. 1 The production program of the enterprise. *Source* Compiled by the authors based on [10]

In the process of studying the theoretical justification of the economic and mathematical model of the production program of enterprises, we found that the main goal of entrepreneurial activity is to make a profit from the sale of goods.

The content of the production program includes the volume and sale of certain products on the economic market, taking into account monthly production.

The formation of the production program of the enterprise depends on the forecast of economic and financial results, which are determined by calculating the economic costs of the organization from the gross income.

Thus, the calculation of the profit of the organization depends entirely on the activities of the enterprise and the effectiveness of the management of managers.

The mathematical model for calculating the profit of an enterprise will depend on the volume of products produced, the larger the volume, the greater the proceeds from the sale of products.

Economic theory shows that the optimal price that increases profit is calculated by the formula:

$$x_{opt} = S_{pr} \times \frac{\varepsilon}{1+\varepsilon}, \tag{1}$$

where Spr—direct unit costs;

$\varepsilon = \frac{dN}{dx} \times \frac{x}{n(x)}$—price elasticity.

We see that the optimal price is obtained by multiplying the direct costs per unit of output by some coefficient, which is an elasticity function independent of fixed costs. The optimal production volume is determined using the demand curve N(x). Price elasticity will not depend on price if the demand curve has the following form.

$$\mathrm{N(x)} = \mathrm{A} \times \mathrm{x}\beta, \tag{2}$$

where A and are constant coefficients, and coincides with the elasticity coefficient.

The analysis of the existing methods of forming the production program of the enterprise, carried out in the study, showed that they do not fully take into account the influencing factors [11]. In our opinion, an integrated approach is required when forming a production program.

This section discusses the complex economic and mathematical models developed by us for the formation of the production program of an enterprise [8, 12]. To build economic and mathematical models, we introduce the notation:

i—a sign of the type of product produced;
k—is the number of the planning period;
x_{ik}—is the price of the i-th type of product in the k-th period, rub.;
S_{ik}—is the direct unit cost of the i-th type of product in the k-th period, rub.;
N_{0ik}—is the production capacity of the enterprise for the production of the i-th type of product in the k-th period.

$$Nik = a_{ik} \times x_{ik} \times b_{ik}, \tag{3}$$

Taking into account the adopted notation, we write the formulas for the annual gross revenue.

$$\beta = \sum_{i=1}^{n} \sum_{k=1}^{L} \times N_{ik} \times x, \tag{4}$$

annual costs

$$U = \sum_{i} \sum_{k} \times S_{ik} \times N_{ik} + U_{const}, \tag{5}$$

and financial result

$$F = V - U = \sum \times \sum [N_{ik} \times x_{ik} - N_{ik} \times S_{ik}] - U_{const}, \tag{6}$$

Based on the obtained formulas, the problem of forming the production program of the enterprise (what to produce, in what quantity and at what price to sell) can be formulated in the form of the following economic and mathematical models.

Model 1. Find $\{x_{ik}\}$ that maximizes the value of the financial result (7 and 8) under the constraints:

- production output in the planned period should not exceed the production capacity:

$$a_{ik} \times x_{ik} + b_{ik} \leq N_{0ik}, \tag{7}$$

- the volume of demand for products in the planning period should not be negative:

$$a_{ik} \times x_{ik} + b_{ik} \geq 0, \tag{8}$$

- not negative variables:

Model 2. Find $\{x_{ik}\}$ that maximizes the value of the financial result (9–12) under the constraints:

- the volume of production in the planning period should not exceed the production capacity:

$$a_{ik} \times x_{ik} + b_{ik} \leq N_{0ik}, \tag{9}$$

- the level of profitability should be equal to the given value R_0 (in percent):

$$\frac{F}{U} \times 100\% = R_o, \tag{10}$$

$$U = \sum_{i} \sum_{k} (a_{ik} \times S_{ik} \times x_{ik} + S_{ik} \times b_{ik}) + U_{const}, \tag{11}$$

– the value of demand for each type of product in the planning period should not be negative:

$$a_{ik} \times x_{ik} + b_{ik} \geq 0, \tag{12}$$

Model 3. Find $\{x_{ik}\}$ that provides the maximum value of the financial result (13 and 14) under the restrictions:

– full use of production capacities:

$$a_{ik} \times x_{ik} + b_{ik} = N_{0ik}, \tag{13}$$

– the quantity demanded must not be negative:

$$a_{ik} \times x_{ik} + b_{ik} \geq 0, \tag{14}$$

Models (1)–(3)—problems of nonlinear mathematical programming [7]. PEC uses the following methods.

If $\{x^*_{ik}\}$—is a solution of models, then the volume of production in the planned period is determined by the formula:

$$N^*_{ik} = a_{ik} \times x^*_{ik} + b_{ik}, \tag{15}$$

With the help of economic and mathematical models, the authors propose a method for determining the need for products.

With the use of PEC, non-linear programming programs, solutions $\{x^*_{ik}\}$ of one of the three proposed models for the formation of an enterprise's production program are found.

The volumes of production of each type of product in the planned period are determined.

$$N^*_{ik} = a_{ik} \times x^*_{ik} + b_{ik}, \tag{16}$$

The relevant economic indicators are considered:

$\beta^* = \sum_i \sum_j N^*_{ik} \times x^*_{ik}$—expected revenue.
$U^* = \sum_i \sum_j N^*_{ik} \times S_{ik} + U_{const}$—expected revenue.
$F^* = V^* - U^*$—expected financial result.
$NP^* = F^*(1 - IT)$—expected net profit, where IT is income tax.
$\frac{NP^*}{U^*} \times 100\% = R^*$—expected rate of return.

Table 1 Demand forecast

Product type	Demand for the i-th type of product (pcs.)	By price (rub.)
Product 1	16	42,426
	18	38,569
Product 2	14	44,725
	16	40,660
Product 3	6	47,586
	7	43,260
Product 4	6	52,228
	7	47,480
Product 5	50	20,339
	67	18,490

Source Compiled by the authors

Let's give an example for an enterprise to form a production program for a month for five types of products, with the following initial data:

$S_1 = 18{,}954$ rub.; $S_2 = 21{,}300$ rub.; $S_3 = 21{,}960$ rub.; $S_4 = 22{,}150$ rub.; $S_5 = 6{,}429$ rub.
$N_{01} = 65$ pcs.; $N_{02} = 65$ pcs.; $N_{03} = 64$ pcs.; $N_{04} = 64$ pcs.; $N_{05} = 258$ pcs.; $U_{const.} = 1{,}000{,}000$ rub.

The demand forecast for the initial types of products is provided by the marketing department operating at the enterprise and is as follows.

4 Conclusion

Thus, the model for building a production program is determined by the financial and economic justification for calculating the production profit of an enterprise for the sale of products. The management of production activities for the formation of effective profit indicators will largely depend on the methods of obtaining the main profit, on the volume of production and sales of products.

References

1. Afanasiev, M. Yu., & Suvorov, B. P. (2003). *Research operations in the economy: Models, tasks, solutions: Proc. allowance* (p. 444). INFRA-M.
2. Sklyarenko, V. K., & Prudnikov, V. M. (2006). *Economics of the enterprise. Textbook* (p. 528). INFRA-M.
3. Volkov, O. I., & Sklyarenko, V. K. (2004). *Enterprise economy* (p. 280). INFRA-M.

4. Garina, E. P., Romanovskaya, E. V., Andryashina, N. S., Kuznetsov, V. P., & Shpilevskaya, E. V. (2020). Organizational and economic foundations of the management of the investment programs at the stage of their implementation. *Lecture Notes in Networks and Systems, 91*, 163–169.
5. Karasev, V. P. (2007). Economic and mathematical model of formation of the production program of the enterprise. In *Interuniversity Collection of Scientific Papers «The Effectiveness of the Functioning of Economic Organizations in Modern Conditions»*, N. Novgorod (pp. 56–62).
6. Andryashina, N. S., Garin, A. P., Romanovskaya, E. V., Kuznetsova, S. N., & Kozlova, E. P. (2020). Analysis of reserves for effective development of production. *Lecture Notes in Networks and Systems, 73*, 403–414.
7. Volkova, O. I., & Devyatkin, O. V. (Eds.). (2007). *Economics of the enterprise (firm). Textbook* (3rd ed., revised and additional, p. 601). INFRA-M. (100 years of the Russian Academy of Economics named after G.V. Plekhanov).
8. Egorova, A. O., Andryashina, N. S., & Kuznetsov, V. P. (2016). Methodology of formation and realization of competitive strategy of machine building enterprises. *European Research Studies Journal, 19*(2 Special Issue), 125–134.
9. Lizunkov, V. G., Morozova, M. V., Zakharova, A. A., & Malushko, E. Yu. (2021). On the issue of criteria for the effectiveness of interaction between educational organizations and enterprises of the real sector of the economy in the conditions of advanced development territories. *Vestnik of Minin University, 9*(1).
10. Kozlova, E. P., Kuznetsov, V. P., Garina, E. P., Romanovskaya, E. V., & Andryashina, N. S. (2020). Methodological bases of the assessment of sustainable development of industrial enterprises (technological approach). *Lecture Notes in Networks and Systems, 91*, 670–679.
11. Romanovskaya, E. V., Kuznetsov, V. P., Andryashina, N. S., Garina, E. P., & Garin, A. P. (2020). Development of the system of operational and production planning in the conditions of complex industrial production. *Lecture Notes in Networks and Systems, 87*, 572–583.
12. Sedykh, E. P. (2019). System of normative legal support of project management in education. *Vestnik of Minin University, 7*(1), 26.

System Factor and Analysis of Economic Security of Organizations

Zhanna V. Smirnova, Victor P. Kuznetsov, Andrey V. Bogatyrev, Svetlana N. Kuznetsova, and Sergey D. Tsymbalov

Abstract This article discusses the theoretical analysis of the economic security of enterprises. The authors carried out a methodological analysis of the economic security management of organizations. On the basis of the analysis carried out, the concepts, goals and objectives of economic security management are substantiated. Purpose: The main ways of efficiency of the planned strategic indicators are considered. The authors identified the main types of threats to economic security. Methodology: The functional components of economic security are determined. It was noted that the management of the enterprise needs to monitor not only the quality selection of personnel, but also at the initial stage they are faced with the task of forming a personnel department from experienced and competent employees. Findings: The types of threats from the information security of enterprises are considered. Taking into account the multiple threats to information in the enterprise, a full-fledged information security must be guaranteed, it must be standardized and constantly under control. Originality: The authors concluded that ensuring economic security is a guarantee of the country's independence, a condition for the stability and effective functioning of society, and one of the important components of the successful functioning of the company. Economic security is one of the important components of the successful functioning of the company. Therefore, the issue of economic security will always be relevant, since ignoring this problem often leads to undesirable results.

Keywords Economic security · Efficiency · Risks · Threats · Enterprises

JEL Classifications R11 · R12 · R58 · Q13 · Q18

Z. V. Smirnova (✉) · V. P. Kuznetsov · S. N. Kuznetsova · S. D. Tsymbalov
Minin Nizhny Novgorod State Pedagogical University, Nizhny Novgorod, Russia
e-mail: z.v.smirnova@mininuniver.ru

S. N. Kuznetsova
e-mail: dens@52.ru

A. V. Bogatyrev
Nizhniy Novgorod Academy of the Ministry of the Interior of the Russian Federation, Nizhny Novgorod, Russia

A. V. Bogoviz (ed.), *Big Data in Information Society and Digital Economy*,
Studies in Big Data 124, https://doi.org/10.1007/978-3-031-29489-1_19

1 Introduction

Today, in order to make correct and effective decisions, it is necessary to have sufficiently complete information about the state of economic security of the organization. Moreover, the main activity of organizations is always associated with many risks, organizations are at risk because they accumulate large cash flows, and this problem affects not only the owners, but also the employees of the organization.

2 Methodology

Given the high importance of this issue, there are a large number of works aimed at studying approaches to ensure the economic and financial (and other) security of business entities. These issues are reflected in the works of Dryagunova [1], Dukhnovsky [2], Sklyarenko and Prudnikov [3], and Zharikov [4].

The main purpose of the study of this problem is to analyze and assess the factors and threats that affect economic security, organizations and develop recommendations for managing the economic security system.

In the course of the study, we selected commercial organizations. The main assessment methods used in studying the level of economic security are: a system of general scientific methods of cognition, system and factor analysis. To calculate the parameters and threshold values of economic security, various indicators of economic security and bankruptcy forecasting models are considered.

3 Results

For the activities of any organization that wants to function successfully, develop, be competitive in the market and maintain sustainability, it is necessary to remember about economic security. In addition, the security of the company affects the level of security of the country as a whole. This works for the reason that the economic tasks facing the state to a greater extent depend on the activity and effective work, it is they who satisfy most of the needs of society.

The definition of economic security is not enshrined in law, but there are opinions of economists on this matter. They define economic security as the absence of danger, the protection of an object from internal or external threats [5].

The definition of the economic security of an enterprise, as a rule, is associated with the ability of an organization to withstand the negative influence of external factors acting on the organization.

The economic security of an enterprise (ESS) is such a state of all types of resources of an organization, in which their effective use leads to the neutralization or reduction of the impact of emerging threats and, as a result, leads to the scientific, technical and social development of the organization and the achievement of strategic goals.

The economic security of a commercial organization depends on the total set of risks. In case of improper management of these risks, threats to the security of the organization may arise [1].

In modern conditions, the role of economic security for the organization is growing, in direct proportion to the growth in the number and variety of threats [2]. In most cases, the authors identify types of ES threats based on a combination of existing factors. At the same time, a detailed classification of threats to economic security, reflecting modern problems, is necessary for the operational provision of ES. Basically, scientists divide threats into external and internal.

The structure of economic security includes many areas of security that must be taken into account when analyzing and assessing the economic security of an enterprise in order to obtain a more complete picture of the economic situation.

Economic security includes the following types (functional components) presented in Table 1.

Many economists believe that it is financial security that is the most important part of the entire economic security of an organization. It represents a state of the financial system of an organization that ensures the efficient use of resources, while counteracting threats or reducing their impact, and at the same time contributes to the achievement of business goals [6, 7].

If serious potential threats are identified, appropriate management decisions should be made, a set of actions should be taken, which the company is reported to take to reduce the level of threats to economic security.

In modern society, among the active actions of informatization, threats in organizations and individual security management structures are increasing.

Table 1 Functional components of EB

Financial	– Efficient use of resources
Personnel	– Effective personnel management
Informational	– Effective information and analytical support for the economic activity of the enterprise
Legal	– Legal support of the enterprise
Property	– Ensuring the physical safety of the employees of the enterprise and the safety of its property
Technical and technological safety	– The degree of compliance of the technologies used at the enterprise for optimizing resource costs

Source Compiled by the authors

In science, there is the most common approach to determine the area of occurrence of threats and risks, they are internal and external factors of economic security. The main risk management methods are:

- risk avoidance;
- acceptance of risk;
- risk transfer;
- risk reduction;
- risk insurance.

The first method is based on the refusal to interact with individuals or companies that may pose a threat. The decision to abandon any risk can be made at the stage of its preparation, as well as in the process of implementing a project in which the company is already participating.

If a company abandons a project in which it is already involved, it will incur additional losses. Therefore, it is important at the initial stage to evaluate everything and make a decision to refuse, so the company can avoid unnecessary expenses by abandoning an unreliable project [8].

This method is considered the simplest, and involves an almost complete reduction in losses, but when using it, it is impossible to make a profit in full.

The next way is to take the risk. This implies accepting the financial consequences of an adverse event. In this case, two options are considered

- risk occurrence (as likely as possible);
- amount of losses in case of risk occurrence.

The Company may accept the risk if it is within acceptable limits or when it is not possible to influence the risk in any other way. However, this does not mean that this risk will be ignored, it will be under constant monitoring. Risk can only be ignored if it is not noticed at all.

In addition, when a company takes on a risk, it accepts the losses associated with it, they must be taken into account. They can be permanent or one-time.

To reduce the consequences of risks, reserve funds are created. They are created in accordance with the company's charter and applicable law. In them, funds are debited at regular intervals [9].

In addition to all this, while taking on risk, the company must have a clear plan of action in case of negative events.

The advantage of this approach is its simplicity and low operating costs. It does not require special knowledge in the field of risk analysis and management.

Risk transfer is another method of risk reduction. Often such a transfer is carried out on the basis of a contract of sale. In these contracts, the transfer is beneficial to both parties. If a commercial company transfers significant risks, then for the host they may be insignificant at all. Typically, the receiver is in the best position to mitigate losses.

There are several ways to transfer risks:

- conclusion of a contract;
- organizational form of business;
- hedging.

The next method of risk management is risk reduction. This method can be achieved in several ways: increasing the accuracy of loss prediction, accumulated experience in this field.

Segregation—minimization of losses in the event of a risky situation by separating assets. Assets can be divided by use, by ownership.

And the consolidation of assets occurs on the basis of the internal growth of the company or on the basis of business centralization, that is, when two or more commercial firms merge. Mergers of firms usually occur in adverse market conditions. It reduces possible losses due to the reduction in the number of units at risk [10].

Among other things, another method is used to reduce losses—diversification. It is based on the principle of separating the company's assets with the subsequent consolidation of possible losses.

This increases the safety of investments by restructuring the investment portfolio. Portfolio risks are reduced due to the multidirectional nature of investments. Portfolios, which consist of risky financial assets, are formed in such a way that if, as a result of unforeseen events, one of the projects turns out to be unprofitable, then other projects can be successful and make a profit. Thus, there are a large number of measures aimed at minimizing the implementation of threats. With competent management in the company, these measures are used, perhaps not all, but only those that are necessary.

4 Conclusion

Of course, the company's economic security system includes not only financial security, but also personnel, information, legal, property, technical and technological security. Each component includes a number of indicators that allow you to reduce risks, respond in time to emerging threats or mitigate the consequences as efficiently as possible. In modern conditions, the role of economic security for the organization is growing, because the number and diversity are increasing [11].

The economic security of a company depends on the total set of risks affecting it (external and internal). If these risks are not properly managed, security risks can arise.

To avoid this, there are groups of indicators that help assess the level of economic security, while each section of it has its own groups of indicators, because in order to get a complete picture of the situation, you should pay attention to all aspects in which the organization operates (financial, personnel, information, legal, property, and technical and technological security). In addition to indicators, there are models that allow you to assess the degree of probability of an organization's bankruptcy [12].

For the safe and efficient operation of the company, to maintain its competitiveness in the market and to achieve its goals, management must be attentive to economic security, monitor the possibility of internal and external risks, try to minimize threats and maintain the stability of the company [13].

References

1. Dryagunova, D. M. (2018). Financial state of the enterprise and its analysis. *Young Scientist, 43*(229), 218–220.
2. Dukhnovsky, S. V. (2020). *Personnel security of the organization: Textbook and workshop for universities* (p. 245). Yurayt Publishing House.
3. Sklyarenko, V. K., & Prudnikov, V. M. (2006). *Enterprise economics textbook* (p. 528). INFRA-M.
4. Zharikov, A. A. (2020). Legal regulation of business security activities: Textbook on the specialty 40.05.04 (Judicial and prosecutorial activity) (p. 148). Development Fund of the Branch of the Moscow State Law Academy Named After O. E. Kutafin in Vologda.
5. Andryashina, N. S. (2014). Modern approaches to the creation of a new product in mechanical engineering. *Vestnik of Minin University, 1*(5), 1.
6. Egorova, A. O. (2014). Analysis of strategic planning at engineering enterprises of the Russian Federation. *Vestnik of Minin University, 1*(5), 4.
7. Sergi, B. S., & Scanlon, C. C. (Eds.). (2019). *Entrepreneurship and development in the 21st century.* Emerald Publishing.
8. Garina, E. P. (2014). Business decisions on the creation of a product in the industry. *Vestnik of Minin University, 1*(5), 2.
9. Garina, E. P., Romanovskaya, E. V., Andryashina, N. S., Kuznetsov, V. P., & Shpilevskaya, E. V. (2020). Organizational and economic foundations of the management of the investment programs at the stage of their implementation. *Lecture Notes in Networks and Systems, 91*, 163–169.
10. Lizunkov, V. G., Morozova, M. V., Zakharova, A. A., & Malushko, E. Yu. (2021). On the issue of criteria for the effectiveness of interaction between educational organizations and enterprises of the real sector of the economy in the conditions of advanced development territories. *Vestnik of Minin University, 9*(1).
11. Popkova, E. G. (2017). *Economic and legal foundations of modern Russian society.* A Volume in the Series E. G. Popkova (Ed.), Advances in research on Russian business and management. Information Age Publishing.
12. Popkova, E. G., & Sergi, B. S. (2018). Will Industry 4.0 and other innovations impact Russia's development? In B. S. Sergi (Ed.), *Exploring the future of Russia's economy and markets: Towards sustainable economic development* (pp. 51–68). Emerald Publishing.
13. Sedykh, E. P. (2019). System of normative legal support of project management in education. *Vestnik of Minin University, 7*(1), 26.

Analysis of the Development of the Strategic Management Models as a Factor in the Development of Oil and Gas Companies in the Transition to Market Relations

Sergei B. Zainullin, Penesiane E. O. G. Seri, Olga A. Zainullina, Wilfried V. R. N'takpe, and Oumar M. Ouattara

Abstract Contemporary international relations are characterized by rapid dynamism and a tendency to further complicate the system of international relations of oil and gas companies, which are acquiring a multidimensional character in all areas—from the economy and finance to the security sector. The mutual dependence of international actors is constantly growing, which requires the search for new forms for their coordinated and coordinated interaction of oil and gas companies. Strategic partnerships are an important foreign policy tool, increasingly used by major countries and integration organizations to improve the effectiveness of oil and gas companies on the international stage. To succeed in foreign markets, oil and gas companies must develop a certain strategy that will allow them to enter a new international market and safely implement their activities in it. The strategy of entering the market of oil and gas companies with a certain product is the most popular strategy of international marketing. Depending on the chosen strategy of the oil and gas company, specific activities are developed. Oil and gas companies often focus on maximum effect without considering risk factors, on minimizing risk without expecting large profits, or on a combination of these approaches.

Keywords Development · Models · Strategic management · Factor · Oil and gas companies · Market relations

JEL Classifications G01 · L1 · M31 · P28

S. B. Zainullin (✉) · P. E. O. G. Seri · W. V. R. N'takpe
Peoples' Friendship University of Russia (RUDN University), Moscow, Russia
e-mail: law_union@mail.ru

S. B. Zainullin
Department of Advertising and Visual Communications, Moscow University for Industry and Finance "Synergy", Moscow, Russia

O. A. Zainullina
Moscow University for Industry and Finance "Synergy", Moscow, Russia

O. M. Ouattara
Ufa State Petroleum Technological University, Ufa, Russia

A. V. Bogoviz (ed.), *Big Data in Information Society and Digital Economy*,
Studies in Big Data 124, https://doi.org/10.1007/978-3-031-29489-1_20

1 Introduction

The development of strategic management was facilitated by the works of I. Ansoff. Influenced by the works of M. Porter, many researchers conducted analytical developments to analyze the space for maneuver that a company has in the gas and oil sector. A well-chosen strategy with a clear mission, goals, and objectives will allow the company to operate effectively in international markets, enabling it to become a leader in the future [1].

The issues of changing the strategy are especially relevant because the world is entering a protracted economic crisis comparable to the Great Depression [2].

2 Methodology

The methodological basis of this research is the classical economic theory and the maximum essential works of Russian and foreign scientists and practitioners withinside the discipline of strategic and international relations. This research used the following methods: cognition, formal and dialectical logic, analysis, synthesis, and theoretical generalization.

3 Results

A strategy is a specific management plan of action to achieve the set goals. It determines how the organization will function and develop and what entrepreneurial, competitive, and functional measures and actions will be taken to ensure that the organization reaches the desired state.

The main reason for oil and gas companies to enter the international market is to maximize profits. However, one factor (no matter how significant its role is) cannot lead to a decisive change in the company's behavior in the gas and oil sector. One way or another, a combination of factors influences the adoption of an important strategic decision.

A review of the economic literature reveals the main reasons (motives) for entering foreign markets or expanding oil and gas activities.

The motives for entering foreign markets are as follows: the nature of the influence of the founding factor (active or reactive) and the possibility of exposure to the factor that causes the appearance of the motive (external or internal).

Active and external factors are as follows:

- The desire to benefit from cheaper resources in foreign markets (raw materials and labor);
- The possibility of obtaining tax benefits or state support for exporters;
- The desire to follow partners or customers who have entered the foreign market;

- Depreciation of the national currency;
- Opportunities for foreign markets.

Active and internal factors are as follows:

- Search for higher profitability of the business;
- The company's desire for growth;
- The desire to obtain economies of scale or additional commercial effects from the use of the advantages of national factors of production and the useful resource ability of the country;
- Increasing the liquidity of the company's assets through the use of foreign currency sources;
- The availability of unique technology or product;
- The desire to benefit from a well-known brand;
- The desire to reduce risks due to the geographical diversification of operations;
- Managers' initiative.

Reactive and external factors are as follows:

- Pressure from competitors in the domestic market;
- Limited capacity (relative saturation) of the domestic market;
- Decrease in demand or the purchasing power of consumers in the domestic market;
- Seasonality of demand in the domestic market.

Reactive and internal factors are as follows:

- Overproduction or availability of unused production capacity.
- Transition to the maturity stage of the product life cycle in the domestic market [3].

A company may have new technological developments in a specialized field. Bearing in mind that the excellent characteristics of the product can subsequently be partially or completely copied by other market participants, most companies in the gas and oil (hydrocarbon) sector respond to business development opportunities abroad and enter the foreign market [4].

The initiative of managers is a motive that reflects the commitment of the management of international activities and the desire to participate in them. The dynamism of managers is probably because they enjoy being part of an international oil and gas company. Moreover, it often gives rise to foreign trips.

However, the initiative of internationalization on the part of managers is often a simple reflection of the desire common to all entrepreneurs to grow and expand the market constantly.

Geographic diversification of operations can be chosen to reduce possible risks if the domestic market has factors that pose a threat to the operations of an oil and gas company (e.g., if the instability of the political situation and imperfect regulatory and legislative framework force the company's management to take into account the danger of possible restrictions on doing business in the national market).

One of the main tools of strategic planning is the strategy of an oil and gas company. In general, the strategy is a fundamental long-term solution for using the international apparatus to achieve the set goals. For a hydrocarbon (gas, oil) company, the strategy is part of its overall strategic development plan; it requires reasonable development methods.

Russian multinationals show some common features in their internationalization strategies for the various industries in which they operate. In energy companies, natural resources are the most important asset that determines a company's competitiveness. Because of the lack of means, competition is oligopolistic: there are only a few large companies. In view of the strategic nature of oil and gas, one of the specialties of this industry is the strong involvement of government at the highest level. In almost all companies, a stake is held by the Russian government. Thus, by owning shares, the government can influence the boards of directors of companies, set rules, or exert this influence by granting drilling and operating licenses. Therefore, political motives for entering foreign markets often dominate economic ones [5].

Thus, companies from emerging economies often begin the internationalization process by gathering resources or critical knowledge, which may be technological know-how, human resources, management skills, global brands, or R&D capabilities. Companies with low levels of R&D tend to focus primarily on their domestic market while seeking to start internationalization from neighboring countries with low cultural distances.

Companies with high R&D spending are entering domestic and international markets almost simultaneously, with overseas affiliates perceived not as mere additions to the domestic market but rather as an important component of a global presence. One of the generally accepted views on this issue is that transnational corporations from developing countries are primarily looking for training opportunities in foreign markets. Simultaneously, companies from developing countries have other motives, such as, for example, fleeing from an unstable home region, which is called "striving for stability" or "escaping from the system."

An analysis of how Russian firms entered foreign markets shows that there are general models of international gameplans continued by enterprises in each of the four groups described above.

Infrastructure is very demanding and cannot be built from scratch, so mergers and acquisitions are often the only solutions. In some cases, international joint ventures help to strengthen innovation and opportunities in R&D. It is important to note that in these four groups of companies, although the existing model constitutes a common logic for internationalization strategies, companies sometimes deviate from this logic because of their unique characteristics.

The international strategies of Russian oil and gas societies are determined by institutional, industry, and internal factors. Institutions impact competition rules, technical standards, innovation development resources, and company growth priorities. Simultaneously, the government plays a major role in developing the export activities of the Russian industry. The influence of the government varies depending on the industry, being especially strong in the fuel and energy complex and taking only an indirect form in other industries [6].

Raw materials, together with the strong dominance of the hydrocarbon industry, are the mainstay of the Russian economy. The share of oil revenues in the total

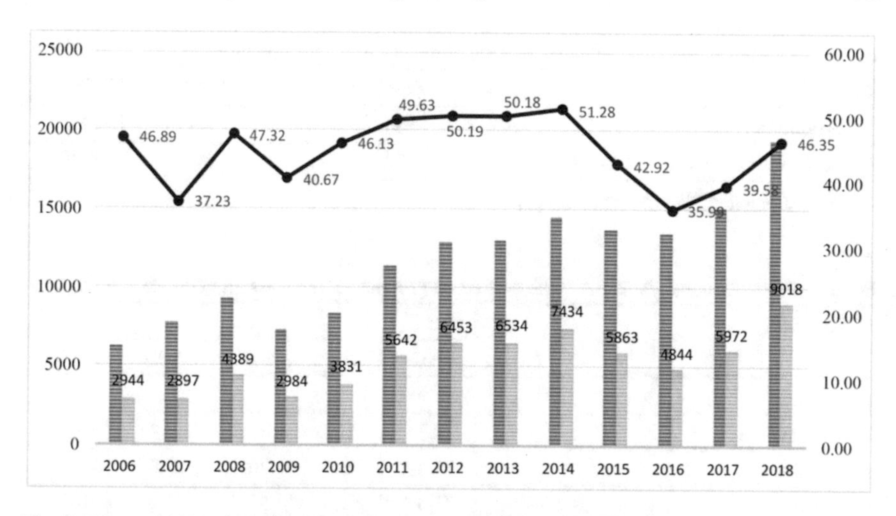

Fig. 1 Share of revenues from oil and gas activities in the total budgetary revenues (in 2007–2020, billion rubles) of Russia. *Source* Compiled by the authors based on [7]

revenues of the federal budget is, on average, about 40%. In 2014–2016, this figure was about 50%; in 2018, it decreased to 35.99 and 46.35% by 2020 (Fig. 1).

The inclusion of these revenues in 2020 will increase the oil revenues of the state budget by 17% (10.5 trillion rubles) compared to the official forecast of the Ministry of Finance.

Rosneft, Lukoil, Surgutneftegas, Bashneft, Gazprom Group, and Tatneft are the six vertically integrated oil companies that dominate the highly concentrated Russian oil market. Among the largest gas and oil companies, the largest contribution to the receipt of budget revenues was made by PJSC Rosneft (2.5 trillion rubles), PJSC Gazprom (1.8 trillion rubles), and Lukoil (1.03 trillion rubles) [8].

Understanding the important role played by the oil industry, the government, represented by the Ministry of Energy of Russia, has been carrying out activities to introduce the latest technologies in the fuel and energy sectors since 2016. From 2017 to 2020, more than 50 projects were considered, ten of which in the gas and oil complex were assigned the status of national projects. In 2020, Rosneft initiated more than 20 innovative projects and received more than 60 patents.

Thus, the government plays a crucial role as a subject of the oil market. For these purposes, the government has adopted a Russian energy strategy for 2035.

To realize Russia's foreign economic interests, it is necessary in the future to ensure a dominant position in terms of energy supplies to world markets.

As of 2020, the Russian Federation ranks third in the world in oil production and sixth in terms of oil consumption. The ratio of oil production and consumption is presented in Table 1.

By the end of 2020, Russian oil production increased by 9 million tons to 555.9 million tons, which is close to the highest absolute production in the Russian Federation since 1988 (570 million tons). This increase in production is linked to the

Table 1 Ratio of production and consumption of oil in Russia, million tons

Index	2015	2016	2017	2018	2019	2020
Oil production	505.1	526.7	534.0	547.6	546.9	555.9
Domestic consumption	125.9	141.3	122.2	138.3	147.1	146.7

Source Compiled by authors based on [7]

Table 2 Production and export of basic petroleum products in the Russian Federation, mln. tons

Mineral oil	Year				
	2016	2017	2018	2019	2020
Production					
Motor gasoline	38.3	39.2	39.9	39.2	39.5
Diesel fuel	77.3	76.1	76.4	76.8	77.5
Heating oil	77.5	71.0	57.0	51.2	48.4
Export					
Motor gasoline	4.2	4.7	4.9	5.2	4.2
Diesel fuel	45.5	45.1	43.7	48.9	54.8
Heating oil and other petroleum products	113.3	115.8	102.2	93.2	91.1

Source Compiled by authors based on [7]

decision of OPEC+ to increase production limits from June 2020, which resulted in Russia's average daily oil production reaching the highest level in October 2020 that Russia has ever achieved in the past (Table 2).

We saw a 3.75% decline in Russian exports in 2012 and a 1.96% decline in 2016. During the rest of the reporting period, oil exports increased steadily compared to last year. This is not only due to the increase in the absolute volume of exports but also because Russian oil consumption has considerably decreased compared to the Soviet period due to the transformation of the Russian economic market, the improvement of oil efficiency, and the replacement of petroleum products by natural gas.

In 2020, exports of petroleum products increased, while negative elements of the industry remain. Thus, if in 2016 Russia produced 78.4 million tons of fuel oil, then by 2020, its production decreased to 48.4 million tons in the composition of production of basic petroleum products, and the share of fuel oil decreased from 40.4% in 2016 to 29.3% in 2020.

The analysis of the processes carried out by the company to enter foreign markets and factors that affect different stages of the company's functioning and intermediate goals allows us to form a sequence of planning stages for entering the foreign market of a company operating in the field of oil and gas.

The sequence of steps in planning the company's strategy for entering the international oil and gas market covers all aspects of its activities at the corporate level.

As mentioned above, at each stage of the activities of a hydrocarbon company, there arise certain risks, which are accompanied by the likelihood of commercial defeat in the market or failure to achieve the set strategic goals and, as a result, economic losses or loss of some resources [9].

The development of an international strategy involves the definition of international support at each stage of planning the process for entering the foreign market of a company operating in the oil and gas sector. To this end, a set of tools is proposed using appropriate matrices to shape an international strategy.

A set of international strategy tools aligned with the process of entering the foreign market of a company operating in the hydrocarbon sector (gas, oil), classification of the process of entering the foreign market of a company operating in the hydrocarbon sector (gas, oil), and then the processes of international activities when a company enters the oil and gas system to the foreign market create opportunities to enter international markets [10]:

1. Analysis of the reasons that prompt the company to enter foreign markets: analysis of macro-and microenvironmental factors, ranking of factors by nature, stability, and degree of influence;
2. Assessing the motivation for oil and gas companies to enter the market of a particular country;
3. Analysis of market conditions: identification of the most important market conditions and factors contributing to access to foreign markets;
4. Intra-brand analysis: analysis of the company's strengths in the hydrocarbons sector as sources of competitive advantages on external links;
5. Formulation of international development goals: setting goals in the foreign market;
6. Setting goals in the foreign market: research of foreign markets, assessment of the attractiveness of the market, and formation of a matrix of the attractiveness of markets;
7. Development of the company's international strategy: development of an international strategy;
8. Implementation of the developed strategy: development of an international complex and programs;
9. Preliminary analysis and control: analysis of the effectiveness of international activities in the process of entering the foreign market.

In the first stages of planning the process of entering the foreign market, it is important to analyze the factors of market conditions. The market environment of any hydrocarbon industry is a significant number of controlled and managed factors, and vice versa: uncontrolled factors that directly affect its activities.

The activities of hydrocarbon industries at every stage of the market environment involve the consideration of inherent factors, which include the peculiarities of legislative regulation of commercial activities, preferences formed by the sociocultural environment, the availability of resources, the competitive environment, the culture of economic and business relations, and others [11].

A distinctive feature of the international regional level is the interaction of state markets with a similar socio-cultural environment, indicators of economic and social development, and approaches to the implementation of foreign economic activity. The regional market environment is quite homogeneous in terms of the national markets it contains in its composition; it should also be noted that contemporary regional markets significantly differ from each other.

However, in the context of developing a strategy for entering the foreign market, it is advisable to consider the global market environment as a set of international regional markets that make up the system of international economic relations. The company's entry into the global market environment has a significant positive impact on its activities and becomes an impetus for improving technologies and products.

The assets that create the most significant advantages to achieve the different international strategic objectives can be identified if and only if we consider the international oil and gas industry competition and its advantages [9]:

- Proprietary technology owned by the company and that can be obtained by others only through new R&D or by acquiring a license from the owner;
- Skills in managing international operations, usually acquired through experience in different countries;
- Access to scarce resources or raw materials;
- Economies of scale (cost reduction);
- Financial economies of scale (access to sources of finance);
- Ownership of a "strong" brand.

After identifying market opportunities to enter foreign markets and analyzing the strengths of oil and gas companies, which can become sources of competitive advantages, the company should compare them with their corporate goals and, as a result, formulate international goals.

4 Conclusions/Recommendations

It is important to assess the degree of integration of oil and gas companies in the international market to develop an effective international strategy. In the hydrocarbon sector, factors such as the type and form of the society and its interaction with international market counterparts, the degree of external economic activity, and the nature of the activities and the development strategies are determined by the degree of integration. In the gas and oil sector, the company's international strategy must include a differentiated set of measures for the different levels of the market environment (national, regional, and global) because each of them has certain characteristics and imposes specific requirements on the world business of gas and oil companies.

According to the abilities of the gas and oil industries, the authors proved the existence of different levels of strategy planning: corporate, commercial, functional, and instrumental levels, at each of which its companies make appropriate strategic decisions. It is established that the strategies of different companies to enter the foreign market provide different options characterized by the ways of entering the country's market, choice of distribution channels, flexibility, and types of control, each associated with certain costs and risks.

To prevent international risks and manage them in a targeted manner, it is necessary to organize a qualitative organization of appropriate information support to develop a strategy for entering a foreign market, which provides for the collection of information on the circumstances of the foreign market environment.

Acknowledgements This paper was supported by the Strategic Academic Leadership Program of the Peoples' Friendship University of Russia.

References

1. Larionova, E. I., Chinaeva, T. I., & Shpakovskaya, E. P. (2019). Analysis of the development of oil and gas industry in present conditions. *Statistics and Economics, 16*(6), 29–36. https://doi.org/10.21686/2500-3925-2019-6-29-36.
2. Zainullin, S. B., Alvarez-Gila, O., Zainullina, O. A., Gómez-Gastiasoro, M. (2022). Comparative analysis of the economic crisis at the beginning of XX century and XXI century in Russia and Spain. In E. G. Popkova & I. V. Andronova (Eds.), *Current problems of the world economy and international trade* (pp. 53–62). Emerald Publishing Limited. https://doi.org/10.1108/S0190-128120220000042006.
3. Pletnev, D. A., Nikolaeva, E. V., & Kozlova, E. V. (2020). Typology of official strategies of large Russian corporations. *Sustainable development of the digital economy and cluster structures: Theory and practice* (pp. 300–321). Polytech-Press.
4. Pushkareva, P. P., Zakharov, G. V., & Klimenko, A. O. (2020). Review and classification of methods and strategies for industrial companies entering foreign markets. *Humanitarian Scientific Bulletin, 7*, 122–130. https://doi.org/10.5281/zenodo.3959173.
5. Rahimov, I. R., & Omarov, M. M. (2020). Key marketing strategies when companies enter international markets. In Yu. V. Kuznetsov, Yu. A. Malenkov, S. V. Sokolova, E. M. Anokhina, V. M. Zhigalov, V. P. Kaisarova, ... O. L. Margania (Eds.), *Actual problems of management: New methods and technologies of management in the regions* (pp. 342–344).
6. Svyatchenko, E. A. (2019). Comparative characteristics of the main forms of company entry into foreign markets. *Economic Development Research Journal, 8*, 48–56.
7. Ministry of Energy of the Russian Federation. (n.d.). *Production of petroleum products*. Retrieved May 7, 2022, from https://minenergo.gov.ru/node/1213.
8. Sidorchak, D. S., & Dyakonova, M. A. (2021). The essence of the company's exit strategy to the international market. *Management of socio-economic development of regions: Problems and ways to solve them* (pp. 146–148). Financial University under the Government of the Russian Federation.
9. Smirnov, V. N., & Novikova, M. S. (2021). Strategies for the exit of companies to international markets. In *Formation of economic stability of regional socio-economic systems: Proceedings of the international scientific and practical conference* (pp. 259–263).
10. Ukolova, E. V. (2020). Specificities of the strategy and tactics of entering the international markets of American companies. *Eurasian Union of Scientists, 5–2*(74), 8–12.
11. Fadeev, A. M., Cherepovitsyn, A. E., & Larichkin, F. D. (2019). *Strategic management of the oil and gas complex in the Arctic*. KSC RAS.

Cooperative Mechanism for the Dissemination and Use of Big Data in the Information Society

Network Organization of Innovation Infrastructure as a Condition for Reducing the Information Entropy of Cooperative Structures

Nadezhda N. Makarova, Vladislav V. Egorov, and Dmitry N. Fetisov

Abstract Digital platforms are acquiring the status of a new trend in automating the network organization of innovative infrastructure for business structures of the agro-industrial complex (AIC), including cooperative associations. Innovative changes lead to an increase in the entropy of the system and a decrease in its stability. The main purpose of this work is to develop the conceptual foundations of the information content of the network organization of the innovation infrastructure of cooperative structures from the standpoint of the entropy approach in the scientific and technological paradigm. The methodological basis of the study is the methods of systemic and comparative analysis, entropy analysis, network theory, and innovations. The article presents a 3D model of the relationship of concepts: "entropy"-"economy"-"information"-"innovation", which allows revealing the directions of development of the digital economy with emphasis on the main derivative of all economic processes—information. The result of the study is a developed flowchart for the development of economic processes and changes in entropy associated with the pooling of resources in the innovation infrastructure of cooperative structures.

Keywords Innovative infrastructure · Business structures · Agro-industrial complex · Entropy · Network · Information · Cooperative structures

JEL Code D81

N. N. Makarova (✉) · V. V. Egorov · D. N. Fetisov
Volgograd Cooperative Institute (Branch) Russian University of Cooperation, Volgograd, Russia
e-mail: yamg@mail.ru

V. V. Egorov
e-mail: vlad-negoro2011@yandex.ru

D. N. Fetisov
e-mail: vkikaffin@yandex.ru

A. V. Bogoviz (ed.), *Big Data in Information Society and Digital Economy*,
Studies in Big Data 124, https://doi.org/10.1007/978-3-031-29489-1_21

1 Introduction

The development and improvement of cooperative processes in the economy and, in particular, in the area of agro-industrial production form the basis for free business and the effective functioning of entrepreneurial structures in agriculture. In the multilevel system of agro-industrial production, an important place is given to associations of participants in this process based on cooperation. Today, consumer cooperatives operating in agricultural production combine the features of commercial and non-commercial organizations. Therefore, attributing them to only one of these forms is incorrect. Therefore, the status of a consumer cooperative must be considered following international and Russian legal norms, because this allows expanding the economic boundaries of the activities of consumer cooperatives and eliminating the duality in the terminological content of consumer cooperation.

Currently, the most important task in the area of organizing agricultural production is the search for scientifically based solutions for the effective functioning and modernization of production processes in cooperative structures based on the widespread use of digital technologies.

The current technological development of different sectors of the Russian economy is accompanied by external shocks, modern global challenges, such as a pandemic. In these circumstances, the Russian economy is undergoing significant qualitative changes that require technological modernization of production and the creation of effective, competitive cooperation, the development of completely new markets. In this case, innovation is the main driver of these changes and an important direction of strategic business development. In accordance with the new paradigm of scientific and technological development, they reflect the permanence of the innovation process that converts knowledge into real achievements in the long term.

In the context of the new scientific and technological paradigm, large-scale restructuring of the economy based on fundamentally new technological solutions takes place, the intensity of innovation activity and the need to introduce advanced innovation with forecasting the results of strategic development in all sectors of the economy are increasing, new open innovation models are being created based on large-scale network interactions.

2 Materials and Methods

Economic transformations in the country are focused on the growth of innovative activity in all basic sectors of the economy, including all cooperative structures. The development of innovative activity in these structures is one of the key problems, the solution of which will not only change the nature of agricultural production but also fill it with a different quintessence and practical content. Therefore, innovative technologies are among the promising directions for the development of agricultural production in Russia. Investments in this aspect are considered long-term and most effective, due to they are aimed at introducing new technological solutions into the activities of cooperative structures [1].

The development of innovations remains one of the most important areas of technological transformation of the economy in terms of its digitalization. The Decree of the President of the Russian Federation No. 203 of 09.05.2017 "On the Strategy for the Development of the Information Society in the Russian Federation for 2017–2030" defines the digital economy as "an economic activity in which digital data, processing of large volumes of data, and the use of the results of the analysis, in comparison with traditional forms of management, can significantly increase the efficiency of various types of production, technologies, equipment, storage, sale, delivery of goods, as well as services".

It should be noted that all areas of business activity in the agro-industrial complex are based on agricultural production, which is highly risky and most vulnerable, and it's impossible to fully structure all business processes there, due to the high level of uncertainty of the external environment.

Strategic management of cooperative structures, which is accompanied by the necessary technological improvements, can only be implemented with the participation of an innovative system of scientific and information support. At the same time, a well-coordinated network of all business entities based on intercompany cooperation should be created: medium and small agricultural, as well as processing organizations, including personal subsidiary farms, larger entities united by the interests of developing innovative infrastructure. Actually, the networks set the spatial coordinates of an effective partnership of economic agents that aim to actualize their innovative potential.

Today, the trend vector for the development of entrepreneurial structures is the formation of a "smart agricultural association (economic entity)", linking all production systems based on a single digital platform, which is positioned as a breakthrough innovation.

The format of the network innovation infrastructure of cooperative formations is a complex dynamic system of interaction between various entities based on the introduction of more advanced technological processes, equipment, and organization of agricultural production with a ratio of information content and entropy of the innovation network. Summarizing all of these, it can be argued that the network form of organizing the innovation infrastructure of cooperative structures is considered as a system that, in turn, constantly accumulates entropy, because uncertainty takes place there.

Quantitative assessment of the level and effectiveness of the development of this structure can be performed based on determining the amount of information contained in it and its entropy (missing information).

Without going into details of the evolution of views on the concept of "entropy" (from the Greek: *entropia*—transformation, rotation), the authors noted, that it was first introduced in thermodynamics by Clausis in 1865 as a measure of irreversible energy dissipation. For reversible (quasi-equilibrium) processes, it was defined as follows [2]:

$$\Delta S = \Delta Q / T,$$

where: ΔS is the entropy change,

ΔQ is the change in heat,

T is the absolute thermodynamic temperature.

Entropy was studied not only in thermodynamics, but it also moved to other systems of scientific knowledge, for example, statistical physics, information theory, theory of dynamical systems, as well as quantum mechanics.

It's known that entropy serves as a measure of the uncertainty of certain structural organizations of the system. The economy or economic system is a controlled system, which sets the goal to achieve a certain structure. An increase in entropy leads to an increase in disorganization in the economic system and an increase in negative dynamics. To prevent this manifestation, the information content of the system from the outside, i.e., from the exogenous environment, is required. As a result, the incoming information contributes to the levelling or containment of the entropy process, which is accompanied by the system overcoming the state of degradation, and the decrease in entropy is a consequence of the decrease in the degree of chaos in the system, the result of the formation and strengthening of order.

In a real economic system, it's advisable to analyze the entropy associated with the concept of information by analogy.

From the position of the theory of tetrads, a relatively stable connection between the concepts of "entropy"-"economy"-"information"-"innovation" arises, which can be represented by the 3D model in Fig. 1.

In this case, the intersection of four planes is considered, forming in three-dimensional space an open box with parallel opposite faces (Fig. 1). The left and right planes are "economy" and "information", respectively. The upper and lower planes represent "entropy" and "innovation".

While crossing each pair of planes, the main directions for the development of the digital economy can be got: "economic entropy", "information economy", "information entropy", and "information innovations". Obviously, information is the main derivative of all economic processes.

Based on the model above and the definition of the digital economy, today, the main factor, in particular in agricultural production, which has the greatest impact on efficiency, is information. From the point of view of a separate structure, an information system is a data warehouse with all its corporate information. Information in this context should be understood as an object of perception, transformation, transmission, and storage of a certain amount of information, which is expressed by successive symbolic designations. It also allows eliminating uncertainty or expanding the end user's knowledge of the world.

So, the approach to information is based on uncertainty and entropy. Brillouin said, "We define information as something different from knowledge, for which we have no quantitative measure" [3]. At the same time, Schrödinger in his work "What is life?" proposed the concept of "negentropy" (negative entropy) in 1943 [4]. He argued that negentropy is a movement towards the ordering of a biological system.

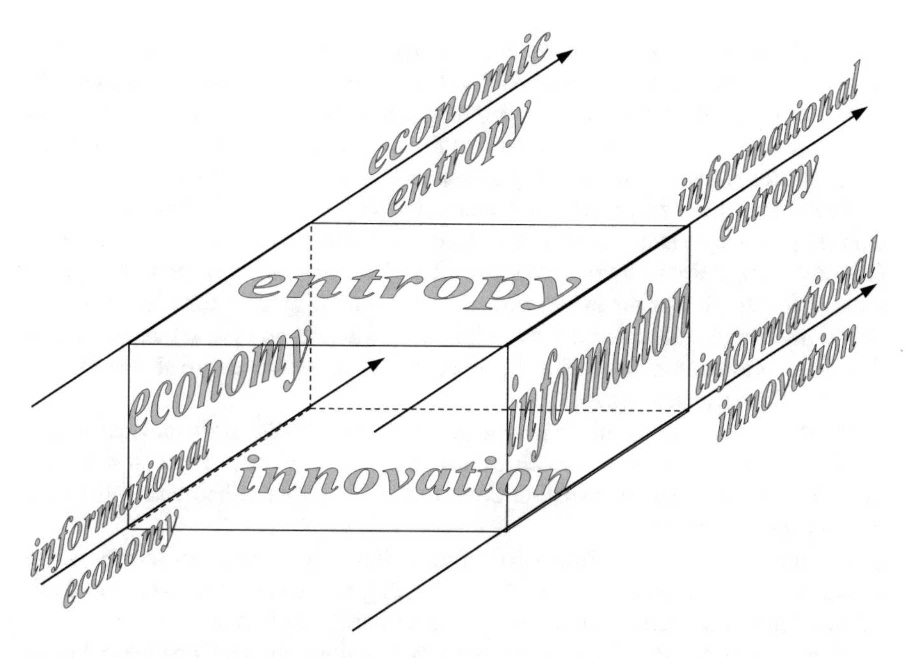

Fig. 1 3D model of the relationship between the concepts of "entropy"-"economy"-"information"-"innovation". *Source* Compiled by the authors

Later, Brillouin established the negentropic principle of information: "information is a negative contribution to entropy" [5].

Thus, entropy acts as a measure of the uncertainty of information, that is, the missing information about the state of the system, which demonstrates the lack of its organization. And new identically related information about the state of the system, which isn't negative entropy,—negentropy—in contrast to it, is a measure of its organization.

Effective development of the economic system correlates with the information component of economic activity.

Digital solutions are increasingly penetrating all segments of agro-industrial production. Information innovation is "advanced information technology" aimed at making it easier to deal with the daily flow of data. Consequently, modern agriculture and related branches of the agro-industrial complex are organized using information technology. The execution of technological operations is accompanied by the collection and prompt processing of large amounts of information. Finally, not all information in complex systems can be recognized. To recognize it, the system needs significant information innovation. This will require more energy and, as a result, more entropy will be produced, which will limit the development of an innovative infrastructure network.

The Shannon entropy [6] serves as the basis for many studies. It's about the production of entropy or the principle of maximum production of entropy. In the absence of external influences on cooperative structures, when a set of microstates of the information system is fixed, economic relations between businesses entities ensure the transition of the system to the state, in which all microstates are equally probable. From the economic point of view, the number of equiprobable states of information systems can be considered as the number of possible implementations of breakthrough innovation used to develop an entrepreneurial structure.

The value of the information contained in the technological methods of agricultural production and management decisions determines the level of knowledge intensity and quality of these categories. Therefore, only information in the conditions of functioning of the information infrastructure (Fig. 2) is the main resource, while the energy and material flow in this infrastructure, in fact, act as an equivalent of the information materialized in them. Financial resources are also one of the varieties of information flows.

In the process of economic functioning, the information infrastructure generates a set of interspecific resources that cause changes in scientific and production activities, as well as directs their information flows through the established channels to the process control system.

For more developed information infrastructures, it's typical to contain large amounts of information structured to varying degrees, which leads to the complication of links between the components of the system functioning.

As a result of the introduction of innovative solutions, the specific physical mass per unit of agricultural production in production systems decreases, which corresponds to the extensive impact of information and can be described by the Hartley formula [7], which represents a logarithmic measure of information to determine the amount of information contained in the message.

If the informational improvement of the technological process determines a decrease in the intensity of energy and substance flows entering the production systems of cooperative structures, a decrease in the cost of material and energy resources per unit of output (work) will be there due to the introduction of innovation. In this case, this is the intensive action of information, reproduced using the Shannon formula (means the average amount of data per one output of the source, and is denoted by the term entropy of the message source) [6].

3 Results

Thus, any process of obtaining information correlates with a change in entropy, and its decrease contributes to an increase in the certainty of the state of the system and its stability. Consequently, the development of innovative infrastructure networks will be a source of negative entropy. Interaction in this infrastructure network of digital platforms contributes to the formation of a digital ecosystem. In this context, a digital platform is an information and communication resource that increases the amount of information processed electronically and reduces entropy in the interaction of all participants in the trajectory of achieving their goals, not discriminating against the interests of other subjects of the innovation network.

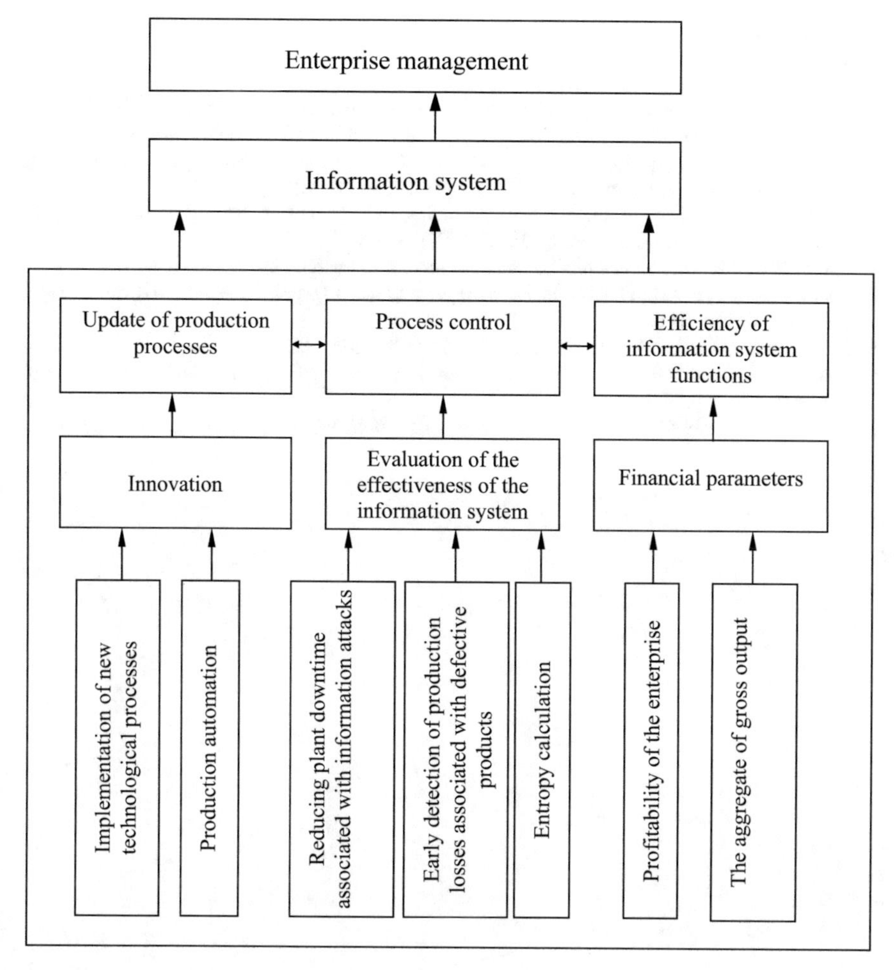

Fig. 2 Block (agricultural enterprise) of the information infrastructure of the network. *Source* Compiled by the authors

4 Conclusion

Undoubtedly, with the introduction of digital technologies and platform solutions in the activities of cooperative structures, their digital transformation can answer such challenges as crises and pandemics, because they allow deploying uninterrupted production while ensuring transparency of processes in conditions of significant restrictions.

References

1. Makarova, N., Timofeeva, G., Shubovich, A., Khalyapina, E., & Ozornnina, E. (2019). Innovation and investment activities in ensuring the sustainability of the functioning and development of agricultural organizations. *International Journal of Recent Technology and Engineering (IJRTE), 8*(3), 848. https://doi.org/10.35940/ijrte.C4034.098319
2. Carnot, S. (1934). Second law of thermodynamics. In S. Carnot, W. Thomson, R. Clausius, L. Boltzmann & M.M. Smolukhovsky (Eds.). GTTI, 312 p.
3. Brillouin, L. (1960). *Science and information theory*. Fizmatgiz, 392 p.
4. Schrödinger, E. (1972). What is life? *From the point of view of physics* (2nd ed). In E. Schrödinger (Ed.). Atomizdat, 88 p.
5. Brillouin, L. (1966). *Scientific uncertainty and information*. Mir.
6. Shannon, K. B. (1963). *Works on information theory and cybernetics*. Publishing House of Foreign Literature, 829 p.
7. Hartley, R. V. L. (1928). Transmission of Information. *Bell System Technical Journal, 3*, 535–563.

Digital Technologies in Accounting and Economics at the Present Stage of Development of Organizations in the Cooperative Sector

Olga V. Ishchenko, Zhanna A. Aksenova, Viktoria V. Saliy, Olga V. Medvedeva, and Anna A. Chernik

Abstract The article presents the characteristic features of the introduction of digital technologies in the modern sector of cooperation, taking into account the trends in the development of the economy. In the context of differences in the possibilities of modernizing the cooperative sector, the authors examine the significance and problems of the development of accounting and analysis. The exhaustion of traditional opportunities for the development of the industry through the introduction of modern digital technologies, which are based on the low cost of labour, is stated. The content, essence, and basic principles of the innovative way of development of the cooperative sector are substantiated. It's shown that the key issue related to the prospects for the further development of the cooperative sector in Russia, increasing the efficiency of using its resource potential is the introduction and use of new automated systems and tools in the context of the digitalization of the economy. Today digitalization is one of the key components in the development of the economy, affecting various spheres of life in modern society. The methodological base of the research includes the theory of accounting and auditing, as well as the principles of blockchain technologies in the aspect of their application in finance. This article provides an overview of the main trends in the development of blockchain technology in accounting and auditing based on foreign scientific articles, as well as an analysis of blockchain technologies in the area of forming the accounting policy of an organization. Conclusions about the degree of applicability of this technology in a modern organization are drawn; the main problems, risks, and benefits from the introduction of this technology are

O. V. Ishchenko · Z. A. Aksenova
Krasnodar Institute of Cooperation (Branch) of Russian University of Cooperation, Krasnodar, Russia

V. V. Saliy (✉)
Kuban State Agrarian University named after I. T. Trubilin, Krasnodar, Russia
e-mail: vlada-2807@mail.ru

O. V. Medvedeva
Kuban State Technological University, Krasnodar, Russia

A. A. Chernik
Krasnodar branch of the Russian University of Economics named after G.V. Plekhanov, Krasnodar, Russia

A. V. Bogoviz (ed.), *Big Data in Information Society and Digital Economy*, Studies in Big Data 124, https://doi.org/10.1007/978-3-031-29489-1_22

reflected. The results of the study are intended for developers of IT technologies and specialists of accounting and analytical profile and should contribute to improving the efficiency of systems for processing and exchanging big data in the area of economics.

Keywords Digitalization · Accounting · Organization economics · Cooperative sector · Accounting systems · Information systems · Data analysis · Blockchain

JEL Codes D85 · C88 · C53 · G10 · F60 · L 86 · M11 · M15 · M40 · M41 · P13 · P17 · P40 · P47

1 Introduction

Digital technologies currently cover all spheres of activity of the modern person; the most changeable environment for the implementation of various automated applications in relation to various processes remains the economy as an area of production, exchange, and consumption of various components, which is a manifestation of economic development.

The digitalization of the economy is acquiring clearer outlines in terms of the scope of the introduction of electronic document management, automation of accounting, Internet banking, and blockchain technologies. Hypothesis about the effectiveness of the formation of a unified register of data on accounting in the cooperative sector of the economy is given. The main characteristics of the concept of the formation of accounting and analytical information systems for the cooperative sector are presented [1].

The introduction of the concept of digitalization into various economic processes in the cooperative sector, the formation of automated accounting and analytical systems for accounting, as well as analytical accounting, began to take place within the framework of expanding the information space in organizations. The automated accounting and analytical information system, which has all the properties of the economic system, as well as a single information language, allows forming a single information space and implementing everything in one software package.

2 Materials and Methods

With regard to accounting, the process of generating an automated data register means a fairly wide range of changes to traditional accounting systems in the further transfer of accounting transactions into electronic form, where blockchain tools will be used. The transition to automated accounting and analytical financial systems with a large set of blockchain elements allows accounting in a new direction. The information space of the organization consists of various arrays of information presented in the form of separate documents stored in the computer memory and portable flash drives [2]. The infrastructure of the modern organization includes:

- Information resources on paper and electronic media;
- Software and hardware and Internet technologies.

Using the term "information resources", the authors mean various accounting documents: contracts, payment documents, invoices for payment, invoices, invoices, various acts of work performed, etc., which determine the composition of the organization's information space.

The structuredness of the information space, which is built based on an ordered data register, is characterized by the orderliness of information for accounting. At the same time, structured information space, where the processes of loading data from accounting entries into the accounting and analytical information system are carried out, will make it possible to determine the links between accounting accounts based on data processing algorithms [3].

While forming a structured data register, it's necessary to observe the principle of consistency of transition processes to ensure the integrity of economic data. Accordingly, the creation of single information space for enterprises in the cooperative sector, based on the data register, will allow for comprehensive automation and digitization of all economic, as well as accounting documents, in which all employees of the financial sector of the organization can work more effectively [4].

In order to form an efficiently and productively operating accounting and analytical information system, information must be sorted by information units, consisting of attributes and indicators, while the composition of them is set by the user. Features must detail the subject area, revealing the functional part of business process optimization, in the process of forming a hierarchy of features reflecting horizontal and vertical relationships in the organization of economic information [5]. In the process of developing an automated accounting information system, attributes-signs can also be single elements or constituent parts of the information space, based on which the corresponding hierarchical links are built. They allow the object to be detailed to the appropriate level [4]. According to this, a blockchain-based data register is formed, in which all transactions are recorded in chronological order and can be viewed by everyone, who has account access. At the same time, the blockchain technology is based on the principle of accounting account, during which the performed operation is recorded twice on two accounts in the form of double-entry.

It's also possible to note the basic principles of using the following blockchain technology, which are the most optimal for accounting in the formation of the data register of the accounting and analytical information system for organizations of the cooperative sector:

1. Decentralization, when all incoming and stored data are stored by all participants included in this circuit;
2. The data is available to all participants of a specific blockchain within the framework of their access rights;

3. Security—all changes with data occurring within the accounting and analytical information system are reflected in the data register records and can be added only by persons who have an account in the register;
4. It's impossible to make changes to the transactions that have already been carried out;
5. All data entering the data register of a certain contour is analyzed and verified by the accounting and analytical information system.

Accordingly, these principles can be applied to form a data register and transfer them to the blockchain technology platform, which will optimize the introduction of accounting:

1. Transactions with counterparties using blockchain technology will make it possible to reconcile settlements and to form transactions on several accounts with the possibility of carrying out a transaction in order to evaluate it for the future.
2. All movements on the assets of the organization will be "transparent", and the functions of the accountant will be reduced to processing accounting data in a single register, which will allow organizing the maintenance of assets by sectors of the register, and after all of these procedures the record is transferred to the accounts of the accounting, as well as an analytical information system.

Thus, the transfer of accounting in real-time to the blockchain platform allows making adjustments to the database, with information on accounting entries, and if these data are transferred to one platform, which is accounting and analytical information system that will be built on based on the register of data of financial and analytical reports received in real-time, this will make it possible to increase the efficiency of the functioning of organizations in the cooperative sector.

The tasks that can be solved based on created accounting and analytical information systems for the cooperative sector, built using blockchain technology, include:

- Preparation of consolidated statements with the provision of summary information in the register of data on the activities of the cooperative sector organization;
- Optimization of financial flows based on the data register;
- Analysis of incomes, costs, as well as fulfilled norms in branches and individual offices;
- Carrying out a comprehensive assessment of the organization's activities in terms of its most essential components;
- Analysis of operational processes and control over the execution of plans, mutual settlements, forecast of the receipt of funds, etc.

Accordingly to this, the cooperative sector, in order to maximize profits and control the movement of funds, must be a combination of the organization into a single chain, uniting the control companies, branches, divisions, departments, and each employee.

The tasks of any accounting and analytical information system are the storage, processing, and analysis of data, which is achieved by the presence of its composition in a data store. It's also necessary to reflect all data sources, linking them into a single integrated model in this data store [6]. The information system architecture must contain the following elements:

1. Means for collecting and primary data processing, containing:
   (a) Data analysis based on information from the local network and web portals;
   (b) Data storage, including data marts, data warehouses;
   (c) Data processing and aggregation;
   (d) Data fixation.
2. Means of implementation:
   (a) Business analysis tools;
   (b) Analytical databases;
   (c) Tools for extracting data from various transactional data sources to transform them into consolidated data.

So, an opportunity is formed to apply modern digital technologies and blockchain technology in accounting.

As it has been noted above, blockchain technology is a software tool with a typical accounting program, only in a simplified form and with a minimum amount of detail. However, this isn't an effective tool, but it allows building new economic accounting and analytical information systems for the cooperative sector.

Due to the blockchain technology wasn't created specifically for accounting and financial control, there are some peculiarities in it, but it complies with most accounting principles. This is because the information posted in the blockchain system must be reliable and truthful, and it's impossible to make any changes into blocks that have been already written, as well as make data substitution [7].

While developing an accounting and analytical information system based on the blockchain technology data register, it's necessary to form each block separately, because it will include the result of making accounting entry based on the documents on the completed operation for which it was carried out, with the introduction of mandatory details for the counterparties of the operation. Consequently, primary documents have to be automatically loaded into the accounting and analytical information system on the condition of an electronic digital signature and with the formation of standard reporting forms, so a single perfect operation will not be missed [8].

Each subsequent block of the data register must also contain information about previous accounting operations, which will prohibit the implementation of additional operations between the blocks that have been already built. If there are cases, in which the logical sequence of operations is violated, increased requirements for the accuracy and reliability of entering data into the registry database may appear.

Sources of unaccounted information, which are entered into the accounting and analytical information system built using blockchain technology in the cooperative sector, can be:

- Results of audits;
- Results of relationships with financial and credit institutions;
- Materials received in the process of interaction with performers;
- Various internal standards, reference books, price lists, etc.

Since all this information must be stored in the organization and be available at any time at the request of the employee, it's necessary to add all additional objects to the system so that there are no "gaps" in the information.

3 Results

Currently, operational databases are present and function in the activities of any organization, containing a lot of unrelated or semi-structured information. All these databases operate based on information systems that have been created at different times by various developers and implemented based on disparate software and hardware. These information systems carry out operational accounting of the results of business operations and transactions in organizations of the cooperative sector, while each organization can operate its system [9].

The introduction of an accounting and analytical information system, made in the form of data register using blockchain technology, is, at the same time, an effective resource for the formation of financial statements, in which the balance sheet for those performed during the reporting period will be visible to all participants in the contour included in the cooperative sector. As a result, there is an absolutization of the network connections of accounting operations, when not only individual people enter data into the register, but "smart" technologies carry out accounting operations and ensure their implementation based on network algorithms to analyze "big data" in single repository [10].

The transaction is carried out only if it's approved by both parties. Each transaction is recorded twice in the same amount for the parties. The fact of economic life is reflected in the same assessment for the debit of one and the credit of another account. This is consistent with the double-entry accounting principle. The information of the system is open and protected at the same time, which ensures its transparency. However, there is a problem with the confidentiality of information. But this is also considered permissible because the users of the chain can have cipher codes, or so-called "keys" to view certain information, i.e., use a private blockchain. The main advantage of implementing a data register is the acceleration of economic transactions with the ability to work without intermediaries in the process of conducting settlements with counterparties [2].

4 Discussion

The developed accounting and analytical information systems containing databases and functioning based on blockchain technology are necessary to track information, which comes from outside and contains intelligence about all operations performed, as well as data entered into a single register [9]. Such registers contain one or more records in the operational database, reflecting the sequence of operators for entering data, moving from one object to another, in a single data store, while reflecting actual data for accounting. They should be characterized by:

- Simultaneous support for a large number of users;
- Short and regular transactions;
- Sufficiently clearly formulated requests;
- Fast response to the request.

Due to the most organizations of the cooperative sector use disparate information systems with different software and hardware to perform all these operations, they can differ within one organization, so, the processing of the request may take place with some delay in time, which is unprofitable for economic object and can lead to sufficient financial losses [1].

Thus, accounting must be carried out in real-time using distributed ledger with the application of blockchain technology, which will serve as the basis for a complete transition to an automated system for keeping it with a reduction in manual labour. As a result, the main advantage will be the abolition of primary accounting, because it will be replaced by the technology of fixing a transaction in the blockchain model in distributed data ledger, and all participants in the cooperative sector will have access to it.

5 Conclusion

It should be said that the use of the "blockchain" technology described in the article can lead to the acceleration of the process of financial and economic activities and accounting of companies in the cooperative sector, as well as bring all accounting, reporting, and control to unified world standards.

In the accounting system, all the basic information necessary for managing the organization is formed, which is subsequently used in other systems for managing the organization. In this case, there are often arise such situations, when information is incomparable with that received from outside, or it can be duplicated in various accounting operations with an unreasonable increase in the used software and hardware.

Thus, it's necessary to assess the feasibility of such a transition and its consequences, taking into account the economic situation of the organization of the cooperative sector of the economy. At the same time, each practical action requires a theoretical justification in the form of the development of a regulatory and linguistic framework for the transition of accounting to the "digital platform" of the blockchain and the development of the plan for the implementation of modern technology in the cooperative sector.

References

1. Saliy, V. V., Ishchenko, O. V., & Aksenova, Zh. A. (2018). Enterprise architecture: study guide. *Krasnodar Cooperative Institute (branch) of the Russian University of Cooperation, 167*.
2. Saliy, V. V., Ishchenko, O. V., & Aksenova, Zh. A. (2018). Main aspects of the organization of management accounting supply and procurement activities of an economic entity. *Bulletin of the Academy of Knowledge, 28*(5), 21–31.
3. Gorlachev, P. V., Kunakovskaya, I. A., Habahu, S. N., Artyushkova, A. Y., & Gribok, N. N. (2021). Development of regional transport infrastructure and its financial mechanisms in Russia. *Studies in Systems, Decision and Control, 316*, 457–465.
4. Guziekova, S. M., & Khachak, S. K. (2013). Neologisms in modern Russian and English languages. Cultural life of the South of Russia. *Publishing House Ultrapress, 2*(49), 67–68.
5. Kamyshansky, V. P., Rudenko, E. Y., Solovyev, A. M., Kolomiets, E. A., & Dudchenko, A. V. (2019). Electronic commerce in the modern economy: Legal aspect. *Advances in Intelligent Systems and Computing, 726*, 904–910.
6. Batov, G. Kh. (2021). Organizational imperatives of the digital economy. http://digital-economy.ru/stati/organizatsionnye-imperativy-tsifrovoj-ekonomiki. Retrieved 10 Oct 2021.
7. Ivanov, A. (2018). How the blockchain will change the accounting department. https://www.klerk.ru/buh/articles/472371. Retrieved 10 Oct 2021.
8. Bondarenko, N. G., Strizhova, E. V., Allalyev, R. M., Smirnov, M. G., & Dudchenko, A. V. (2018). History and main trends in the development of copyright. *Journal of Advanced Research in Law and Economics, 9*(1), 41–47.
9. Saliy, V. V., Ishchenko, O. V., Bush, V. G., Gladysheva, E. G., & Abyzova, E. V. (2021). Accounting and analytical systems as an integral element of contemporary accounting. *Studies in Systems, Decision and Control, 316*, 739–746.
10. Gogolin, S. S., & Fadeeva, E.Yu. (2018). Increasing the efficiency of managerial decision-making in logistics based on the use of information and analytical systems. *Bulletin of the Academy of Knowledge, 29*(6), 111–117.

Social Planning in the Planning System of Consumer Cooperation

Vladimir V. Dudukalov, Galina N. Dudukalova, and Raisa V. Kalinicheva

Abstract The purpose of the study is to establish approaches to identifying the relationship between the planning processes and the conditions of economic activity of an operating cooperative. The article analyzes the possible problems and prospects of planning the activities of cooperation (including social planning), promoting and ensuring the integrated development of rural areas, effective economic activity of consumer cooperation.

Keywords Cooperation · Cooperative · Consumer society · Social planning · Planning principles · Strategic plans

JEL Classifications J54 · O21 · P13 · P11 · P21 · R13 · R58 · J54 · O21 · P13 · P11 · P21 · R13 · R58

1 Introduction

Despite attempts to slow down globalization, it continues to be the main direction in the development of human civilization at the present time. The process of strategic planning of consumer cooperation activities is becoming increasingly developed, and integration processes are at work in business.

The trend towards the development of the system of planning the activities of consumer cooperation is transformed from the methods and principles of current

V. V. Dudukalov · G. N. Dudukalova (✉) · R. V. Kalinicheva
Volgograd Cooperative Institute (Branch) Russian University of Cooperation, Volgograd, Russia
e-mail: gdudukalova@ruc.su

V. V. Dudukalov
e-mail: vdudukalov@ruc.su

R. V. Kalinicheva
e-mail: rkalinihteva@ruc.su

A. V. Bogoviz (ed.), *Big Data in Information Society and Digital Economy*,
Studies in Big Data 124, https://doi.org/10.1007/978-3-031-29489-1_23

planning, characteristic of the economic processes of the last century, to the development of strategic planning of the activities of consumer societies and their unions.

Consumer cooperation in Russia has been conducting active economic activities for the benefit of its shareholders and the population for many years, acting as a kind of counterbalance to globalization, defending the interests of people at the local level.

2 Methodology

During the research, the authors used a dialectical methodological approach based on the main laws of dialectics: the law of unity and struggle of opposites, the law of negation of negation, the law of the transition of quantitative into qualitative changes. The authors used the empirical method of observation, the theoretical method of comparative analysis and the historical method in their scientific research.

The most important trends in the development of management in the XXI century should include the processes of globalization and increasing uncertainty for development.

The process of globalization tends to displace consumer cooperation. For many years there has been a struggle for a political and economic system that would give maximum well-being to the greatest number of people. The market economy has shown greater efficiency compared to the directive one.

At the same time, studies of consumer cooperation activities conducted by Fedulova [1] confirm that it is market conditions close to the nature of consumer cooperation that contribute to the fact that some cooperatives provide high efficiency through the use of scientific and technical achievements in their activities, cooperative interaction as a tool of economic and trade cooperation, including at the international level. The vast majority of cooperative organizations demonstrate binding contractual relations, improving the quality of goods and services and stable financial condition.

However, the market constantly differentiates countries, peoples and individuals by the level of wealth, which is the basis of constant social tension. At present, the strict requirement of making a profit prevails and its fulfillment is sometimes achieved at the expense of the requirements of ecology, culture, morality and other elements of human existence that do not give immediate profit.

3 Results

In this controversy, according to the authors, there are not so many systemic entities that turned out to be viable and would meet the requirements of both "old" and "new" Russia. And consumer cooperation stands out among them since it is a cultural heritage, which by its very nature is a socially oriented system.

The research of consumer cooperation activities conducted by Podobed [2] confirms that the philosophy and rich experience of consumer cooperation activities presuppose the development of a strategy for the development of cooperation, the definition of a strategic vision as the result of a deep analysis of the external environment, its opportunities and threats, in order to ensure flexibility and adaptability.

In these conditions, from the point of view of the authors, the planning of the activities of cooperative organizations is of particular importance. Scientific planning of consumer cooperation activities, and as a result of its implementation, the development of strategic solutions contributes to improving the efficiency of consumer cooperation organizations, reducing the degree of uncertainty in the future.

Akoff [3], a well-known American specialist in planning, considers strategic planning, which he calls the design of the future, as one of the most complex types of human mental activity.

At the same time, the consumer cooperation system has gained a lot of experience in planning economic activities over the almost two-century history of development. It is especially important to emphasize the experience of planning the activities of consumer cooperation in the period of a directive economy.

There are several planning principles that form the basis of consumer cooperation plans. One of the basic principles assumes at the initial stage of the planning process a broad participation in it, first of all, shareholders, management and employees of consumer cooperatives with the mandatory participation in this process of the teaching staff of cooperative educational organizations. According to the authors, it is the personal participation of shareholders and employees in the planning process that creates the conditions for its successful implementation in practice, successful completion of tasks that are developed independently and therefore more understandable than those that are sent "from above".

Another principle that manifests itself in the planning of cooperative organizations is the principle of continuity, given that it is an iterative process in which all current plans are developed taking into account the implementation of the past and the fact that they will serve as the basis for making plans in the future. In fact, the continuity of planning is manifested.

The system of consumer cooperation presupposes the presence of different levels of management at the level of consumer societies and unions of consumer societies, respectively, planning processes are coordinated at different levels of management, which corresponds to the principles of coordination and integration of plans of both consumer societies and their unions vertically and between consumer societies horizontally.

As for one of the most important principles of planning, which is cost-effectiveness, it is mandatory in consumer cooperation and manifests itself in the fact that the cost of drawing up a plan is less than the effect that cooperators receive when executing it.

The creation of conditions in consumer cooperation organizations for the implementation of the developed plans also confirms compliance with one of the basic principles of planning.

All the above management principles are applied at all levels of management in the system of consumer cooperation, in consumer societies and in unions of consumer societies, as universal principles of planning. At the same time, specific planning principles are applied and operated at various levels of consumer cooperation, for example, in consumer societies it is important to adhere to the bottleneck principle, according to which production output is calculated taking into account the possibilities of equipment whose productivity is the lowest one. The principle of scientific planning is more typical for unions of higher-level consumer societies.

Given the multilevel nature of the consumer cooperation system, plans are formed at all levels of cooperation activities, and their integration and coordination at different levels provides an understanding of planning problems and planned solutions, they are not independent of others. The Centrosoyuz of the Russian Federation, regional consumer unions and consumer societies ensure the unity of forms of planning.

Studies show that the rapid changes that occur in the external environment and within consumer societies and their unions often do not allow planning processes to keep up with them, taking into account certain inertia, bureaucracy and time spent on planning calculations, and often the lack of planning departments in cooperative organizations.

In the first half of the last twentieth century, consumer cooperative organizations mainly used current plans, which was possible in the conditions of the predicted slow changes taking place in the external environment and the predictability of demand in the internal environment of cooperatives.

The second half of the last century, when the speed of changes in the external environment began to increase, in addition to current plans, cooperative organizations began to apply a system of medium- and long-term planning. In the practice of planning consumer societies and unions, targeted programs, budgets, profit plans were also used.

The end of the twentieth century is characterized by the rapidity and unpredictability of changes occurring in the internal environment. This situation has actually made it impossible to work on current and medium-term plans and has led to the transition of consumer societies and their unions to strategic planning, i.e. planning future activities based on mathematical models, management based on strategic decisions that involve taking into consideration changes in current operational plans.

In consumer cooperation, it is customary to divide plans based on the time horizon into long-term (plans-goals), medium-term in the form of programs and short-term operational plans.

The history of the Russian consumer cooperation has accumulated extensive experience in planning the social development of cooperative organizations. The purpose of such plans is to determine the prospects for the social development of the team.

In the new economic conditions, the role of socio-economic development plans is especially great when the authoritarian methods of economic management have been replaced by methods of economic management that stimulate all employees to reveal their abilities.

The peculiarity of the labor collectives of consumer cooperation is their small number. Therefore, social development plans are being drafted here, starting with consumer societies. The plan for social development of the collective is the final section of the overall socio-economic development plan of a consumer society or a union and is based on indicators of the trade, production and financial situation of the organization.

The plan for social development of the collective is a promising social program for the activities of a cooperative organization. It creates real conditions for scientific management of social phenomena and processes in the main link.

The main purpose of planning the social development of cooperative organizations is the development of such an interconnected system of planned measures, the implementation of which will ensure the successful progressive development of the collective, the effective fulfillment of its tasks, the creation of conditions for the best possible satisfaction of the spiritual and material needs of the members of the collective, the improvement of their moral traits and the education of their conscious attitude to work and to society.

Long-term plans in the consumer cooperation system aimed at solving strategic tasks are traditionally made for five to ten, and sometimes even more years.

As an example, the "Concept of development of the consumer cooperation system for 2017–2021" should be cited.

Consumer cooperation, as a socially oriented system, supplies rural residents with products, basic necessities, provides household services, in other words, contributes to improving the living conditions of the rural population. Currently, cooperation from individual economic communities, which united small and medium-sized producers, has gradually transformed into a system that has become an integral part of rural social infrastructure.

Consumer cooperation, increasing the well-being of rural residents, solving their social and domestic problems, helps to reduce social tension in rural areas.

For the villagers, the supply and delivery of bread, the maintenance of remote outlets, the sale of goods on credit and at a reduced price to rural residents-shareholders, the creation of opportunities for vocational education to shareholders and employees, the provision of school meals. All these consumer cooperation activities are carried out within the framework of a social mission.

Rural residents are suppliers of milk, vegetables and other agricultural products to cooperative organizations for processing, which at the same time contributes to the development of private subsidiary farms. This in turn provide employment for the population in subsidiary farms and processing enterprises and generates cash incomes of the rural population.

Consumer cooperation organizations operate in 89 thousand settlements in 71 regions of the country. Consumer societies and their unions, together with state authorities, solve various social problems of the population and the corresponding territories. Currently, consumer cooperation, fulfilling its social mission, can and should make a significant contribution to solving the problem of unemployment and increase the employment of the population with socially useful work. The directions of development and the social role of consumer cooperation have an impact on the process of forming a social economy focused on citizens' motivation for labor participation in the production process, fair distribution, exchange and consumption of material goods.

Integrated results based management of social phenomena and processes in the collectives of cooperative organizations has traditionally been carried out through the development and implementation of social development plans. The concept provides for the intensification of work in the elaboration of plans for the social development of cooperatives and monitoring their implementation.

Gorbachev [4] in his work notes that consumer societies have certain reserves to provide jobs for residents of rural settlements, which is reflected in the plans of organizations.

At the same time, a special place in the plans is given to the forms and methods of work and relationships with shareholders, whose number has been noticeably decreasing in recent years, as well as the number of the rural population as a whole. The aging of the shareholders also poses a certain problem.

Currently, there are not many young people in the cooperative. Research shows that motivation is needed to join the consumer society. The first place among the motives has always been social ideas, there is no reason to say that the situation has changed.

A. O. Gorbachev in his dissertation [4] argues that it is necessary to take into account a number of conditions that can become the basis for work on the involvement of the population in the cooperative movement:

- people should acutely become aware of all the unfavorable consequences of their isolated position as a consumer;
- they should have sufficient motivation to deal with these adverse circumstances by associating with other consumers;
- people should believe that their association can significantly improve their living conditions;
- joining a cooperative should not cause any difficulties for a person, especially bureaucratic ones;
- it requires a skillful organization of business, goodwill, the ability to convince, knowledge of the basic principles on which consumer cooperation is built.

4 Conclusion

According to authors such as Dudukalov et al. [5], there is a large scope for activity for departments engaged in organizational work among the public, which should restructure their work—leave their offices, "be closer to people", become propagandists of ideas and principles of cooperation, convince people of the advantages of a cooperative association.

By doing this, they will make a great contribution to the realization of the social mission of consumer cooperation.

Today and in the foreseeable future, it is necessary to shift the focus to young people in the work of involving the population in the cooperative movement. To do this, it is important to create school cooperatives so that people from a young age know the advantages of cooperation, its social significance. These tasks can also be implemented in the process of creating social cooperatives for childcare, assistance to the disabled and elderly, single pensioners, household management, animal care, grazing, and so on.

It is a misconception that hard times will pass, life will get better, society in the XXI century will be rich and happy, and the need for consumer cooperation will disappear. An analysis of the current situation in the world indicates that the global market economy (with all its advantages) will not change its nature. Most likely, economic activity in the context of globalization will become even more complicated and the need to protect the interests of people will constantly increase.

Consequently, the social orientation of the planning of the activities of cooperative organizations and the content of their main motto "Union—force" acquire special significance in the work of consumer societies "with the help of people, for people and with the participation of people."

References

1. Fedulova, E. V. (2008). Organizational and economic aspects of the formation and functioning of consumer cooperation organizations. Candidate's dissertation. https://www.dissercat.com/content/organizatsionno-ekonomicheskie-aspekty-formirovaniya-i-funktsionirovaniya-organizatsii-potre/read. Retrieved 23 Dec 2021.
2. Podobed, V. M. (2001). Organization of consumer cooperation management in the region. Candidate's dissertation. Moscow. https://search.rsl.ru/ru/record/01002287339. Retrieved 23 Dec 2021.
3. Akoff, R. L. (2002). Akoff on management. The series "Theory and practice of management". https://lib.sale/menedjmenta-osnovyi/akoff-menedjmente-seriya-teoriya-praktika.html. Retrieved 22 Dec 2021.
4. Gorbachev, A. O. (2003). *Development of public relations in consumer Cooperation organizations in Russia. Extended abstract of candidate's dissertation.* Moscow University of Consumer Cooperation.
5. Dudukalov, V. V., Dudukalova, G. N., & Chekina, M. V. (2016). Strategy of development of consumer cooperation of Russia—New vectors of development.: theory and practice [Text]. In *Proceedings of the scientific-practical conference "Cooperative sector of the economy in the innovative development of Russian society"* (Volgograd, May, 9, 2016) (pp. 45–51). Volgograd Scientific Publishing House.

Automation of Accounting and Settlement Operations as a Form of Digital Transformation of the Cooperative Sector of the Economy

Elena N. Makarova, Regina V. Skachkova, Danara A. Takhtomysova, Elena S. Fedotova, and Anna I. Aukina

Abstract The relevance of the research topic is determined by the direct relationship between the parameters of the automation of accounting and settlement operations and the increase in the quality of the implementation of accounting procedures at a cooperative trade enterprise. As a result, rationally organized accounting of settlement transactions with suppliers and contractors will assist in achieving the planned financial performance indicators and other targets of the organization. The paper aims to study the features of the formation of accounting procedures in terms of settlements with suppliers in the context of digital transformation of the consumer cooperation system and consider practical possibilities for their automation. During the research, the following methods were used: observation method, comparison method, calculation method, balance method, as well as an integrated approach that ensures the uniform application of methods at all stages of the study. The study yielded several results. First, the paper reflected the peculiarities of bookkeeping of accounting and settlement operations in relation to incentive payments of commercial nature. Second, the directions of using the possibilities of digitalization during the optimization of accounting procedures in the activities of a trade enterprise of the non-profit cooperative sector of the national economy were formulated.

Keywords Economy · Accounting · Settlements with suppliers · Digitalization · Consumer cooperation

JEL Classification M41

E. N. Makarova (✉)
Volga Region Cooperative Institute (Branch) of Russian University of Cooperation, Engels, Russia
e-mail: makarova1811@mail.ru

R. V. Skachkova · D. A. Takhtomysova · E. S. Fedotova · A. I. Aukina
Yuri Gagarin State Technical University of Saratov, Saratov, Russia

A. V. Bogoviz (ed.), *Big Data in Information Society and Digital Economy*, Studies in Big Data 124, https://doi.org/10.1007/978-3-031-29489-1_24

1 Introduction

The current situation on the retail market in the segments that are of interest to the subjects of cooperative trade is marked with several factors transforming the traditional trading environment. In the context of the stated topic, in addition to the factors of trade business consolidation in general and the dominance of organized network capital in some regional markets, we can note such ambivalent phenomenon as high transportability of sales transactions at the level of the individual participant of trade relations. Market transparency increases the level of trust in the chain of trade transactions. Simultaneously, it imposes more stringent requirements for the execution of transactions [1].

The conditions of market competition urge participants in the competitive market to find ways to improve their competitiveness. One of the areas of market optimization is the relationship between suppliers and buyers. The importance of their interaction is determined by the fact that any of the market participants in its business activities acts as a supplier and a buyer in relation to their counterparties. In any of these qualities, the market participant can provide the economic partner special terms of delivery of goods, payments, and additional services based on the importance of the partner enterprise in its economic circuit and its long-term prospects. In other words, the key economic partner has the right to expect special conditions in economic relations.

To disclose the research topic stated in the abstract, the authors set the following research tasks:

(1) To study the composition and importance of accounting and settlement operations of a trading company;
(2) To consider the specifics of reflecting the methods of sales promotion and the promotion of supplier goods in the accounting procedures of the purchasing company;
(3) To propose measures to automate accounting and settlement operations of trade enterprises.

2 Methodology

The current situation in the established practice of methodological and informational support of accounting procedures in terms of accounting and settlement operations of trade enterprises is marked with insufficient elaboration of some aspects. They primarily affect the issues of automation and objective reflection of new operations that have not been previously used and are dictated by the current market practice.

Among the operations in the field of commodity-money relations with trading partners and manufacturing enterprises that supply commodity products for resale, we can highlight the echelon system of reducing costs of trade organization by building special market relations with large and stable functioning suppliers. As a regular customer of such suppliers, the retailer can optimize costs by gaining access to their bonus programs and regular discounts from the standard market price [2].

The new practice of market interaction in the relations between counterparties implies that the related accounting and settlement operations must be solved based on the current requirements by methods of digital transformation of traditional economic activities. In this regard, the organizational and methodological issues of accounting of settlements with suppliers of products and documentation of accounting and settlement operations must be solved at a new level of automation based on digital platforms and advanced information technology [3].

Until recently, accounting work on calculating various incentives (discounts, premiums, and bonuses) was performed using a standard set of office computer programs. This process was marked with increased labor intensity and unproductively occupied excessive working time, which was ultimately reflected in the violation of the planned timelines for providing processed reporting information to internal and external recipients. The result of the negative impact was the adoption of improper management decisions and a decline in profitability targets.

A specialized program for accounting discounts and markups can help a contemporary accountant in a trading company overcome the disadvantages of traditional office programs. One of these specialized programs is the computer program "1C: ERP-Enterprise Management." The program allows the accounting department of a commercial organization to conduct accounting and settlement operations using broader digital and methodological tools. The new digital platform focuses on accelerating business processes and is an automated system for enterprise resource planning. From a structural point of view, this program is divided into constituent elements (blocks). Each block accumulates the necessary forms of primary documents and the corresponding forms of analytical reports.

3 Results

The study of accounting registration of operations on settlements with suppliers of the cooperative organization showed that the use of marketing tools is a sufficiently effective form of long-term encouragement of the prospective counterparty in the commercial turnover of manufactured products and, particularly, trade in consumer goods. Among the variety of its forms and methods, let us make a methodical emphasis on some advanced means of stimulating sales activity, namely:

1. Bonus payments;
2. Discount system;
3. Bonus programs.

The unifying point of all of these stimulating measures is their incentive nature. They allow buyers to save their own money and help increase the size of assets.

To objectively reflect operations on the receipt of incentive remuneration from trading partners of a higher wholesale level in the accounting system of the buyer company, it is proposed to allocate separate sub-accounts to the account 60 "Settlements with suppliers and contractors," including the following:

- Subaccount 2 "Bonuses received," which is supposed to consider the bonuses of the wholesale and retail levels;
- Subaccount 4 "Bonus received by stores," designed to account for bonuses at the retail level only.

These measures will contribute to a clearer control of the dynamics of mutual settlements with suppliers and make the process of monitoring the situation more transparent. It is also highly desirable to document all agreements with counterparties on the provision of bonuses, premiums, and other incentive payments. This can be done, for example, in the form of additional agreements to the main contract. In this case, the main details of the additional agreement will be as follows:

- Calculation basis for the incentive payment;
- Percentage of premium, discount, or bonus;
- Ways to receive the incentive payment;
- Terms of payment and terms of granting.

A promising form of relationship between partners in the circulation of goods is the so-called transmitted discount. As a means of sales promotion, transmitted discounts arise at the wholesale and retail levels. The parties to the transmitted discount are as follows:

- Manufacturer or wholesaler,
- Retail organization.

In terms of content, a transmitted discount is a reduction in the retail price of the product at the manufacturer's expense. From an accounting point of view, the use of transmitted discounts can lead to complications and failures in the pricing for the manufacturer's products. The procedure for granting, accounting system, control procedures for this type of discount, and other forms of incentives for resellers is regulated by Federal law "On the principles of state regulation of trading activities in the Russian Federation" (No. 381-FZ) [4].

Transmitted discounts are becoming more widespread in Russian trade practice as a method of market promotion of products. Their popularity is due to their potential effectiveness and practicality for the trade enterprise, including the following:

- Expansion of customer base;
- Increase in turnover;
- Improved financial performance.

A current approach to objectively verify the amount and order of accrual of a broadcast discount and other means of sales promotion is the use of an online cash register. The prospects of applying online cash register are proved by the practice of commercial activity of progressively managed trade organizations. The use of online cash registers allows one to quickly provide the enterprise-supplier with receipts indicating the selling price of goods subject to a discount or incentive measure of the appropriate type. Additionally, online cash registers also allow one to conduct refunds. In this case, the advantage of online cash registers is the independence of the return on the sale of goods, which promotes trust in the relationship between counterparties. Thus, the integral advantage of using online cash registers is a significant acceleration of procedures for checking settlements between counterparties in the accounting system of the trade organization.

The following entries may be applied in the accounting system of a trading company in the registration of accounting and settlement transactions in respect of discounts and commercial incentives:

- Debit 62 "Settlements with buyers and customers" sub-account 1 "Settlements with buyers in rubles";
- Credit 90 "Sales" sub-account 1 "Revenue" is used by the accounting department of the trade organization for a full reflection of the amount of sales in monetary terms, considering the discount received from the supplier or the manufacturer;
- Debit 76 "Settlements with different debtors and creditors" sub-account 6 "Settlements with debtors and creditors";
- Credit 91 "Other income and expenses" sub-account 3 "Other income" is applied to determine and accrue the amount of debt in accordance with the granted translatable discount.

The type of specific compensation affects the reflection of entries, namely:

- Debit 60 "Settlements with suppliers and contractors" sub-account 1 "Settlements with suppliers in rubles";
- Credit 76 "Settlements with different debtors and creditors" sub-account 6 "Settlement with debtors and creditors"—to reflect the situation of debt repayment by the seller through mutual set-off;
- Debit 51 "Current account";
- Credit 76 "Settlements with different debtors and creditors" sub-account 6 "Settlement with debtors and creditors"—to reflect the situation when the seller repays debts using a settlement account by means of money transfer;
- Debit 41 "Goods";
- Credit 76 "Settlements with different debtors and creditors" sub-account 6 "Settlement with debtors and creditors" – to reflect the situation of debt repayment by the seller by transferring bonus products.

An ambivalent aspect of providing incentive discounts in commercial practice is limiting or constraining promotional budgets. This applies to cash and quantity discounts. The problem is that disregard for the discount limits can provoke losses on a deal with a particular supplier. Naturally, a supplier is not ready to compensate for the excess of the pre-agreed level of compensation. Accordingly, the loss will have to be covered from own funds, which cannot be considered a rationally incurred expense. To counteract unprofitable work, it is necessary to set up the correct accounting of these procedures in the accounting system of the organization. Such a measure could be the opening of an analytical sub-account to account 76 in the context of each supplier, which would allow tracking the completeness and timeliness of discounts reimbursement.

To accurately conduct accounting and settlement transactions with respect to incentive payments, the buyer of commodity products should use a special form of the contractual relationship. It must contain the following basic details:

(a) Name of the incentive payment: discount, premium, and bonus;
(b) Information about the supplier that granted the discount, premium, or bonus;
(c) Basic estimated cost;
(d) Value-added tax;
(e) Amount of the incentive payment;
(f) Number and name of the appropriate account used to record the incentive payment;
(g) Term of the discount, premium, or bonus.

This form is filled out jointly by such enterprise departments as the commercial department, financial department, and accounting department. Among the primary documents generated for the automatic accrual of incentive payments, we must highlight the act of calculating the incentive payment and the invoice. The following postings will be created automatically:

- Debit 60 "Settlements with suppliers" sub-account 2 "Bonus received";
- Credit 91 "Other income and expenses" sub-account 3 "Other income," reflecting the fact of accrual of bonus payment;

or another possible option:

- Debit 60 "Settlements with suppliers" sub-account 4 "Bonus received by stores";
- Credit 91 "Other income and expenses" sub-account 3 "Other income," reflecting the fact of accrual of incentive payment for retail trade.

It is advisable to conduct control measures for the completeness of the amount of accruals and the correctness of their distribution in the accounts based on the report on the incentive payments with a periodicity of once a month.

Automation of the system of accounting and settlement operations in respect of incentive payments to the intermediate consumer has the following main features:

1. Ensuring the needs of synthetic and analytical accounting information and reporting documentation;
2. Adaptability or the ability to adjust the main types of documents used in the accounting process, the chart of accounts, forms of primary documents, varieties of analytical procedures, and standard postings;
3. Fast printing of primary documents and automation of the process of forming tax registers.

4 Conclusion

In conclusion, we should note that the result of the digital transformation of the accounting and settlement transactions based on integrated application of advanced electronic platforms and computer programs will reduce the labor intensity of these operations, increase accountant productivity, and provide high-quality information for accurate and effective management decisions.

References

1. Kachkova, O. E., Demina, I. D., Krishtaleva, T. I., Kosolapova, M. V., & Alferova, E. Y. (2019). Building the concept of the control-oriented accounting system. *International Journal of Civil Engineering and Technology, 10*(2), 1830–1837.
2. Rykhtikova, N. A., Anisimov, E. Y.,Evdokimov, S.Yu., Ivanova, E. V., & Lebedeva, O. E. (2018). Improvement of enterprise financing system in unstable economic environment. *The Journal of Social Sciences Research, S3*, 298–303. https://doi.org/10.32861/jssr.spi3.298.303
3. Takhtomysova, D. A., Fedotova, E. S., & Makarova, E. N. (2021). Methods and main stages of expert study of accounts receivable in enterprises of road construction sphere. In E.G. Zhulina (Ed.), *Current trends and prospects for digital transformation in the economy and business* (pp. 92–106). ISRDPC.
4. Russian Federation. (2009). Federal Law "On the principles of state regulation of trading activities in the Russian Federation" (December 28, 2009 No. 381-FZ, as amended July 2, 2021). Moscow, Russia.

Formation of Digital Skills of Personnel in the Cooperative Sector of the Economy in the Face of New Challenges

Evgeniya A. Sysoeva, Elena G. Kuznetsova, and Tatyana E. Shilkina

Abstract Industry 4.0 is the central topic of numerous publications by foreign and Russian scientists, most of which are conceptual in nature. The current labor market requires the formation of new competencies and skills demanded in the digital economy. Digital reality dictates the need to develop new competencies—digital skills. Knowledge of computer technology, the use of software, and programming have become essential. For most organizations worldwide, the COVID-19 pandemic has accelerated the digitalization and adoption of digital ways of working. The development of new forms of employment (remote and combined) using digital technology will contribute to positive changes in the labor market. Based on these trends in the digital transformation of current society, the research aims to examine the digital skills of employees in the cooperative sector of the economy and their demand in the labor market in the context of Industry 4.0. The authors use descriptive statistical methods to process official data from the Federal State Statistics Service of the Russian Federation and the Higher School of Economics. The research indicates that the transformation of the labor market and forms of employment requires the development of new competencies demanded in Industry 4.0. In the digital economy, employment opportunities increasingly depend on a person's digital skills and competencies. This paper is one of the first comprehensive studies of the changing labor conditions, working environment, and the emergence of new competencies in Industry 4.0. This research contributes to the problem of studying digital skills as a component of human capital and can be the basis for further research in this subject area.

Keywords Digital skills · Competencies · Fourth industrial revolution · Industry 4.0 · Digital transformation · Labor market

JEL Classifications D20 · D25 · O30

E. A. Sysoeva
National Research Mordovia State University, Saransk, Russia

E. G. Kuznetsova · T. E. Shilkina (✉)
Saransk Cooperative Institute (Branch) of Russian University of Cooperation, Saransk, Russia
e-mail: tekuznetsova89@mail.ru

A. V. Bogoviz (ed.), *Big Data in Information Society and Digital Economy*,
Studies in Big Data 124, https://doi.org/10.1007/978-3-031-29489-1_25

1 Introduction

The transformations occurring during the Fourth Industrial Revolution (Industry 4.0) are characterized by a significant scope and degree of improvement of society. It is the era of advanced technologies based on the use of information and communication. The distinguishing features of Industry 4.0 are speed, scale, and transformative ability.

Russia is taking on new challenges of the global development of humanity, actively participating in the new technological order. By 2030, it is planned to achieve the digital maturity of key sectors of the economy and social sphere [1].

The conditions of the COVID-19 pandemic have particularly sharply raised the question of the full or partial transition of the personnel to remote work and the inevitability of using digital services and programs for organizations of the cooperative sector of the economy. The current situation dictates the need for the digital transformation of these companies and the availability of digital skills and competencies of personnel for the successful and timely fulfillment of tasks.

The research relevance is due to the fact that in the context of the digital transformation of society, the ongoing technological revolution contributes to alterations in the labor market as the most sensitive indicator of the occurring changes. Labor and technology have been closely interrelated historically. Advanced technologies are radically changing the nature of work, automating it in many industries and professions. According to experts, Industry 4.0 will lead to a change in the labor market structure and the displacement of a person from the sphere of employment associated with performing mechanical, monotonous work and solving typical tasks. The widespread use of digital and information technologies will increase the share of intellectual and creative work [2]. Simultaneously, these challenges can cause the mass release of low-skilled labor. Thus, it is necessary to conduct comprehensive research in the field of the demand for digital skills of personnel in the labor market.

The paper aims to study the digital skills of personnel of companies in the cooperative sector of the economy and their demand in the labor market in the conditions of Industry 4.0.

Based on this goal, the following research objectives were set, which determined the logic and structure of the research:

- Analyzing the development of the digital environment in Russia;
- Studying the composition of skills and professional competencies in the context of the development of the digital economy;
- Identifying the digital skills that are most in-demand for most vacancies in the context of Industry 4.0.

2 Methodology

Discussions about the impact of information technology on the labor market can be divided into two groups. The first group indicates that the use of robots, artificial intelligence, intelligent systems, and algorithms in business life will eventually lead to mass unemployment, increasing poverty, and social tension. According to another group, the development of technology will help increase labor productivity.

The materials of the research conducted by E. Erer and D. Erer confirm the existence of a close relationship between the labor market and technology. According to scholars, technological development increases labor productivity and creates new jobs only in industries requiring specific skills, negatively affecting the employment of low-skilled workers. In general, Industry 4.0 contributes to changes in the employment structure [3].

A. A. Patrushev and S. V. Bespalyy also considered the emerging threat to specialists with low and medium qualifications with the development of automation. The scholars noted that most people are not ready for such changes: they lack the digital competencies necessary in the current conditions [4].

The study of E. G. Popkova and K. V. Zmiyak on the priorities of training digital personnel for Industry 4.0 is of particular interest. According to the results of the experiments, it was found that the level of robotization of socio-economic systems of the countries worldwide will be very low and will not cause an increase in unemployment until 2022. The scholars concluded that Industry 4.0 would develop under the scenario of moderate automation and robotization while maintaining the dominance of human labor in most business processes and economic spheres. Communication with people will become the basis for the activities of digital personnel, and social competencies will become a priority for them [5].

According to E. V. Gutorova, under the influence of digital technologies, the profile of many professions is significantly transformed, modifying the methods of labor organization. The so-called digital and non-cognitive skills are the most important for employment [6].

The significance and relevance of digital skills in the context of digital transformation were the basis of the study conducted by C. Hughes, L. Robert, K. Frady, and A. Arroyos. A shortage of skilled workers and a technical skill gap make it challenging to recruit and retain a high-tech workforce. The scholars focused on solving problems related to personnel recruitment, professional development, and best management practices in the current conditions [7].

The potential impact of Industry 4.0 on human resource management was explored by L. B. Liboni, L. O. Cezarino, C. J. C. Jabbour, B. G. Oliveira, and N. O. Stefanelli. The scholars dwelled on the features of employment, job profile, and requirements for qualifications and skills of the modern workforce [8].

The methodological basis of the research was the methods of descriptive statistics, which allowed processing and systematizing empirical data in the studied subject area through the main statistical indicators and visually presenting them in the form of graphs and tables. Popular job search sites were analyzed to identify the digital skills demanded among employers in the context of Industry 4.0. In the course of the research, causal relationships between the studied phenomena of digital transformation of modern society were established, on the basis of which specific conclusions were formulated within the problem under consideration.

The official data of the Federal State Statistics Service of the Russian Federation (Rosstat) [9] and the Higher School of Economics were applied in the research [10].

3 Results

The Global Connectivity Index, developed by Huawei in 2013, is used to compare the development of the digital environment on a global scale. This index reflects the progress of the world's largest countries in the transition to digital technologies; it shows the relationship between the level of investment in information and communications technology infrastructure and economic growth based on information and communication technologies. The index is calculated based on 40 indicators reflecting the degree of development of countries and the influence of five main technological growth factors accelerating the digital transformation of the country's economy [11]. Depending on the number of points scored, countries are divided into three groups:

- Leaders (more than 65 points);
- Catching-up countries (40–65 points);
- Newcomers (less than 40 points).

For several years in a row, Russia has been among the group of catching-up countries in terms of the value of the global network interaction index. In 2019, Russia took 41st place with an indicator of 49 points (one point higher, but two places lower in the ranking compared to 2018). According to the authors of the rating, the catching-up countries are in advantageous positions because their return on investment in information and communications technology infrastructure is growing exponentially. In these countries, the growth drivers are government initiatives to expand networking, increase the speed of broadband internet access, and create conditions for the development of cloud services [1].

High indicators of the use of digital technologies in Russia are recorded in organizations of the financial and business sectors of the economy, with the former leading in the spread of network technologies (e.g., broadband internet, cloud services) and the latter leading in the use of technologies that change business processes (e.g., enterprise resource planning systems, radio-frequency identification [RFID] technologies, such as automatic identification of objects by radiofrequency tags). The dynamics of the use of digital technologies in organizations by sector in Russia (as a percentage of the total number of organizations in the relevant sector) for 2015–2019 was presented in the statistical collection of the Higher School of Economics "Indicators of the Digital Economy, 2020" [10]. Currently, RFID technologies are attracting more users, and their number is increasing faster than ever. So far, this technology is

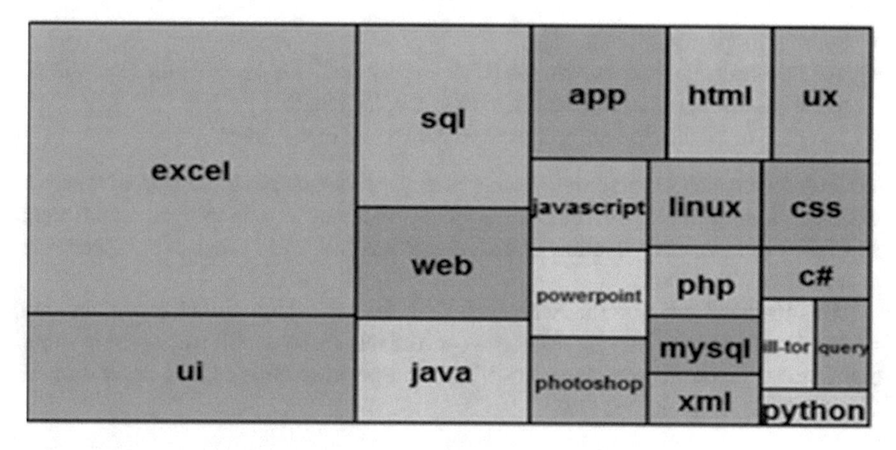

Fig. 1 Digital skills most in demand. *Source* Compiled by the authors

in little demand, but it has a huge potential for developing the Internet of Things. The demand for cloud services is growing in the social sphere: every fourth organization uses them. Regarding the availability of broadband Internet access, the social sphere is comparable to the business sector.

Until recently, the skills required by employees were conditionally divided into two levels: hard skills and soft skills. However, modern digital reality dictates the need to develop a new block of competencies—digital skills [12] that are key for employment and successful career growth.

Exploring the most popular Russian job search websites (e.g., Head Hunter, Rabota.ru, and Superjob.ru) allowed us to identify the digital skills demanded for most vacancies. According to the study, employers impose requirements on applicants of different professions for the possession of digital skills (Fig. 1).

Most employers require solid skills in working with the following:

- Spreadsheets, visualization, and data analysis tools in Microsoft Excel;
- An interface for transferring information between a human user and hardware and software components of a computer system (user interface [UI]);
- A programming language to create, modify, and manage data in a relational database managed by a corresponding database management system (structured query language [SQL]);
- A system for accessing interconnected documents on various computers connected to the internet (Web);
- A general-purpose object-oriented programming language (Java).
- Future applicants are also required to possess the following skills:
- The hypertext markup language [HTML];
- A general-purpose scripting language for developing web applications (Hypertext Preprocessor [PHP]);
- A formal language for describing the appearance of a document (web page) (Cascading Style Sheets [CSS]);

- Python programming language;
- The skill to design user interaction (User Experience [UX]) and use the capabilities of the multifunctional Adobe Photoshop graphics editor;
- The skill to create PowerPoint presentations, and many others.

When determining the personnel strategy, today's companies in the cooperative sector of the economy focus on the cumulative set of competencies of employees of different specializations, which will allow one to form teams with a full range of competencies necessary to solve the task.

HR specialists are sure that with each year, the job profile will change more and more often, and those who are now ten years old will change their work profile about eight times. The change of activity will occur especially often at the intersection of disciplines or technologies [13].

4 Discussion

In a globalized and dynamically developing world, digital technologies have spread rapidly. The economy, labor market, and society are changing at an unprecedented rate. Digital technologies are one of the main factors stimulating growth and increasing productivity, competition, and innovation potential. In the labor market, these technologies threaten existing professions, especially those involving performing monotonous tasks. However, digital technologies open up opportunities for creating new professions, mainly related to the digital economy. There are jobs where people increasingly interact with digital tools and artificial intelligence.

The COVID-19 pandemic contributed to the development of a trend of remote employment among companies in the cooperative sector of the economy. In a crisis, switching to remote work allows one to save money and avoid mass layoffs. The development of new forms of employment (remote and hybrid) using digital technologies can result in positive changes in the labor market. The creation of rules and procedures for the transition to remote work, the consolidation of relevant norms in labor legislation, and the development of the necessary information infrastructure will increase the resilience of the economy to potential crises.

5 Conclusion

Industry 4.0, which has no analogs in the history of humanity in terms of scale, scope, and complexity, is rapidly transforming the entire world.

The COVID-19 pandemic has accelerated digitalization and the introduction of digital ways of working for most organizations worldwide. The active and widespread digitalization of business processes has led to the fact that a certain set of basic skills of working in the digital space becomes not just a competitive advantage but a necessary requirement for many professions and specializations. Thus, digital skills are one of the components of human capital that largely determine the possibility of employment.

The transformation of forms of employment and the labor market requires forming new competencies demanded in the digital economy. With the development of the information society and the digital economy, professions are becoming more complex. The requirements for specialists are changing. First, digitalization saves time to solve more complex and creative tasks. Second, the requirements for the level of qualifications and a set of competencies are increasing.

Therefore, as a fundamental consequence of Industry 4.0, there are transformations concerning the essence of the work performed. Employment opportunities are increasingly dependent on a person's digital skills and competencies. This situation will entail the need for a complete restructuring of the personality, and each person will face the task of mastering new skills and competencies.

References

1. Abdrakhmanova, G. I., Vanyushina, M. D., Vishnevsky, K. O., Gokhberg, L. M., Gribkova, D. E., Demidkina, O. V., & Utyatina, K. E. et al. (2021). *Trends in internet development: Readiness of economy and society to function in a digital environment*. Analytical report. HSE University. https://issek.hse.ru/mirror/pubs/share/536491500. Retrieved 9 Dec 2021.
2. Emelianenko, E. E. (2019). The fourth industrial revolution: New opportunities for the labor market. In *Collection of selected articles on the materials of scientific conferences of the State Research Institute "National Development"* (pp. 229–231). GNII Natsrazvitie.
3. Erer, E., & Erer, D. (2020). Industry 4.0 and its role on labor market: A comparative analysis of Turkey and European countries. In B. Akkaya (Ed.), *Agile business leadership methods for Industry 4.0* (pp. 85–102). Emerald Publishing Limited. https://doi.org/10.1108/978-1-80043-380-920201006
4. Patrushev, A. A., & Bespalyy, S. V. (2020). Digital competencies and skills in modern conditions of economic development. *Information Technology. Problems and Solutions, 2*(11), 88–93.
5. Popkova, E. G., & Zmiyak, K. V. (2019). Priorities of training of digital personnel for industry 4.0: Social competencies vs. technical competencies. *On the Horizon, 27*(3/4), 138–144. https://doi.org/10.1108/OTH-08-2019-0058
6. Gutorova, E. V. (2020). Digital skills and competencies: EU experience. socio-economic development of organizations and regions in the digital economy. In *Proceedings of the international scientific and practical conference* (pp. 116–121). Vitebsk, Belarus.
7. Hughes, C., Robert, L., Frady, K., & Arroyos, A. (2019). Middle-skill-level employees and technological environments. In *Managing technology and middle- and low-skilled employees (The changing context of managing people)* (pp. 13–28). Emerald Publishing Limited. https://doi.org/10.1108/978-1-78973-077-720191002
8. Liboni, L. B., Cezarino, L. O., Jabbour, C. J. C., Oliveira, B. G., & Stefanelli, N. O. (2019). Smart industry and the pathways to HRM 4.0: Implications for SCM. *Supply Chain Management, 24*(1), 124–146. https://doi.org/10.1108/SCM-03-2018-0150
9. Federal State Statistics Service of the Russian Federation. (2020). *Regions of Russia. Socio-economic indicators: Statistical collection.* Rosstat. https://rosstat.gov.ru/storage/mediabank/LkooETqG/Region_Pokaz_2020.pdf. Retrieved 9 Dec 2021.
10. Abdrakhmanova, G. I., Vishnevsky, K. O., Gokhberg, L. M., Demidkina, O. V., Demyanova, A. V., & Dranev, Yu. (2020). *Digital economy indicators: Statistical collection.* HSE University. https://issek.hse.ru/mirror/pubs/share/387609461.PDF. Retrieved 9 Dec 2021.

11. Kokh, L. V., & Kokh, Yu. V. (2019). Analysis of existing approaches to measurement of digital economy. *St. Petersburg State Polytechnical University Journal. Economics, 12*(4), 78–89. https://doi.org/10.18721/JE.12407
12. Brolpito, A. (2019). *Digital skills and competence, and digital and online learning*. European Training Foundation. https://www.etf.europa.eu/sites/default/files/2018-10/DSC%20and%20DOL_0.pdf. Retrieved 9 Dec 2021.
13. Kochetkova, L. N., & Kozlova, M. A. (2017). Fourth industrial revolution: Social transformations and new human requirements. Relevant problems and prospects of development of radio engineering and infocommunication systems. In *Proceedings of the III international scientific and practical conference* (pp. 444–449.). MIREA—Russian Technological University.

Production Optimization Model for Cooperative Farming

Angelica A. Nikitina, Irina N. Girfanova, Alsou F. Mukhamedyanova, Flyuza A. Tukayeva, and Ilyusa M. Khanova

Abstract In this work, the authors proposed options for cooperating farms according to the principle of inter-farm cooperation, and assessed the effectiveness of the functioning of unions and associations of farms. Small forms of business operating in the agro-industrial complex are the most vulnerable and unprotected in conditions of monopolization of industrial resource supply, existing competition in the market, it is most difficult for them to adapt and function in conditions of competition with large agricultural producers. In this regard, inter-farm cooperation, the unification of small businesses into unions and associations will help strengthen the position of farmers and gain their market share.

Keywords Cooperation · Farms · Unions · Efficiency · Production optimization · Specialization · Resource potential · Production capacity · Individual entrepreneurs

JEL Classifications Q12 · R1 · R2

1 Introduction

In the context of the transformations taking place in the agrarian sector, the problems of the formation of new forms of management that meet the requirements of a market economy acquire special relevance. The formation of small business in agriculture occupies an important place, they combine various forms of farms that provide a trend for the development of a market economy. The activity of farms is an entrepreneurial

A. A. Nikitina (✉) · I. N. Girfanova · A. F. Mukhamedyanova
Bashkir Cooperative Institute (Branch) of the Russian University of Cooperation, Ufa, Russia
e-mail: aa_nikitina@mail.ru

I. N. Girfanova
e-mail: irina13091970@mail.ru

I. M. Khanova
Ufa University of Science and Technology of the Bashkir State University, Ufa, Russia

F. A. Tukayeva
Ufa Law Institute of the Ministry of Internal Affairs of the Russian Federation, (FGKOU VO UYI of the Ministry of Internal Affairs of Russia), Ufa, Russia

A. V. Bogoviz (ed.), *Big Data in Information Society and Digital Economy*, Studies in Big Data 124, https://doi.org/10.1007/978-3-031-29489-1_26

activity aimed at making a profit. The development of inter-farm cooperation as a form of interaction between entrepreneurs in the course of their activities and being one of the most accessible means contributes to the deepening of the level of specialization of production, an increase in its concentration.

At one of the first stages of the creation of inter-farm cooperatives, it is necessary to resolve such issues as the substantiation of the composition and structure of the association; selection of production capacity and location; determination of the powers of the management, as well as all members of the cooperative; determination of the size of production and the term for the provision of certain types of resources; the size of financial investments of a capital nature.

2 Methodology

Materials of financial and economic reporting of the Ministry of Agriculture and Food of the Republic of Bashkortostan, the Local Agency of the Federal State Statistics Service for the Republic of Bashkortostan, the materials of the authors served as sources of information during the study. In the study of issues of inter-farm cooperation, methods and techniques of modern science, such as analysis and synthesis, experiment and economics-mathematical modeling based on the definition of dynamics, interconnection, interdependence were used.

Works of Buranbaev and Sabirov [1], Dashkovsky [2], Khanova et al. [3], Nikitina [4, 6], Mukhametzyanova et al. [5], Zhilina and Mukhamedyanova [7], Rusina [8] were studied when writing the article.

3 Results

Economic and mathematical models that most of all reflect the most significant aspects of the specialization of farms, the emerging economic relationships in the performance of their assigned tasks, a full-fledged production process were built to substantiate the economic efficiency of creating inter-farm cooperative associations [3].

The presented model which unites farms, their organizational structure is block-based. The production conditions of each farm are reflected in a separate block. The unifying block describes the restrictions on the total volume of production and sale of agricultural products by farms, as well as the conditions for their cooperation on the use of available production resources. The organizational structure of inter-farm cooperation in the optimization model consists of 8 blocks (Fig. 1).

The main goal of optimization is the totality of all sectors of individual inter-farm organizations and farms, the search for reserves to increase production efficiency and improve financial results.

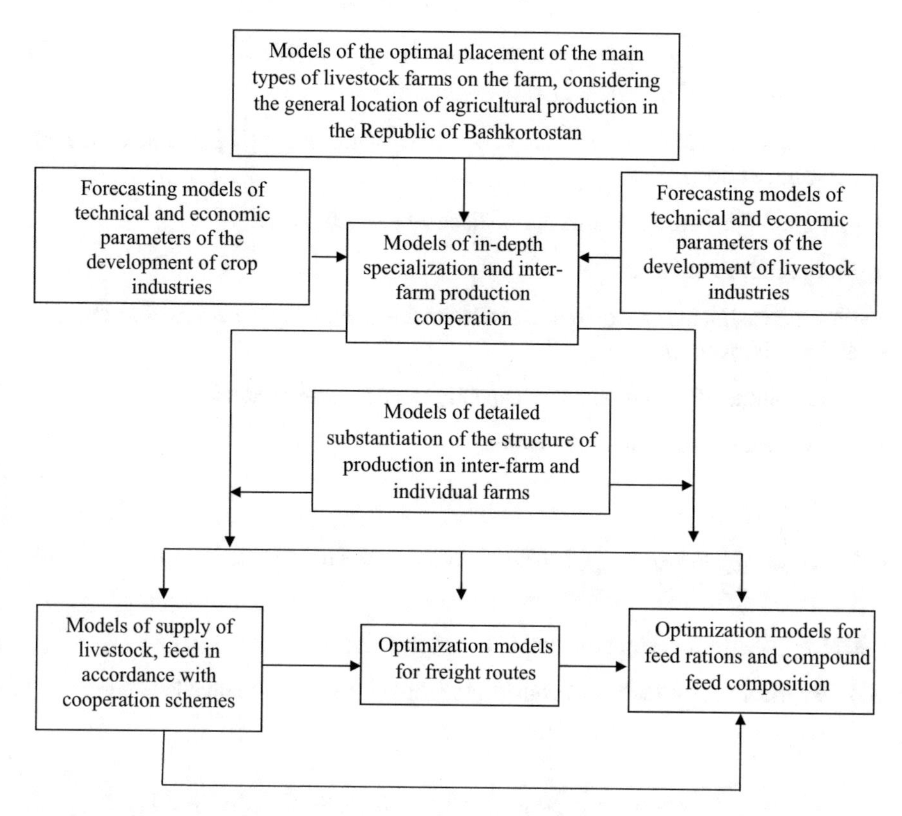

Fig. 1 Optimization model of the structure of the organization of an inter-farm cooperative. *Source* Developed by Nikitina [5]

The main functional criterion F_X is the improvement of financial results (profit maximization):

$$F_X = \sum_{l \in L} \sum_{j \in J_1 \cup J_2} c_{jl} x_{jl} \to \max \tag{1}$$

where c_{jl}–financial results;

x_{jl}—the value of the gross production of the jth agricultural industry in the l—th enterprise (J_1—the variable values of the plant growing industries; J_2—the variable values of the livestock industries; $l \in L$, where L is the number of farmers and personal farmsteads);

The indicators were defined as limitations:

(1) By the area of agricultural land of the ith type of crops in the lth enterprise

$$\sum_{j \in J_1} a_{ijl}x_{jl} - \overline{x}_{il} + x_{il} = A_{il} \quad (i \in I_1, l \in L) \tag{2}$$

where a_{ijl}–coefficient of consumed resources of the ith type per unit of the jth agricultural sector;

I_1–the maximum available area at the disposal of each farmer;

x_{il}–cooperative land area;

$\overline{x}_{il}$–for increasing the area of agricultural land due to plowing during the development of an inter-farm cooperative;

A_{il}–the volume of own production capacity of each farm production;

(B) By the amount of labor resources:

$$\sum_{j \in J_2} a_{ijl}x_{jl} + \sum_{j \in J_2} a_{ijl}x_{jl} - \overline{x}_{il} + x_{il} \leq A_{il} \quad (i \in I_2, l \in L) \tag{3}$$

where I_2—restrictions on the number of employees;

(C) By the need for feed ($\overline{x}_{il}$) considering the inter-farm exchange between farms:

$$-\sum_{j \in J_1} v_{ijl}x_{jl} + \sum_{j \in J_2} a_{ijl}x_{jl} - \overline{x}_{il} + x_{il} \leq 0 \quad (i \in I_1, l \in L) \tag{4}$$

where v_{ijl}–production volumes of finished products of the ith type;

I_3–restrictions on types of feed.

The economic and mathematical model makes it possible to evaluate the options for optimizing the combination of agricultural production branches in the inter-farm cooperative "Kirillova" of the Ufa region of the Republic of Bashkortostan.

The production potential of the "Kirillovo" inter-farm cooperatives was made up of land plots and property shares of former members of the "Kirillova" collective farm leased by organized farms. Industrial buildings and structures, agricultural machinery and vehicles were later used jointly, on the basis of an organized inter-farm cooperative.

The lease agreement for the property complex stipulates that it is transferred for temporary use with the option of redemption in stages. If the leased property is redeemed, then the lease contract is considered closed from the date of full redemption. The lessee can buy the property either in full or in parts within five years prior to the expiration of the lease. According to the terms of the agreement, the payment for the use of the property share in the inter-farm cooperative is charged in the amount of 10% of the volume of production, and farmers must annually increase the volume of gross production by at least 5%. In accordance with the additional agreement of the parties, the rent from the use of the property complex is further directed to the renewal of fixed assets and their annual repair.

The amount of rent is adjusted during the term of the lease agreement in accordance with the issued regulations on the revaluation of fixed assets and other material values included in the property complex.

An assembly of members of an inter-farm cooperative may require termination of the lease agreement with the farm when using and operating the leased property complex in violation of the terms of the lease agreement or for other purposes.

Increasing the efficiency of dairy farming, sheep breeding and vegetable production in suburban farms must be carried out in accordance with changing economic conditions based on the priority use of all factors that ensure high efficiency. The project of the inter-farm production cooperative "Kirillova" was developed to achieve high production efficiency. The research results are presented in Table 1.

Option 1 involves the organization of production focused on dairy farming. Dairy products are perishable, but the presence of nearby large settlements located at a distance of 10–20 km will make it possible to sell milk (whole) in bottling, in the nearest markets of the villages of Shaksha and Iglino, as well as in Ufa at selling prices–40 rubles per liter.

Option 2 provides for the delivery of milk to milk processing plant at a price of 25 rubles per liter. The distance of transportation from the farms of the peasant (farmer) economy to the receiving point of the dairy plant is 2 km.

Option 3 is calculated considering the organization of production of farms on the basis of partnership and inter-farm cooperation: F 1–F 2 and F 3–F 4. The sale of livestock products will be carried out at retail outlets in Ufa at a price of 40 rubles per liter.

As a result of solving the economic and mathematical problem according to the options presented, an assessment of the efficiency of production and sale of dairy products produced both by individual farms and by an inter-farm cooperative in general, considering various options for organizing the sale of finished products, was given. The optimization model made it possible to assess the financial result of the inter-farm cooperative (Table 1). Thus, an inter-farm cooperative can make a profit in the amount of 10.6–37.5 million rubles, including each farm–from 2.6 to 10.6 million rubles.

The most profitable option for organizing production for farms with the maximum economic effect in the form of high profitability is the implementation of the 3rd option. With this option, production costs, as well as costs associated with the delivery of finished products to consumers are reduced due to the concentration of production. Reducing costs contributes to an increase in profit from the sale of products. Farms receive most of the proceeds from the sale of milk, which is explained by the year-round production and the proximity of sales markets. The production of crop products is associated with the seasonality of production and does not ensure the receipt of proceeds evenly.

To increase the efficiency of production and sale of milk in farms based on the inter-farm cooperatives of Kirillov's farms, it is necessary to organize milk processing

Table 1 Options for organizing production in the Kirillova Inter-farm Cooperative, Ufa District

Indicators	Option 1				Total IFC Kirillova option 1	Option 2				Total IFC Kirillova option 2	Option 3		TSotal IFC Kirillova option 3
	F 1	F 2	F 3	F 4		F 1	F 2	F 3	F 4		F 1 – F 2	F 3 – F 4	
Number of employees, people	12	15	17	17	61	12	15	17	17	61	27	34	61
Labor costs per man-hour	28,632	36,981	42,489	40,201	148,303	28,668	37,401	42,489	40,201	148,759	72,710	82,599	155,309
Arable land, ha	298	339	470	450	1557	298	339	470	450	1557	637	920	1557
Hayfields area, ha	17	19	56	47	139	17	19	56	47	139	36	103	139
Pasture area, ha	75	75	127	112	389	75	75	127	112	389	150	239	389
Livestock of cows, head	93	115	114	107	429	93	116	114	107	430	236	222	458
Livestock of young cattle for fattening, head	8	30	30	30	98	8	30	30	30	98	38	60	98
Number of horses, head	3	3	2	2	10	3	3	2	2	10	6	4	10

(continued)

Table 1 (continued)

Indicators	Option 1				Total IFC Kirillova option 1	Option 2				Total IFC Kirillova option 2	Option 3		TSotal IFC Kirillova option 3
	F 1	F 2	F 3	F 4		F 1	F 2	F 3	F 4		F 1 – F 2	F 3 – F 4	
Livestock of ewes, head	4	4	5	5	18	5	5	5	5	20	4	4	8
Sales of products, c													
Winter rye	600	330	330	330	1590	600	300	330	330	1560	500	600	1100
Winter wheat	578	300	300	300	1478	578	300	300	300	1478	500	600	1100
Spring wheat	400	100	100	100	700	400	100	100	100	700	200	400	600
Barley	200	200	200	200	800	200	150	200	200	750	300	400	700
Oats	200	200	200	200	800	200	200	200	200	800	300	400	700
Cabbage	150	150	–	–	300	150	150	–	–	300	300	–	300
Milk	2775	3351	5402	5051	16,579	2773	3394	5402	5051	16,620	6977	10,489	17,466
Cattle meat	20	75	75	75	245	20	75	75	75	245	95	150	245
Sheep meat	0,088	0,088	0,11	0,11	0,396	0,11	0,11	0,11	0,11	0,44	0,088	0,088	0,176
Wool	1	1	1,25	1,25	4,5	1,25	1,25	1,25	1,25	5	1	1	2
Sale of young animals 1 year of birth	84	81	80	86	331	84	81	80	86	331	231	155	386

(continued)

Table 1 (continued)

Indicators	Option 1				Total IFC Kirillova option 1	Option 2				Total IFC Kirillova option 2	Option 3		TSotal IFC Kirillova option 3
	F 1	F 2	F 3	F 4		F 1	F 2	F 3	F 4		F 1 – F 2	F 3 – F 4	
Cost of products sold, thousand rubles	2538	2766	3864	3664	12,832	2546	2774	3874	3674	12,868	5250	7542	12,792
Sales proceeds, thousand rubles	9025	10,220	15,163	14,320	48,728	5220	5638	7743	7383	25,983	20,808	29,510	50,318
Profit from product sales, thousand rubles	6487	7454	11,299	10,656	35,896	2674	2864	3869	3709	13,115	15,558	21,968	37,526
Rent for the use of land and property shares, thousand rubles	126,9	138,3	193,2	183,2	641,6	127,3	138,7	193,7	183,7	643,4	262,5	377,1	639,6
Profitability level, %	255,6	269,5	292,4	290,8	279,7	105,0	103,2	99,9	100,9	101,9	296,3	291,3	293,4

Source Developed by Nikitina [5]

and bottling, which will allow it to be supplied to the market in a higher quality, and, accordingly, at a higher price.

Farms F 1 and F 2 sell milk (whole) in the markets of Ufa, the villages of Iglino and Shaksha, where they are in high consumer demand, as noted above. Therefore, the organization of processing and bottling of milk will expand sales markets and conquer new niches. The acquisition and launch of milk processing and packaging facility in an inter-farm cooperative will contribute to an increase in revenue from milk sales by 2.7 times, and, accordingly, to an improvement in financial results.

4 Conclusion

The study made it possible to assess the economic efficiency of production cooperative formations and to prove the feasibility of organizing inter-farm associations.

Practical implementation of the methodology proposed by the authors will increase the production of private farms, as well as regional production indicators in general.

Inter-farm cooperatives will help strengthen the position of farms, as well as protect the interests of farmers in the market, increasing the competitiveness of their products.

Acknowledgements The authors express their gratitude to the Deputy Minister of the Ministry of Agriculture and Food of the Republic of Bashkortostan for the opportunity to freely use scientific and practical information on the development of the cooperative movement in Russia, to the Chairman of the Council of the Tsentrosoyuz of the Russian Federation, to the Chairman of the Council of Bashpotrebsoyuz for assistance in the development of scientific research in the field of the cooperative movement.

Conflict-of-Interest Notification I, the author of this article, bindingly and explicitly declare of the partial and total lack of actual or potential conflict of interest with any other third party whatsoever, which may arise as a result of the publication of this article. This statement relates to the study, data collection and interpretation, writing and preparation of the article, and the decision to submit the manuscript for publication.

References

1. Buranbaeva L.Z., Sabirova Z.Z. (2018). Agricultural consumer cooperation: revival and development prospects. In *Modern cooperation in the system of sustainable development goals. Materials of the International Scientific and Practical Conference* (pp. 64–67). Russian University of Cooperation. Moscow.
2. Dashkovsky I. (2019). Cooperation on paper. Why there are no European-sized cooperatives in Russia, working paper. Agricultural engineering and technology. https://www.agroinvestor.ru/regions/article/31727-kooperatsiya-na-bumage. (data accessed 15 Oct 2021).
3. Nikitina A.A. (2017). Analysis of the efficiency of the functioning of peasant (farmer) households considering the regional characteristics of the Republic of Bashkortostan: monograph.—Orenburg, p. 188.

4. Nikitina A.A. (2021). Traditions and innovations in modern science and education: theory and best practice. Collective monograph. Monograph code: MON-62, Publisher: ICNP "NEW SCIENCE", Publisher: International Center for Scientific Partnership "New Science", Petrozavodsk, pp. 269–286.
5. Nikitina A.A., Khanova I.M., Dubinina E.V., Sibagatullina R.M., & Nurlygayanova A.M. (2021). Innovative directions of assessing the regional resource potential of small forms of management functioning in the field of agriculture in Bashkiria, working paper. *Economy and Entrepreneurship, 1*(126), 498–505
6. Nikitina A.A., Mukhametzyanova E.R., Yuldybaev B.R., Knyaginina G.V., & Nurlygayanova A.M. (2021). Regional resource potential of small forms of farming in the field of agriculture of the Republic of Bashkortostan Regional, working paper. *Economy and management: electronic scientific journal, 1*(65), 20
7. Rusina, A. A. (2020). The role and significance of farming activities abroad/A. A. Rusina//Bulletin of the faculty of land management of the St. *Petersburg State Agrarian University, 6*, 89–91.
8. Zhilina E.V., Nikitina A.A., & Mukhamedyanova A.F. (2021). Directions of development of small and medium-sized businesses in Russia considering foreign experience, working paper. *Bulletin of BIST (Bashkir Institute of Social Technologies), 1*(50), 29–37

Innovative Approaches to Managing the Costs of Production Cooperatives in Modern Conditions

Svetlana L. Blau, Victoria V. Varfalovskaya, Elena V. Khomutova, and Nadezhda F. Shchukina

Abstract The article substantiates the need to apply innovative approaches to managing the costs of production cooperatives in modern conditions. The authors conducted a detailed review and critical analysis of approaches to cost management. Existing approaches to cost management are rethought and supplemented with elements of GAP analysis and economic crisis management in support of reducing the cyclicality of economic systems and increasing the resilience of production cooperatives to it. The scientific novelty and theoretical significance of the results consist in the fact that they formed a systemic methodology for managing the costs of production cooperatives in an unstable market environment. The practical significance of the results obtained in the research lies in the fact that the proposed scientific and methodological recommendations for cost management can improve the adaptability of production cooperatives to the dynamically changing market environment. Management implications of the author's developments is that they allow increasing the effectiveness of financial management of production cooperatives.

Keywords Production cooperative · GAP analysis · Operational analysis · Break-even point · Profitability threshold · Fixed costs · Variable costs · Marginal income · Operating leverage

JEL codes D24

S. L. Blau (✉)
Russian University of Cooperation, Mytishchi, Russia
e-mail: sblau@yandex.ru

V. V. Varfalovskaya · E. V. Khomutova · N. F. Shchukina
MIREA–Russian Technological University, Moscow, Russia
e-mail: varfalovski@mail.ru

E. V. Khomutova
e-mail: khomutova.e.v@yandex.ru

N. F. Shchukina
e-mail: nada0973@mail.ru

A. V. Bogoviz (ed.), *Big Data in Information Society and Digital Economy*,
Studies in Big Data 124, https://doi.org/10.1007/978-3-031-29489-1_27

1 Introduction

An urgent need to improve the technology for managing their resources arises to overcome the negative consequences of the economic shock caused by the pandemic and its consequences, to ensure the survival and stabilize the development of commercial organizations. At the same time, in a crisis economy, the problem of cost management, which largely determines the competitiveness of a commercial organization and its viability, is significantly exacerbated.

This problem does not lose its relevance in a steadily developing economy. A crisis of a certain number of organizations is a normal phenomenon of a market economy, when non-viable members of the population of economic entities are crowded out along with the growth and development of competitive economic entities [3, p. 74]. This is confirmed by the fact that a significant part of commercial organizations is unprofitable even during the period of relatively stable economic development. Thus, according to the Federal State Statistics Service of the Russian Federation (Rosstat), in the last pre-pandemic year, the share of unprofitable organizations was 28.1% [10, p. 83]. The problem of improving methodological approaches to cost management, without losing its relevance during a period of stable economic development, becomes especially acute in a crisis and post-crisis economy, considering the role of costs in the formation of the profit of commercial organizations.

In a stable economy, the main goal of any commercial organization is to increase the capitalization of the company and the welfare of its owners. The main goal of a production cooperative is to ensure its growth and development, as well as to increase the welfare of its shareholders, considering that it is a corporate for-profit organization. These goals are achieved through the distribution of its profits, part of which is directed to the development of the cooperative's activities, and the other is distributed among its members.

In a crisis economy, the goals of management are different—it is to maintain the solvency and financial stability of organizations, to ensure the break-even of their activities. At the same time, the main task of management is to prevent the bankruptcy of a commercial organization, to ensure its survival and further development.

The need to develop innovative approaches to managing the costs of production cooperatives is also justified by the processes of informatization and digitalization of economic management processes. So, Mandych and Bykova believe that "all over the world, the new paradigm of economic development is characterized by an increase in the role of the production of high-tech products, intellectualization of resources, the expansion of the international market for intellectual property objects as an independent sector of the world market, as well as relying on innovative sources of growth" [8, p. 80]. In these conditions, there is an urgent need to develop a technology for cost management, an adequate information economy and based on modern management technologies. This is what determined the topic of this article, devoted to the study of the possibilities of improving the cost management of a production cooperative on the basis of modern management technologies.

2 Methodology

The innovative approaches to cost management, contained in the article, are based on modern methods of operational analysis using *GAP* analysis tools and crisis management technologies.

In theoretical terms, this work is based on the works of domestic scientists on anti-crisis management: Alferov et al. and others [1, 2, 5, 7, 11–13], as well as on the author's research on the optimization of cost management.

The main goal of a production cooperative, like any commercial organization, is to ensure its growth and development, as well as to increase the welfare of its owners (shareholders). The most important condition for achieving these goals is to ensure consistently profitable work. Due to the profit, which is directed to the development of its activities, economic growth is ensured. The other part of the profit is distributed among the members of the cooperative, contributing to an increase in their well-being.

In the context of the global economic crisis caused by the coronavirus pandemic and its consequences for the effective management of business processes, it is necessary to apply modern management technologies and appropriate tools.

At the level of a commercial organization, it is advisable to use the methodological apparatus of *GAP* analysis (gap—space, break) as a tool. In our opinion, "its essence consists in assessing and analyzing deviations of the actual values of the monitored indicators from the target (planned) ones in the development of corrective actions to eliminate deviations (strategic gaps)" [4, p. 203].

In the event that the result of the *GAP* analysis shows that it is necessary to take measures to prevent losses due to the projected decrease in sales, it is advisable to use operational analysis, which implements the management focus on ensuring the break-even operation of a commercial organization, maintaining its solvency and making a profit in volume necessary for further development. The operational analysis toolkit allows finding the optimal balance between variable and fixed costs, price and sales volume.

In particular, the operational analysis makes it possible to determine the profitability threshold (break-even point), that is such a volume of sales of the production of a production cooperative at which it no longer suffers a loss, but does not yet begin to make a profit. Such a volume of sales will not ensure the development of this organization and the growth of the well-being of its member-shareholders, but will allow it to maintain its vital activity and solvency in a critical situation and in the short term to avoid bankruptcy.

Let us denote variable costs per unit of sales a, and fixed costs b to determine the break-even point. With a sales volume of x units and a unit sales price P, the equation for determining the break-even point has the following form:

$$Px = ax + b \tag{1}$$

This is the break-even point:

$$x = b : (P - a) \tag{2}$$

Consequently, the profitability threshold (break-even point) is defined as the quotient of dividing the sum of fixed costs by the marginal income per unit of output [6, p. 272].

Let us determine the condition for the break-even economic activity of a commercial organization using the operational analysis apparatus.

At the break-even point, the profit volume is $R = 0$. Therefore, the equality holds:

$$Px - ax - b = 0 \tag{3}$$

In crisis conditions, the primary task of any commercial organization is its survival and prevention of bankruptcy. In this regard, the determination of the break-even point, when a commercial organization is able to fulfill its obligations to pay mandatory payments and fulfill monetary obligations, but does not receive profit, has a certain practical significance.

The equation for determining the volume of sales x, providing a profit in the amount of $R > 0$ has the following form:

$$R = Px - ax - b. \tag{4}$$

Consequently, the condition for the break-even activity of the organization, which consists in making a profit $R \geq 0$, is expressed by the inequality (5):

$$Px - ax - b \geq 0 \tag{5}$$

The main goal of a commercial organization in a stable economy is to maximize profits. Therefore, at each stage of economic activity, its goal can be formally described by the function (6):

$$R = Px - ax - b \rightarrow max \tag{6}$$

Let us note that the net profit earned by a production cooperative is usually spent on reinvestment in the development of the company and distribution among its members. These payments may be temporarily delayed in times of crisis and lack of sufficient profits in order to maintain the viability of the commercial organization. Such measures will increase profits directed to development, avoid bankruptcy and keep the company.

The volume of sales predictably falls in a crisis economy, when the effective demand of the population decreases, and its economic activity also decreases in the event of a pandemic.

The volume of sales, which ensures the receipt of profit on a certain type of product in the planned volume of R units, can be determined by the formula (7):

$$x = \frac{b + R}{P - a}. \tag{7}$$

In the context of the economic crisis, a negative trend towards a decrease in this indicator is likely to form. At the same time, it can fall below the profitability threshold, which will lead to a loss based on the results of the organization's activities. Therefore, a financial strength margin (the difference between the volume of sales of products and the threshold of profitability) acquires special importance in a crisis economy. The value of this indicator makes it possible to assess the possibility of a relatively safe decrease in the volume of sales of products within the boundaries of the break-even point [10, p. 83].

A decrease in the value of this indicator is an alarm signal. If the value of this indicator decreases and becomes critically small, then it is necessary to immediately make management decisions aimed at correcting the situation. Therefore, it is advisable to regularly monitor the margin of financial strength. Its value shows how much it is possible to reduce the volume of sales of the organization's products without reaching the break-even threshold.

The main role in the choice of management decisions to maintain the financial strength at the proper level belongs to the indicator of marginal income *(Px-ax)*, which is primarily intended to cover fixed costs, and its balance represents profit for the period [9, p. 12].

If the result of the analysis shows that it is necessary to take measures to prevent a loss due to the projected decrease in sales, it is advisable to use the effect of operating leverage (E_{ol}), which is measured as the ratio of the marginal income to the profit of the production cooperative:

$$E_{ol} = \frac{Px - ax}{Px - ax - b} \tag{8}$$

The high operating leverage effect leads to the fact that even a small change in sales volume can lead to a significant change in profit. Thus, an increase in the effect of operating leverage leads to an increase in the risk of loss of profit due to lower sales volumes.

3 Discussion

We note that if the economy is healthy, it is advisable to have a high operating leverage effect. In this case, the profit will grow at a faster pace than the sales volume. In a crisis economy, it is advisable to take measures to reduce the value of this indicator. This will reduce the risks of a critical decrease in profits in the event of a drop in sales, and, thereby, will help maintain the solvency of the production cooperative and prevent its bankruptcy.

In order to reduce the effect of operating leverage in an economic crisis, it is advisable to take measures to reduce the share of fixed costs in the structure of the organization's costs. This can be realized through such anti-crisis measures as partial disinvestment of non-current assets, reduction of management personnel, transfer of part of fixed costs to variables. For example, the production staff of a cooperative can be converted to piecework wages. All this will reduce the strength of the operating leverage and reduce the risk of a loss as a result of a decrease in sales in crisis conditions.

In general, the use of innovative approaches to cost management based on modern methods of operational analysis, *GAP* analysis tools and anti-crisis management technologies can contribute to the survival of a production cooperative in crisis conditions, to successfully overcome the consequences of the economic crisis and ensure its further development through cost optimization, and, as a result, an increase in profits.

In particular, the application of the methods of *GAP* analysis will strengthen the orientation of the management process towards achieving the set goals. Operational analysis will ensure that management is focused on the breakeven operation of the production cooperative, maintaining its solvency and making a profit in the amount necessary for its further development and meeting the needs of its shareholder members. The use of tools for the threshold of profitability and the margin of financial strength in the process of managing the costs of an organization will help prevent insolvency and bankruptcy of the organization, which is especially important during a period of economic crisis.

4 Conclusion

The use of innovative approaches to managing the costs of production cooperatives based on modern methods of operational analysis using GAP analysis tools and anti-crisis management technologies allows:

- Setting a threshold for the profitability of an organization, defining the edge of its financial security, crossing which leads to insolvency and bankruptcy;
- Determination of the break-even condition and the target function of the economic activity of the production cooperative;
- Monitoring the financial strength of the organization, which makes it possible to assess the value of an additional relatively safe decrease in the volume of sales of products;
- Regulation of the size of the operating leverage in such a way as to reduce as much as possible the loss of profit of the production cooperative with a decrease in sales in the context of the economic crisis;
- Timely assessment of the deviation of monitored profit indicators from target values, determination of their causes and development of anti-crisis management actions to reduce the share of fixed costs in the cost structure.

The use of the innovative approaches to cost management discussed in the article will increase the efficiency of management decisions aimed at achieving the main goal of management in the context of the global economic crisis—to prevent the bankruptcy of a production cooperative, to ensure its survival and further development. The use of modern management technologies during the period of stable economic development will increase the efficiency of resource use, thereby increasing the competitiveness of the cooperative.

References

1. Alferov, V. N., & Filyaeva, A. E. (2019). Financial sustainability of the organization: Theoretical aspects of analysis and management. *Problems of Modern Economics, 4*(72), 57–60.
2. Blank, I. A. (2004). Financial security management of the enterprise. Kiev, Ukraine: Nika-Center Elga.
3. Blau, S. L. (2004). Management of the resource potential of the region from the standpoint of food security. Moscow, Russia: Information and Implementation Center Marketing.
4. Blau, S. L., & Yusupova, S. Y. (2018). Methodological approaches to improving the management of state corporations. *Management Sciences in the Modern World, 1*(1), 200–204.
5. Borodach, Yu. V., & Belyaeva, A. V. (2016). Enterprise management using anti-crisis management tools. *Economy Management Innovations, 6*(6), 6–15.
6. Khomutova, E. V., & Shchukina, N. F. (2021). Operational analysis as a tool for managing a commercial organization in an economic crisis. In Priority directions of innovative activity in industry: collection of scientific articles of the III international scientific conference (pp. 270–275.). Kazan, Russia.
7. Kovan, S. E., Kryukova, O. G., Ryakhovskaya, A. N., Alferov, V. N., Berezin, K. A., & Plaksin S. Yu. (2014). Transformation of anti-crisis management in modern economic conditions. Moscow, Russia: INFRA-M.
8. Mandych, I. A., & Bykova, A. V. (2019). Trends of innovative and investment development of high-tech enterprises. *Russian Technological Journal, 7*(5), 79–92.
9. Nikolaeva, O. Y., & Shchukina, N. F. (2020). Profit management of a commercial organization based on controlling tools. *Economic Science Issues, 6*, 10-13.
10. Repushevskaya, O. A., & Khomutova, E. V. (2021). Using operational analysis as a management tool for a commercial organization in a crisis economy. *Russian Journal of Management, 9*(2), 81–85.
11. Ryakhovskaya, A. N., & Kovan, S. E. (2019). Anti-crisis management as a scientific direction in the system of management sciences. *Economic Sciences, 174*, 109–116.
12. Savina, M. V., & Zhilchenkova, V. V. (2018). Features of cost management during the period of anti-crisis management of an industrial enterprise. *Current Trends in the Development and Prospects for the Introduction of Innovative Technologies in Engineering, Education and the Economy, 4*(1), 215–220.
13. Ushanov, P. V. (2010). Anti-crisis management as a new management paradigm. *Effective Anti-Crisis Management, 1*(60), 66–79.

The Impact of Digitalization of the Economy on the Cooperation of Economic Entities in the Post-pandemic Period

Veronika R. Akaeva, Yana F. Narshirvanova, and Natalya V. Savintseva

Abstract According to the authors, the digital transformation of the global economic system, which has affected all areas of business and sectors of the economy, has led to the formation of a new business model aimed at creating a single financial and economic space without borders, contributing to new forms of obtaining added value and "digital dividends" by combining various technologies (for example, cloud technologies, sensors, big data, 3D printing), as well as the development of markets for goods and services, labor reserves and capital through transformations at all social levels. The authors believe that all of the above opens up expanded opportunities for organizing and doing business and allows increasing the potential for creating radically new products, services and innovative business models focused on sustainable business development in the new conditions of digitization of the economic system. In this regard, the paper explores key approaches to the definition of the term "digital transformation of business." The trends of business digitalization and, accordingly, the factors that are inhibitors and drivers of the development of a new business model of cooperation and cooperation of modern organizations were identified. In the process of analysis, the authors determined the vector of development of business models in the context of the digital transformation of the global economic system.

Keywords Digital transformation · Digitalization of the economy · New business model · Intercompany interaction · Cooperation · Sustainability and business development · COVID-19

JEL Classifications F15 · F63 · G3 · M1 · P11

V. R. Akaeva (✉)
Institute of Management, Economics and Finance, Department of Marketing, Kazan Federal University, Kazan, Russia
e-mail: r-akaev_80@mail.ru

Y. F. Narshirvanova · N. V. Savintseva
Faculty of Economics and Customs, Department of Economics and Management, Kazan Cooperative Institute (branch), Russian University of Cooperation, Kazan, Russia
e-mail: yana.nashirvanova@mail.ru

A. V. Bogoviz (ed.), *Big Data in Information Society and Digital Economy*,
Studies in Big Data 124, https://doi.org/10.1007/978-3-031-29489-1_28

1 Introduction

The transformations on the introduction of advanced technologies that began in 2020 led to the emergence of a new technological order, with the widespread use of cyber-physical systems through the digital transformation of the global economic system [6]. In this regard, the "digital state" has been developed, within its framework, programs to support the digitalization of economic sectors and government operations have been initiated [18]. The blurring of the boundaries of service markets, the strengthening of external factors influencing the formation of new forms of cooperation and coordination of economic entities are observed: virtual companies and business networks are actively developing, a single financial and economic space without borders is being formed, contributing to new forms of value added and "digital dividends", as well as the development of markets for goods and services, labor reserves and capital [12, 15]. All of the above confirms the relevance of the topic of this article. In this connection, its purpose is to study the trends of digital transformation of business and to determine the vector of development of "digitized" organizations. The set goal forms block of tasks to be solved: (1) to reveal the conceptual approaches to the definition of the term "digital business transformation"; (2) to identify key business digitalization trends; (3) to determine the vector for the development of business models in the context of the digital transformation of the global economic system.

2 Methodology

As part of writing this article, we used the following methods: (1) content analysis of scientific publications of domestic and foreign authors, indexed in the Russian Science Citation Index, presented in free access on the Internet; (2) critical analysis of the existing approaches to the definition and study of the problems posed in the article. In studies on digitalization, it was revealed that most of the studies pay special attention to the search for relationships between innovative activity and the level of inter-firm interactions [3–5, 11, 13, 16, 17, 19] to strengthen positions in international competition. In our article, we will focus on these areas as well.

3 Results

Before proceeding to the analysis of the modern business model that has emerged in the context of the digital transformation of the global economic system, we would like to draw attention to what, in fact, such a process as business digitalization is and what it is aimed at. Digital modernization refers to the process of changing the model of organization of the socio-economic system. That is, digitalization involves serious, profound changes in society and the economy. It creates fundamentally new markets (high-tech), new practices of social interaction (social networks and e-commerce), and government regulation (e-government). In terms of cooperation, the digital modernization of economic systems offers expanded opportunities for joint investment, risk sharing, knowledge sharing, and resource sharing [2, 6]. Simultaneously, the consideration of the opportunities presented by digital transformation requires a flexible and responsible approach to cooperation. This requires a transformation of business management practices.

In accordance with the above, digital transformation of business should be understood as the integration of digital technologies in all areas of business, which radically transforms how it functions and how it reproduces customer value, creates benefits for its own well-being and staff, as well as organizations with which contracts are concluded [1]. In general, digitalization is a deep business transformation at all levels of value creation based on the use of digital technologies, optimization of business processes and management in order to simplify interaction with consumers, employees, suppliers and other stakeholders. Therefore, digitalization makes modern organizations think about the feasibility of ways to organize their own business, change the current corporate strategy and the need for a comprehensive study of new opportunities for its development at an early stage.

The COVID-19 pandemic has become a direct catalyst for the development of digital transformation projects. This is explained by the fact that the problem of the survival of organizations in the face of increased competition and instability was acutely manifested during this crisis period. Therefore, the ability of organizations to quickly adapt to post-pandemic supply chain disruptions and rapidly changing customer expectations has become critical, and spending priorities reflect this reality. According to the World Data Corporation (IDC) Digital Transformation Expenditure Committee for May 2020, spending on the digital transformation of business practices, products and organizations continues "…at a sustained pace despite the challenges posed by the COVID-19 pandemic." IDC predicts that global spending on digital transformation technology and services will grow by 10.4% in 2020 to 1.3 trillion dollars. This compares to a 17.9% growth in 2019, "…but remains one of the few highlights in a year of sharp declines in overall technology spending…," IDC notes [8].

In this regard, the expansion of digital services became imperative when it became impossible to interact with customers and partners in an analog way. Hundreds of company managers around the world say that the coronavirus pandemic has had a significant impact on their companies' digital transformation efforts. In this regard, many organizations have revised their strategic development priorities and developed new adaptation strategies. In the past, there has been reluctance to decide to innovate business processes involving the virtual holding of important business meetings or the transition to a fully electronic workflow in some companies. The coronavirus pandemic has made it possible to adjust the organization of business processes of companies in all sectors of the economy and to the interaction between them. The accelerated modernization supported the evolving business goals of improving business continuity and resiliency by increasing the flexibility associated with operations and business processes. Business continuity or sustainability during 2021 increased by 21% points over the previous COVID-19 rating. The desire to improve business continuity is associated with the need to stabilize and develop a competitive market position.

Another key business goal is to increase the flexibility associated with operations and business processes, here, there has been an increase of eight percentage points since the outbreak of COVID-19. Organizations are looking for super-efficiency through digitization as it allows business processes to be digitized, and there is an opportunity to automate those processes more effectively through the application of new rules and artificial intelligence.

Two other, more important priorities—each of which has increased by five percentage points—increase the potential for new business proposals designed to accelerate innovation and use corporate data to reveal new business and operational ideas. Together, the emphasis on agility, innovation and analytics demonstrate the growing interest of leaders in constant change to quickly understand the changes taking place in the market.

Now companies must learn from the 2019 business transformation experience and turn it into updated strategies. The improvements replicated in the leading companies help inform about this process, because the initiatives to transform the business model in them were successful before the coronavirus pandemic and remained highly effective after it intensified. For example, some leaders use continuous integration or continuous delivery to modernize processes, the process of accelerating deployment and continuous improvement of new applications to meet changing business requirements, adhere to lean manufacturing methodologies, turn to agile methodologies for activities inside and outside information technology. More than half of them report that agile practices support the project management processes in their organizations while encouraging steps such as building cross-functional teams that drive transformation through a series of iterative steps.

CI/CD methods and guidelines are being given significant attention by leading digital companies globally as a resource for speeding up workflows, because focusing on speeding up processing is one of the most important areas that organizations need to focus on today. At the same time, it should be noted that the modern modernization of technologies requires a multifaceted approach, because, most often, it involves the transformation of the technologies used as part of this process. This connection exists because outdated technologies are one of the main obstacles to accelerating business processes, and optimized (digitized) systems allow organizations to interact more effectively with each other in the digital economic space through video conferencing, data exchange and content creation, and accordingly the creation of new business models. Leaders' focus on collaboration is not surprising, considering the ubiquity of the remote workforce caused by the 2019–2020 pandemic. To expand analytics capabilities, 41% of firms plan to invest in cloud analytics platforms, which were already a top priority before the pandemic and are steadily growing in importance. Because analytics, including artificial intelligence, are the tools to improve and transform a business—from calculating a customer's lifetime value to ensuring that a company offers the right product for a particular customer. In our opinion, such analyzes are very important for companies seeking to increase the level of competitiveness in the context of business digitalization. Based on expert assessments, the efficiency of companies that have digitized their business processes has increased by 13 percentage points. This data shows that a growing number of executives believe that effective internal services are the basis for customer retention, which is critical to maintaining profitability, given the costs associated with attracting new customers [8].

Let us consider in more detail how the relationship between business structures and other organizations is built in the context of the digitalization of the economy. The digitalization process is gradually changing the way of doing business, forms of cooperation, increasing the ability to manage the interaction between business partners, establishing closer business ties through a clear definition of the role and responsibilities assigned to each of them. As it was revealed, business cooperation began to transform from a traditional form into multilateral cooperation aimed at more comprehensively meeting the needs of the digital age. Partners invest in digital assets focused on relationships, technology alignment and competency development. The benefit of digitalization is that digital knowledge sharing processes increase transparency and efficiency in the use of data.

One of the main factors determining the involvement and activity of digital modernization of enterprises is their openness and globalization. Large corporations, especially multinational corporations that rely heavily on their international operations, are more likely to use digital innovation. The reason is not only that they have more opportunities to do so but also that they are subject to stronger market incentives. In the pursuit of global competitiveness, these corporations cannot stay away from global innovation and technological trends. In contrast, enterprises that focus on local markets and local (in a region, town, or village) competitiveness are somewhat isolated from global trends. Digital technology and advanced telecommunications infrastructure are less available to them. This requires managers of the companies in question to manage change more flexibly and comprehensively when implementing digital business modernization strategies [18]. Innovation policies that promote diffusion are aimed at helping test new digital applications, for example, by setting up test stands and normative "sandboxes." Innovative initiatives also facilitate the early deployment of advanced digital technologies. For this purpose, they help innovators access cutting-edge tools and expertise (for example, in the field of artificial intelligence or supercomputing) [7].

4 Discussion

The competitive advantage in the process of business digitization is the result of mutually adapted inter-firm relations and the joint contribution of partners, which allows these companies to develop together, generating rents based on the digitization of relationships [14]. Firms that integrate their resources in a particular way may provide a better advantage than competing firms that are unable or unwilling to do so, although competition between firms may still be the general rule. At the same time, it should be noted that the implementation of integrated products and services can be successful only when organizations deploy them together in close cooperation. In this regard, it is worth noting the possibility of using additional resources, assets associated with specific relationships, knowledge sharing procedures and effective management throughout the entire supply chain as key factors that determine and form interorganizational competitive advantages. In our opinion, these components of business digitalization have significant potential for understanding how inter-firm relations are being transformed through digital servitization. In fact, these determinants can generate relationship rents or "excessive" profits jointly generated within the exchange relationship that cannot be generated by any of the companies individually and can only be created through the joint contribution of specific alliance partners. At the same time, access to additional resources and opportunities is considered the initial basis for creating a partnership. In this case, the marginal return on the partner's resources increases when there are resources from the other partner [4, 9]. Also, "…the greater the proportion of synergy-sensitive resources owned by alliance partners, which together increase the degree of value, rarity, and difficulty of imitation of resources, the greater will be the potential for their creation."

The advantage of cooperation is the acceleration of the pace of digital modernization and its fullest coverage of local enterprises. However, contrary to current perceptions that this advantage is due to the increased availability of digital technology [20], it is based on knowledge integration and collaborative team building. This is due to the fact that local businesses do not have enough resources to master and establish effective use of digital technology as large and multinational corporations do on their own. In contrast, local businesses need to exchange experiences and ready-made application solutions.

Cooperation allows creating and transferring know-how on the use of digital technology in all business processes, from management to production and sales. Cooperation also makes it possible to select the most effective practices for each individual market, the boundaries of which are defined by product and geographic criteria. In addition to investing in relationship-specific assets, partners are establishing the knowledge-sharing processes and procedures necessary for digital easement. Partners must agree on the mechanisms for managing their partnership in order for cooperation to be successful, because this is critical to creating the opportunity to extract the maximum benefit from the use of digital assets and to form an effective digital knowledge exchange procedure. Thus, the balance of control and flexibility of managing the formed new business model between partners is corrected [20].

As the relationship develops, partners establish trust-based relational governance to focus on win–win improvements. As partnerships mature and digital services become more complex, governance is increasingly based on trust without tight upfront controls or tight monitoring of transactions. For example, they highlight the digitalization capabilities that providers must develop in order to interact and co-create value with their customers, intelligence, connectivity and analytics capabilities. In terms of digital servitization, the more digital intensity increases within a company's business strategy, the more likely that its scaling options will be based on partnerships with other companies through shared digital assets. For example, customers' factories need to be connected to the supplier's digital architecture using a compatible communications network to guarantee the availability of machines and plants [20].

The greatest effect of synergy in cooperation is achieved in the strategic cooperation of enterprises. The long-term orientation of cooperation creates an atmosphere of mutual trust in the enterprises. This ensures the transition from the pursuit of their own separate goals to the establishment and achievement of joint goals for the development of an integrated business. Simultaneously, a high degree of independence of cooperating business structures can be maintained. Mutual trust also makes a more serious transformation of value chains possible, with suppliers and buyers making common changes in production and distribution: from a unified electronic document management system to an industry standard of robotization. Thus, a key feature of strategic cooperation is the joint management of change in mutually beneficial business areas [11].

5 Conclusion

In the context of the digitalization of the global world system, there is the creation of new digital markets and business models based on digital technologies, often adjacent to traditional sectors. While new business models are emerging in the economy, the scale and potential of these trends vary by sector. In some cases, these business models can replace traditional ones [7]. In this regard, digitized organizations are more adapted to today's economic realities and more attractive for investment than organizations that do not have these capabilities [7, 10].

In general, it is worth noting that digital transformation involves the use of advanced technologies to radically increase the productivity or reach of enterprises. In this regard, leaders in all industries are leveraging digital advances such as analytics, mobility, social media and smart embedded devices, as well as improving the use of traditional technologies such as ERP to transform customer relationships, internal processes and value propositions. Thanks to digitalization, the following improvements in business development are observed. First, data is becoming a key input to innovation. Second, the innovative activity of businesses ensures their digital competitiveness, allowing them to expand the scale of their activities. Third, the pace of the innovation process is speeding up, the stages of which are becoming even more integrated through digital technology. Fourth, innovation is becoming systemic: it no longer involves individual business processes but entire business models and cooperatives [7].

In this regard, the key performance indicators of "digitized" organizations show that global enterprises see the benefits of their digital transformation efforts in terms of improving the efficiency of the entire business system through investments in the latest digital tools and in process modernization [21].

References

1. Akaeva, V. R. (2020). Digital logistics as an innovative tool for managing business processes in a modern organization. In *Digital region: Experience, competencies, projects: Collection of articles of the III International Scientific and Practical Conference dedicated to the 90th anniversary of the Bryansk State Engineering and Technology University* (pp. 47–51). Bryansk, Russia.
2. Akaeva, V. R. (2020). The impact of digital transformation on the business environment. In *Digital region: Experience, competencies, projects: Collection of articles of the III International Scientific and Practical Conference dedicated to the 90th anniversary of the Bryansk State Engineering and Technology University* (pp. 52–55). Bryansk, Russia.
3. Etzioni, A. (1965). *Political unification: A comparative study of leaders and forces*. New York, NY: Rinehart and Winston.
4. Fritsch, M. (2003). Does R&D-cooperation behavior differ between regions? *Industry and Innovation, 10*(1), 25–39.
5. Fritsch, M., & Lukas, R. (1999). Innovation, cooperation, and the region. In D. B. Audretsch & R. Thurik (Eds.), *Innovation, industry evolution and employment* (pp. 157–181). Cambridge University Press.
6. Gryaznevich, V. (2021). *Our time has come: How a new technological order is being created*. Retrieved from https://www.rbc.ru/spb_sz/03/01/2021/5ff097a39a7947903446c7e8. Accessed 14 Oct 2021
7. Guellec, D., Paunov, C., & Planes-Satorra, S. (2020). *Digital innovation: Cross-sectoral dynamics and policy implications*. Retrieved from https://www.oecd-ilibrary.org/sites/ee2a2c2f-en/index.html?itemId=/content/component/ee2a2c2f-en. Accessed 13 Oct 2021
8. Harvard Business Review Analytic Services. (2021). *Accelerating transformation for a post-COVID-19 world*. Retrieved from https://www.redhat.com/rhdc/managed-files/so-accelerating-transformation-post-covid-19-analyst-material-f28723-202105-en.pdf. Accessed 16 Oct 2021
9. Hess, A. M., & Rothaermel, F. T. (2011). When are assets complementary? Star scientists, strategic alliances, and innovation in the pharmaceutical industry. *Strategic Management Journal, 32*(8), 895–909.
10. International Institute for Applied Systems Analysis (IIASA). (2018). *Digitalization will transform the global economy*. Retrieved from https://previous.iiasa.ac.at/web/home/resources/publications/IIASAPolicyBriefs/pb20-web.pdf. Accessed 13 Oct 2021
11. Kamalaldin, A., Linde, L., & Sjödin, D. (2020). Transforming provider-customer relationships in digital servitization: A relational view on digitalization. *Industrial Marketing Management, 89*, 306–325.
12. Kuzovkova, T. A., Salyutina, T. Yu., & Sharavova, O. I. (2019). The impact of digital development on the transformation of the organizational and methodological apparatus of statistics and the economy of infocommunications. *Age of Quality*, *2*, 106-119.
13. Medvedeva, V. R. (2010). Improving the efficiency of resource saving management of a petrochemical complex enterprise using automated systems. *Bulletin of Kazan Technological University, 9*, 787–789.
14. Milgrom, P., & Roberts, J. (1995). Complementarities and fit strategy, structure, and organizational change in manufacturing. *Journal of Accounting and Economics, 2–3*, 179–208.
15. Nashrvanova, Y. F. (2021). Trends in the formation of social and labor relations in the context of digital transformation of the economy. *Russian Journal of Management*, *9*(1), 196-200.
16. Nikulina, I. E., & Burets, Yu. S. (2016). Formation of interregional cooperation in the field of innovation based on the diagnosis of gaps in the innovation process. *Economics and Entrepreneurship, 12*(2), 346–349.
17. Pyke, F., & Sengenberger, W. (Eds.). (1992). *Industrial districts and local economic regeneration*. ILO.
18. Song, W. (2007). Regionalization, inter-regional cooperation and global governance. *Asia Europe Journal, 5*(1), 67–82.

19. Timchenko, K. Y. (2018). Forms of intercompany interaction. *Modern Innovations, 5*(27). Retrieved from https://cyberleninka.ru/article/n/formy-mezhfirmennogo-vzaimodeyst-viya. Accessed 5 Oct 2020
20. Torre, A., & Rallet, A. (2005). Proximity and localization. *Regional Studies, 39*(1), 47–59.
21. Westerman, G., Bonnet, D., & McAfee A. (2014). *The nine elements of digital transformation.* Retrieved from https://sloanreview.mit.edu/article/the-nine-elements-of-digital-transformation. Accessed 16 Oct 2021

Applied Application of the Apparatus of Formal Mathematical Models in Software Development in the Cooperative Sector of the Economy

Ludmila A. Gaynulova, Alfira M. Akhmedova, Guzel Z. Khabibullina, Irina V. Zhazhneva, and Elena S. Shchigortsova

Abstract The digitalization of society as a whole and the widespread introduction of advanced information technologies based on the application of the apparatus of formal mathematical methods and models affect all areas of life in Russia, in particular, the cooperative sector of the economy. The development and implementation of software (SW) for use in the cooperative sector of the economy is an important task from the point of view of practice for consumers of agricultural products and producers. The article deals with the application of some of the formal mathematical models, for example, the hierarchical data model and the relational data model presented in the works of (Gainulova et al. in. Application of Formal Models in the Applied Software of the Simulator "Driving School-Driving". Cooperation and Sustainable Developmen, pp. 311–315, 2022 [6]; Georgiev et al. in Studies in Systems, Decision and Control. Springer, pp. 89–97, 2021 [7]; Potashev et al. in Studies in Systems, Decision and Control. Springer, pp. 299–308, 2021 [10), in the cooperative sector of the economy, while creating digital projects.

The applied use of algorithms of the apparatus of mathematical modelling, in particular, based on a hierarchical data model and a relational data model, is considered The implementation of this approach is presented in the form of application software for the digital project "Products of agricultural cooperatives".

L. A. Gaynulova (✉)
Kazan Cooperative Institute (Branch) Russian University of Cooperation, Kazan, Russia
e-mail: lagainulova@bk.ru

A. M. Akhmedova · G. Z. Khabibullina · I. V. Zhazhneva
Kazan (Volga region) Federal University, Kazan, Russia
e-mail: alfira233@yandex.ru

G. Z. Khabibullina
e-mail: hgz1980@rambler.ru

I. V. Zhazhneva
e-mail: iraira90@bk.ru

E. S. Shchigortsova
Kazan State Power Engineering University, Kazan, Russia
e-mail: 1996yulia@gmail.com

A. V. Bogoviz (ed.), *Big Data in Information Society and Digital Economy*,
Studies in Big Data 124, https://doi.org/10.1007/978-3-031-29489-1_29

The digital project “Products of agricultural cooperatives” is intended for producers and consumers of agricultural products and includes original application software for the operation of the thematic site, mobile application of the site, bot, database, including developed original queries in the SQL language, etc. The software makes it possible to solve problems of universal type and specific problems of the subject area, in particular, the problems of the cooperative sector of the economy. The software of the digital project “Products of agricultural cooperatives” allows meeting the ever-growing requirements for solving informal problems that arise due to promoting the products of agricultural cooperatives. Such informal problems include the type of “what happens if” problems, the main meaning of which is to switch from clear formal algorithmic structures that determine the solution of the problem to logical programs based on expert systems that are used to diagnose the current state in the availability area, promoting the products of agricultural cooperatives. One more important meaning is also to predict the development of the situation of the cooperative system and to manage this cooperative system.

Keywords Digitalization · Digital project · Applied software · Formal mathematical models · Cooperative sector of the economy · Products of agricultural cooperatives · Hierarchical data model · Relational data model

JEL Classifications L86 · M15 · C80 · D58

1 Introduction

The digitalization of society as a whole and the widespread introduction of advanced information technologies based on the applied application of the apparatus of formal mathematical methods and models affect all areas of the life of the Russian Federation, in particular, the cooperative sector of the economy. The development and implementation of software (SW) for use in the cooperative sector of the economy is an important task from the point of view of practice for consumers of agricultural products and producers. The article deals with the application of some of the formal mathematical models, for example, the hierarchical data model and the relational data model presented in the works of [6, 7, 10], in the cooperative sector of the economy when creating digital projects. The applied use of algorithms of the apparatus of mathematical modelling, in particular, based on a hierarchical data model, is considered. The implementation of this approach is presented in the form of application software for the digital project “Products of agricultural cooperatives”.

2 Materials and Methods

Innovative processes in the cooperative sector of the economy are an actual reality today. Innovation has affected the entire system. The number of potential consumers of farm products is growing. Real farm products produced by a person, not within the framework of agricultural holdings have organoleptic advantages. However, the lack of funds and theoretical knowledge, price disparity, etc. lead to violations in the technology of manufacturing farm products, as well as violations of the requirements of Sanitary and Epidemiological Rules and Norms and State Standard.

Approaches to each other of producers and consumers of agricultural products are actual tasks now. Practice shows that this can be achieved using evidence-based and practice-oriented methods [1, 2, 5, 11]. Digitalization in this area gives qualitative results.

The digitalization of the cooperative sector of the economy sets the goal of developing and implementing digital projects for the development of a wide range of relevant demanded areas in this area.

To achieve the stated goal, the authors will solve several groups of problems. The first group—to explore electronic platforms of similar subjects; identify consumer interest in various types of farm products; to select effective practical algorithms for checking the quality of farm products; to study algorithms for assessing the economic efficiency of the implementation of farm products; to design and create a database in this area; to design and develop a website and its mobile application in this area; to develop activities to promote the site and its mobile application; to calculate the effectiveness of the proposed activities; to identify the risks of creating a digital project "Products of agricultural cooperatives", etc.

The main priority areas and methods of the second group of tasks will be to improve the functionality of the site and its mobile application, improve the theme of the chatbot, expand and refine the content of the database, etc.

All electronic sites taken for research and analysis don't meet the modern requirements of buyers. Each site, in fact, doesn't provide conditions for the full sale of agricultural products. If there are conditions for the sale, there are no conditions for the full sale of agricultural products on the market. The studied electronic platforms are also tied to different social networks.

Visible errors of the studied electronic platforms were taken into account in the process of creating the digital project "Products of agricultural cooperatives". So, conditions were made for the introduction of new functions and opportunities for sellers and buyers due to the creation of the digital project "Products of agricultural cooperatives". And the promotion of agricultural cooperatives and their products was designed.

For the seller and the consumer, the functional content of the site varies [3, 8, 11]. In the process of registering for the seller, one function is issued: Registration in the site system as a seller; Access to social networks (advertising and promotion of products and cooperatives); Adaptability for mobile devices (application); Ability to create an extensive customer database; Logistics conditions for delivery and payment methods; Condition of communication between cooperators and cooperatives.

For the buyer, the functions of the site change: Registration in the system as a buyer; Website and application for creating a consumer basket; Possibility of self-delivery or home delivery; Different payment methods; Quality control.

However, certain features of the site remain the same for the seller and the buyer: Notes; Forum; Commodity fullness of each seller.

The functional content of the site directly depends on the services provided by the site, on the consumer basket of a potential buyer and on suggestions for improving the functionality of the users of the site.

It's necessary to constantly improve the work of the bot to optimize sales. Sellers have to process a huge number of requests, but not every visitor is ready to become a client. The chatbot qualifies the user and shares information to make a decision, collects contacts and transfers the "warmed up" client to the operator. The bot works with clients 24/7. And, if the client doesn't promptly receive a response, the likelihood of going to competitors is very high. The bot is also available at any time and from any device, which means that customers will always be able to get up-to-date information.

The solution of the tasks set is presented in the form of application software of the digital project "Products of agricultural cooperatives".

For mock implementation, hierarchical and relational data models were chosen. The hierarchical structure is a set of elements interconnected according to certain rules. Objects connected by hierarchical relationships form a directed graph. The relational data model is a logical data model, which can be an application to data processing tasks.

3 Results

The digital project "Products of agricultural cooperatives" is intended for consumers and producers of agricultural products and includes:

1. Original application software based on the use of such formal models as hierarchical data model, relational data model;
2. Hardware-software block, including a computer peripheral device, etc.

The software makes it possible to solve problems of universal type and specific problems of the subject area, in particular, the problems of the cooperative sector of the economy.

The software of the digital project "Products of agricultural cooperatives" allows meeting the ever-growing requirements for solving informal problems that arise in the process of promoting the products of agricultural cooperatives. Such informal problems include the type of "what happens if" problems, the main meaning of which is to switch from clear formal algorithmic structures that determine the solution of the problem to logical programs based on expert systems that are used to diagnose the current state in the availability area, promoting the products of agricultural cooperatives; to predict the development of the situation of the cooperative system; to manage this cooperative system.

All of these different solutions to questions and problems can be combined by the application software to simulate the best situation.

4 Discussion

The algorithmic solution of the problems of the first group of the second section is described as an approach to the representation of behaviour in the form of a hierarchy of processes of a special type—finite processes. Each final process and its relationship in the hierarchy are described in natural language. The solution of the second group of problems is based on the concept of a process. This model is built, taking into account the user's actions and describing the directive and request-response processes [4, 7, 9, 10].

As a subject of discussion, the authors put forward the need to study the effectiveness of using other formal models for solving the groups of problems presented in the second section of the article. These models include models based on semantic networks; models based on the theory of formal languages; tensor models; models based on game theory; models based on functional Petri nets; probabilistic models.

5 Conclusion

The article considers the approach of applying formal mathematical models, in particular, the hierarchical data model and the relational data model presented in the works of [6, 7, 10], in the cooperative sector of the economy. The applied use of algorithms of the apparatus of mathematical modelling based on a hierarchical data model and a relational data model is considered. The implementation of this approach is presented in the form of application software for the digital project "Products of agricultural cooperatives".

References

1. Abdrakhmanov, A. L., Abdrakhmanova, L. V., & Shchigortsova, E. S. (2019). Modern international monetary system: Possibilities of using crypto currencies as alternative money settlements. *Scientific Review: Theory and Practice, 9*(12), 1868–1878.
2. Abdrakhmanova, L. V., & Shchigortsova, E. S. (2021). World economy: Current development trends in connection with the spread of COVID-19. *Scientific Review: Theory and Practice, 11*(4), 1179–1189.
3. Akhmedova, A. M., & Zhazhneva, I. V. (2018). Study of the role of PR activities in improving the competitiveness of travel agencies. Growth potential of the modern economy: opportunities, risks, strategies. Materials of the V International Scientific and Practical Conference (Moscow, November 22, 2018). In: A.V. Semenova, M.Y. Parfenova, L.G. Rudenko (eds.). Moscow University named after S.Yu. Witte, pp. 88–92.
4. Burdinov, K. A., Gainulov, E. R., & Karpov, A. I. (2016). Synthesis of ACS by a video camera installed on a multicopter UAV. Scientific progress—the work of the young scientists. *Volga State University of Technology, 3–4*, 77–79.

5. Fatkhullina, N. K., Shamsutdinova, V. V., Shchigortsova, E. S., Vakhidova, Z. R., & Khasanova, L. R. (2022). Problems of development of insurance cooperation in Russia. *Cooperation and Sustainable Development*, 909–915.
6. Gainulova, L. A. Akhmedova, A. M., Khabibullina, G. Z. Matrenina, O. M., & Nigmedzyanova, A. M. (2022). Application of formal models in the applied software of the simulator "driving school-driving. *Cooperation and Sustainable Development*, 311–315.
7. Georgiev, V. O., Biktimirova, K. S., Gaynulova, L. A., Akhmedova, A. M., & Kurmankulova, N. Z. (2021). The research on the application of formal mathematical models in industry-oriented development. frontier information technology and systems research in cooperative economics. studies in systems, decision and control (Vol. 316, pp. 89-97). Springer.
8. Khabibullina, G. Z., Akhmedova, A. M., Makletsov, S. V., Khairullina, L. E., & Raisovna, K. A. (2019). On the Effectiveness of the training programs in the learning process. *Journal of Research in Applied Linguistics, 10*(Special Issue), 590–597.
9. Kondrakov, N. P. (2017). Fundamentals of small and medium-sized businesses: A practical guide. In: N. P. Kondrakov & I. N. Kondrakov (Eds.), INFRA-M, 446.
10. Potashev, A. V., Potasheva, E. V., Ahmedova, A. M., & Gaynulova, L. A. (2021). Mathematical modeling of economic processes in the activities of cooperative organizations. frontier information technology and systems research in cooperative economics. *Studies in Systems, Decision and Control, Springer, 316*, 299–308.
11. Smolentseva, L. V., Gainulova, L. A., Akhmedova, A. M., Khabibullina, G. Z., & Yunusova, G. R. (2022). Application of the information system 1C: Enterprise in the cooperative sector of the economy. *Cooperation and Sustainable Development*, 933–939.

Geographical Latitude Index as an Indicator of Responsible Cooperation of Entrepreneurship Based on Ecological and Socially-Oriented Clusters in the Digital Economy

Nina V. Khodarinova, Valeriy L. Shaposhnikov, Taisiya N. Sidorenko, Svetlana G. Boychuk, and Olga V. Romanova

Abstract The aim of the study is to create an invariant index of responsibility for business cooperation based on environmental and socially oriented clusters in the digital economy. Modern monitoring is an information mechanism of socio-economic management, actively applying invariant indicators in relation to the sphere of entrepreneurial activity (Loiko et al Scientific Journal of Kuban State Agrarian University, 129(05):1382–1406, 2017). In this article, the authors propose to consider a monitoring invariant indicator, the geographical latitude index, as an indicator of the results of the activities of responsible business cooperation based on environmental and socially-oriented clusters in the digital economy. The authors also proved that this indicator has a differentiating capability to identify the most socially responsible business cooperatives, companies, enterprises, organizations, creative teams working in environmental and socially-oriented clusters. The issues of ecological (green) and socially-oriented economy in the conditions of digital transformation are considered. The cluster mechanism is justified as universal in relation to various territories and regions for the implementation of the fundamental principles of responsible cooperation of entrepreneurship. The fundamental methods of this research are methods of set theory, clustering of economics, and the scientific foundations are modern concepts of monitoring as an information and analytical mechanism of socio-economic management.

Keywords Monitoring · Geographical latitude · Index · Responsible cooperation · Cluster Products

N. V. Khodarinova (✉) · V. L. Shaposhnikov · O. V. Romanova
Krasnodar Institute of Cooperation (branch) of Russian University of Cooperation, Krasnodar, Russia
e-mail: hodarinovanina@mail.ru

V. L. Shaposhnikov
e-mail: shaposh.vl@mail.ru

T. N. Sidorenko · S. G. Boychuk
Russian University of Cooperation, Mytishchi, Russia

A. V. Bogoviz (ed.), *Big Data in Information Society and Digital Economy*, Studies in Big Data 124, https://doi.org/10.1007/978-3-031-29489-1_30

JEL Classifications Q00 · Q01 · Q29 · Q30 · Q38 · Q39 · Q50 · Q56 · Q59 · R10 · R15 · R19 · Z13 · Z19

1 Introduction

In the digital economy, with the use of modern information technologies and artificial intelligence, careful monitoring is carried out not only of the activities of business cooperatives, but also an assessment of their effectiveness in terms of corporate responsibility for the environment and social well-being of employees.

Corporate responsibility is a system of principles of business cooperation that considers the interests of the local community, taking responsibility for its impact on the environment and social well-being of all subjects of the public sphere of its geographical location. This system goes beyond the generally accepted legislative framework and assumes that business cooperation independently takes additional measures to effectively improve the environment, and not just eliminates negative impacts on nature and is engaged in improving the quality of life not only of cooperative workers and their families, but also of the local community and society as a whole.

An effective mechanism of social management in business cooperation presupposes the existence of criteria and diagnostics of business processes and internal and external features of business processes affecting its activities, as well as algorithms and rules for their interpretation. The authors of the article also share the opinion that when conducting this study, it is necessary to consider several parameters in order to avoid the presence of one-sided indicators, or even to allocate one of the many indicators as a priority and adjust it to the normative values.

The incorrectness of the activity lies in the search for artificial improvement of corporate responsibility indicators. In practice, there are many examples that prove the facts of falsification of performance indicators of enterprises, companies, cooperatives reflecting their achievements in various fields of the economy. It is necessary to develop indicators that are unchanged in relation to the sphere of activity of the business cooperation, so that, on the one hand, their use would improve the management of the cooperation, and on the other hand, these indicators could not be arbitrarily adjusted when determining the effectiveness of the business cooperation in terms of ecology and social activities.

Based on the above, the methods and experience of managing the work of responsible business cooperation show that the task of developing indicators that are unchanged in relation to the field of activity, resistant to arbitrary adjustment, is relevant. The problem of the study is to solve the problem of finding an indicator that remains unchanged in relation to the sphere of activity of the business cooperation, objectively reflecting the responsibility of the work of the business cooperation and resistant to arbitrary adjustment. The present study is aimed at finding and substantiating the invariable, regardless of the field of activity, the index of responsibility of the results of the work of the business cooperation.

In the digital economy, the task of assessing corporate responsibility is becoming more and more interesting. Widely applied territorial development projects that aim at social and environmental transformations that take into account the interests of the local community. Unlike a number of foreign countries, in Russia, social investment of entrepreneurship is defined as a contribution to the transformation of the ecology and social life of the territories where enterprises and companies that are part of the business cooperation are located and where generations of cooperative workers live. Therefore, for example, the district consumer society involves local communities, local authorities and its employees and founders in the process of ecological and social transformation of the territories in which it carries out its work. In addition, the cooperation actively participates in the preparation of projects and development programs and in their implementation. Business cooperation is interested in participating in environmental and social transformations in both environmental and socially-oriented clusters, as this allows it to participate in their development and implementation of its own business projects. This imposes obligations on the cooperation for its impact on the nature of the territories and the local community, and, consequently, it must report to society for its actions and, in order to avoid controversial issues, keep records of them.

Foreign practice shows that a number of guidelines and standards have been developed for this. For example, the ISO 1400 Environmental management Standard, etc. The most interesting, according to the authors of the article, is the FTSE Group index—FTSE4Good. The securities of companies with this index, which are regularly placed on the London Stock Exchange, are absolutely liquid. This certainly encourages companies to engage in social responsibility.

Working locally, entrepreneurship takes care of fulfilling the instructions of the local community in terms of ecology and social security, which allows the cooperation of entrepreneurship to avoid problems with local authorities and ensure the loyalty of the local community to the business of entrepreneurship. The activities of foreign companies that position themselves in the business community as socially responsible organizations are of interest. At the same time, it should be noted that they strongly orient entrepreneurship towards the development of social responsibility. The participation of companies in social responsibility helps them to participate in stock trading, and, consequently, to increase their capitalization. Moreover, corporate social responsibility enhances the image of companies in the market.

However, it should be noted that in the conditions of the digital economy, unfortunately, there are no invariable indicators of responsibility for the environment and social support, regardless of the sphere of activity, which allow differentiating the attitude to such important activities of companies and enterprises of various forms of ownership. This also applies to the activities of the business cooperation.

2 Methodology

Research of assessments of the responsibility of business cooperation, clustering of the economy, methods of mathematical modeling.

3 Results

The authors of this article believe that the index of geographical latitude of responsibility of the results of the activities of the business cooperation is objectively an invariable indicator, regardless of the sphere of activity of the business cooperation. In the context of the article, responsibility is considered to be the results of the activities of the business cooperation, which in one form or another have the recognition of society and the business community.

Following the methodology of the study of the responsibility of entrepreneurship cooperation, the authors of the article investigate the problem of invariant indicators of responsibility in the digital economy, considering the method of clustering the economy. The authors believe that clusters are a specially organized economic space in which business cooperation works and in which a synergy effect is achieved, manifested in both ecological and socially oriented clusters. This helps to solve the problem of the responsibility of business cooperation, which is important for local communities and the environment. Clusters can be identified as a geographical space for the implementation of economically significant tasks using local resources and involving the population of the cluster territories in the creative process [2]. In the digital economy, with the continuous process of creating and applying artificial intelligence, business cooperation has the capability to continuously coordinate business entities in solving large-scale economic and environmentally important tasks within the framework of environmental and socially-oriented projects.

The activity of the business cooperation in the territories of its presence is characterized by the possibility of exploring local natural resources and developing modern high-tech businesses. An important role in solving these problems is played by highly qualified entrepreneurs, the efficiency of using available capital, their own entrepreneurial energy and vision of the prospects for applying innovations in these territories, as well as mutual understanding of cooperation and the local community, including local authorities.

Business cooperation consists of business entities. Therefore, the most effective is the unification of business structures aimed at solving common business problems in a particular territory and jointly ensuring its ecology and social protection of the local population. The responsibility of the business cooperation, therefore, has a specific geographical reference. Considering the cooperation of entrepreneurship as a socio-economic phenomenon and issues of motivation of the subjects of cooperation, it can be confidently stated that entrepreneurship is the source of the most a durable and flexible framework of the local national economic model of the economy. At the same time, it has responsibility for local territorial environmental management and provides social support practically to themselves, since they live in this territory. Business cooperation, which unites business structures of various forms of ownership and geographically belonging to a single territory, participates in the processes of environmental safety and social security in accordance with territorial affiliation and thereby participates in the formation of responsibility at a specific geographical latitude. The subjects of business cooperation have legal and economic independence, and therefore independently make decisions on participation in environmental and socially-oriented clusters. However, the management of the cooperation is carried out by the board. It is the management board that decides on the participation of the business cooperation in economically significant territorial projects, in the implementation of which its members will be involved.

The Board of the Business Cooperation forms a planned development, interregional ties, works on creating an innovative and IT infrastructure system in the digital economy that implements investment and innovation partnership. Therefore, considering the responsibility of the business cooperation in the most sensitive issues of ecology and social responsibility for society, depending on geographically different territories where the cooperation conducts its business and where it influences their development, environmental situation, socially-oriented local community, according to the authors of the article, it is necessary to use the example of ecological and socially-oriented clusters. The development of new modern technological directions of entrepreneurship in the field in the digital economy will reduce inequality in the development of territories. Considering the responsibility of business cooperation in the management of local natural resources, and socially-oriented projects on the ground, it is necessary to take into account their impact on the development of the territory. Socially-oriented projects are implemented by the cooperation of entrepreneurship working in a specific territory, where business structures representing the interests of different business areas, united by a single purpose, the main component of which is to ensure the economic development of the territory of business functioning, take part in their implementation. Ecological and socially oriented clusters carry out not only regional development, but also solve the tasks of social security and environmental safety, increasing the responsibility of entrepreneurship through the work of modern enterprises and businesses in the territory in which they operate.

So, let us turn to the consideration of the index of geographical latitude of responsibility for the results of business cooperation. In accordance with the approach common in Russia, when social investments of business cooperation are defined as a contribution to the systemic transformation of the social life of territories [1], i.e. thereby assume social responsibility for the transformation of the ecology and social life of a geographically defined territory of the location of clusters in which the corresponding business cooperation operates. The index of geographical latitude of social responsibility of business cooperation is calculated as follows: $\mathbf{Q} = \mathbf{F(W)} \cdot \mathbf{I_K/S}$. Here: **F(W)** is a function reflecting the number of localities geographically included in the ecological and socially-oriented clusters in which entrepreneurship cooperation is represented and social investments of cooperation take place as a contribution to the systemic transformation of the social life of settlements; **Ic** it is the average value of social investments in one locality geographically situated on the territory of the presence of entrepreneurial cooperation within clusters; **S** is the area (in km^2) of the territory on which the settlements are located. When a cluster is located in a single locality, then **S** is its area (in km^2). When considering the location of the cluster in two settlements, the area **S** (in km^2) is considered numerically equal to the sum of the areas of localities. The proposed index of geographical latitude of responsibility of the results of the activity of the cooperative of entrepreneurship is universal, i.e. it is invariant with respect to the subject area, the sphere of activity of the cooperative and shows the volume of investments of the cooperative of entrepreneurship per unit area of a geographically defined territory of its presence. Obviously, the greater this ratio, the more social investments the cooperative has made in the territory of its presence, therefore, the higher its social responsibility. According to the authors of the article, the use of the proposed index of geographical latitude of responsibility of the results of the activity of the cooperative of entrepreneurship gives a number of interesting indicators: the indicator of the participation of the cooperative in contributing to the environmental safety of the territory of presence, the indicator of the participation of business entities in social support of the local population, the indicator of the dynamics of the responsibility of entrepreneurship, etc.

4 Discussion

Currently, in the digital economy, the use of artificial intelligence helps not only to control the environmental and social aspects of production and economic activities, but also to manage them effectively. This is due to the digital transformation of business processes of entrepreneurship, ensuring the comprehensive implementation of environmental requirements and solving their own economic problems. The economic benefits of high indicators of the geographical latitude index of responsibility can be expressed in the following:

- sales growth and improved market position;
- increase in labor productivity;
- building the image of business cooperation;
- formation of customer loyalty;
- formation of the involvement of the founders and employees of the cooperation in the work processes;
- increase in the cost of business cooperation by increasing the assessment of its reputation;
- facilitating access to investments;
- priority for public procurement of budgets of different levels;
- weakening of control by state bodies.

The indicators of the geographical latitude index of the responsibility of the cooperative of entrepreneurship influence the question of the increasing importance of the social responsibility of entrepreneurship in the process of competition in local markets and, especially, in the process of dominance in certain market niches, when it is necessary to consider the activities of the cooperative in the territories entrusted to it through the criteria of environmental safety and ground to solving these issues, the higher its responsibility, and therefore the more social investments it will make locally. This will immediately affect the value of the geographical latitude of responsibility index, which is a signal that in a particular territorial entity, the business cooperation is making efforts aimed at improving socially-oriented areas of activity and meeting the environmental safety requirements of business processes carried out by the cooperation. Significant attention to the dynamics of changes in the index of geographical latitude of responsibility on the part of the business cooperation will allow entrepreneurship to better navigate the processes related to environmental standards of the business processes carried out, as well as pay due attention to socially-oriented projects in their development and implementation. All this together will allow entrepreneurship to achieve a serious effect in the implementation of its own business processes and receive the deserved attention of both the business community and the local establishment.

5 Conclusion

The application of the geographical latitude index of the responsibility of the cooperative of entrepreneurship proposed by the authors of this article in solving the problem of preserving the ecological and social balances of the territory of the presence of the cooperative of entrepreneurship and the place of residence of the population employed in the cooperative is relevant, because people's health depends on its solution. The ease of use of the geographical latitude index of the responsibility of the business cooperation makes it possible to increase the transparency of the indicators of the social responsibility of entrepreneurship and its participation in the environmental and socially-oriented development of the territories of its presence and, most importantly, the index is invariant with respect to the spheres of activity of the business community.

In the modern context of the digital economy in ecological and socially-oriented clusters, conditions are being formed for the optimal balanced use of all types of resources necessary for the successful entrepreneurial activity. Business cooperation has the capability to use resources such as energy, finance, information, artificial intelligence, information technology and infrastructure. This allows the cooperation of entrepreneurship to develop modern technologies, efficient production projects based on scientifical solutions for the use of the resource base of the territories of presence and thereby ensure economic growth, employment of the population, improving the level of well-being and quality of life. The geographical latitude index, as an indicator of the responsibility of the business cooperation, will allow for the effective development of clusters, in which cooperation operates, taking into account the contribution of entrepreneurship to solving territorial environmental and social problems and increase mutual trust within the cooperation.

References

1. About the Social Investment program (2021). Gazprom Neft. https://rodnyegoroda.ru/about-program. Data accessed 08 Sep 2021.
2. Babkin, A.V., Novikov, A.O. (2016). Cluster as a subject of economy: essence, current state, development. *Scientific and Technical Bulletin of St. Petersburg State University of Economic Sciences*, 1(235), 9–29. https://doi.org/10.5862/JE235.1
3. Loiko, V.I., Romanov, D.A., Kushnir, N.V., Kushnir, A.V., & Shaposhnikov, V.L. (2017) Scree plot method as a basis of solving metrological tasks in social and humanitarian knowledge fields (on the example of the objectives of economics, pedagogics and sociology) *Scientific Journal of Kuban State Agrarian University*, 129(05), 1382–1406. https://doi.org/10.21515/1990-4665-129-099

Case Study on the Application of Big Data in the Digital Economy of the Eurasian Economic Union

Automation of Customs Operations as an Element of Creating a Single Point of Acceptance of Extended Advance Notification About the Importation of Goods into the Customs Territory of the Eurasian Economic Union at Air Border Crossing Points

Valentina S. Arsenteva, Maryana V. Arkhipova, Elena A. Yakushevskaya, Victoria B. Gorbunova, and Ekaterina A. Levit

Abstract The paper aims to conduct a comprehensive review of the issues and problems associated with the provision of advance notification (AN) on goods intended to be imported into the customs territory of the Eurasian Economic Union (EAEU) by air. The introduction of new administrative technologies in the activities of the Federal Customs Service of Russia is aimed at optimizing the timing of customs operations and customs control, including those spent at air checkpoints. Special attention is paid to AN to reduce the time allocated for customs operations when importing goods into the customs territory of the EAEU. The primary legal acts regulating AN of goods intended to be imported into the customs territory of the EAEU by air are considered and analyzed. The authors emphasize the need to hold the persons concerned liable for failure to submit or untimely submission of advance notification to the customs authority. The research identified gaps in the activities of customs authorities in the application of AN on goods intended to be imported into the customs territory of the EAEU by air and proposed ways to address them.

V. S. Arsenteva (✉) · M. V. Arkhipova · E. A. Yakushevskaya · E. A. Levit
Kaliningrad Branch of the Russian University of Cooperation, Kaliningrad, Russia
e-mail: v.s.arsenteva@ya.ru

M. V. Arkhipova
e-mail: marhipova@ruc.su

E. A. Yakushevskaya
e-mail: e.a.yakushevskaya@ruc.su

E. A. Levit
e-mail: e.a.levit@ruc.su

V. B. Gorbunova
Kaliningrad State Technical University, Kaliningrad, Russia
e-mail: viktoriya.gorbunova@klgtu.ru

A. V. Bogoviz (ed.), *Big Data in Information Society and Digital Economy*,
Studies in Big Data 124, https://doi.org/10.1007/978-3-031-29489-1_31

Keywords Advance notification (AN) · Air transport · Information technology · Responsibility

JEL Classifications F02 · F13 · K33

1 Introduction

In today's market environment, international cooperation is developed under the influence of many factors that can create its competitive advantage in the current, medium, and long term [6].

The integration of the world economy brings new demands on customs. It is also facilitated by the globalization of business, which ensures the development of world trade. The main requirement of any change is the introduction of advanced digital technology.

Customs administration is no exception; it must follow the world's leading practices [8]. The business actively notes the benefits of the digitalization of customs operations and the desire to provide a comfortable environment for the formation of that very business.

The interaction between participants in foreign economic activities and customs authorities has already reached a whole new level. The newly adopted Development Strategy of the Customs Service of the Russian Federation is calculated up to 2030.

It sets a major goal for the entire customs system: the organization of a completely new, versatile, reconfigurable, and smart customs service, communicating with artificial intelligence, building information bridges between external and internal partners, introducing a minimum of obstacles for business, but being the most effective for the country [6].

To achieve this voluminous goal, 23 target benchmarks were formed, including automation and digitalization of all activities of customs authorities of the Russian Federation. The International Customs Forum, held in October 2019, discussed topical issues of the new interaction between business and customs authorities. The business expects customs to provide faster services and shorter customs clearance times, i.e., an improved customs logistics mechanism.

Throughout the existence of the EAEU, the Federal Customs Service of Russia has been actively participating in the creation of a national "single window" mechanism, the development and modernization of which ensures the synchronization of single window systems of foreign countries throughout the customs territory [3].

Previously, the authors noted [5] that mutually beneficial and equal cooperation with third countries is possible only through enhanced cooperation with regional integration associations and international organizations, participation in the implementation of global initiatives; it is a key area of the activities of the EAEU.

The institution of advance notification (AN) is a relatively young instrument of customs control. For the first time, the global community has proposed the use of information about goods, cargoes, and vehicles before they cross the customs border in the SAFE Framework of Standards. The main objective of the SAFE Framework of Standards is to promote trade facilitation worldwide, to apply a uniform approach to various aspects of customs activities, and counter international terrorism.

In accordance with Standard 6 of the SAFE Framework of Standards, authorized bodies must request advance notification electronically in a timely manner to assess risks properly.

Over the past 10–15 years, preliminary notification has transformed from one of the elements (ways) of interaction between customs authorities and participants in FEA into an independent institution of customs affairs.

Based on the standards of the World Trade Organization and the World Customs Organization, the EAEU customs authorities are now faced with the task of modernizing processes of the unified automated information system, capable of accepting extended AN that can be used by control authorities at border crossing points. This will further accelerate the development of foreign trade activities of the EAEU members.

To maximize results in this area, customs authorities need to actively promote mandatory AN in electronic document management. This obligation should apply to customs authorities, other state controlling bodies, and all stakeholders involved in the movement of goods across the customs border.

2 Materials and Method

All data used in the research was obtained from publicly available sources, including official websites of international organizations and state executive authorities, bulletins, research articles, etc.

3 Results

According to clause 28 of Article 2 of the EAEU Customs Code, advance notification (AN) is information (presented in electronic form) on goods intended to be moved across the EAEU customs border, means of international transportation carrying such goods, time and place of arrival of goods on the EAEU customs territory, and passengers arriving on the EAEU customs territory [2].

Mandatory AN on goods transported by air was introduced in 2017. A year later, since July 1, 2019, the decision of the Board of the Eurasian Economic Commission [1] introduced mandatory AN into the practical work of customs authorities.

The main purpose of AN is enshrined in Article 11 of the EAEU Customs Code. The main purpose is to obtain information on goods planned to be introduced into the territory of the EAEU. This information is needed to assess risks and make preliminary decisions on the choice of objects and forms of customs control.

The legislator requires the following obligatory information when submitting AN:

- Information about the subject of the information;
- Number of the document confirming inclusion in the register of customs representatives;
- Data on the carrier of goods;
- Aircraft data;
- Data on the presence/absence of prohibited items on board;
- Data on goods in accordance with the transportation documents.

The above information allows customs officials to determine the possible risks in advance and select a set of control elements [7].

As in any other mode of transport, AN of customs authorities in air transport has the same positive and effective result [9]. Primary, the time spent by customs authorities and participants in foreign economic activities on organizing and carrying out customs procedures is reduced.

It is also worth noting that the availability of pre-filed documents does not guarantee the absence of verification. Even if the information is submitted in advance, customs officials have the right to conduct a full inspection of the goods being transported and the vehicle itself. However, such inspections are currently regulated by the rules and regulations of a specially designed risk management system.

Due to the development of technology, there are many ways in which the participants in foreign economic activities can provide the necessary information to authorized officials through a personal account of the FEA participant on the web portal or by using specialized software.

Various mechanisms, equipment, and software are used to ensure correct and competent interaction in the process of AN by all parties to the FEA. Otherwise, everything would become too cumbersome.

In general, the technology of AN on the territory of EAEU members is approximately identical for all types of vehicles [4]. However, we will consider a scheme that is specifically related to air transport (Fig. 1).

Let us consider this scheme in a little more detail, breaking it down by stages:

1. Submission of a package of documents to the Central Information and Technical Customs Directorate of the Federal Customs Service of the Russian Federation;
2. Automatic verification of the submitted package of documents for compliance in terms of quantity and quality with the existing regulatory and legal documents of national and international levels;
3. Providing a participant of foreign economic activity with a unique identification number for the transported goods;
4. Receipt by a customs official of the number generated in the unified automated information system, as well as all necessary basic and additional information;
5. Comparison of data previously received in electronic form and data provided in real-time.

In the case of a discrepancy between the data listed in step five, the customs authority determines the degree of non-compliance and the level of threat and,

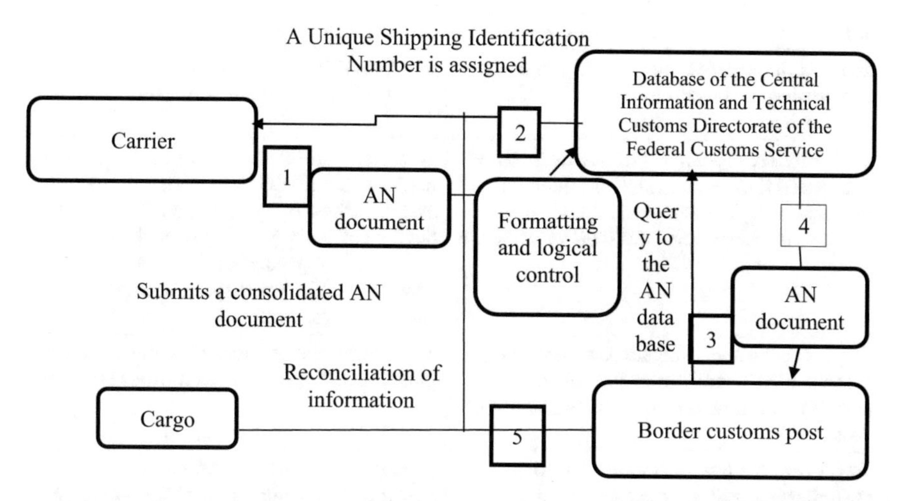

Fig. 1 Sequence of prior notification for goods transported by air transport. *Source* Compiled by the authors

applying risk management, makes one or another decision that directly impacts the outcome of a particular shipment.

The field of freight transportation in the former Soviet Union is a very conservative industry; that is why full-fledged digitalization has only now begun when most areas and organizations have optimized their activities thanks to new equipment and new technologies.

As for the airfreight industry, there are only a few steps that should be taken to fit into the modernized structure:

- To get a digital signature;
- To draw up a contract with the information operator for the uninterrupted transfer of information between the subjects concerned;
- To connect the internal information system to the general information system on the rights of a participant in FEA with modifications and amendments, if necessary.

The information obtained through AN can be used not only by customs authorities but also, based on the "single window" mechanism, by other public authorities authorized to conduct other types of state control, such as phytosanitary control.

The general rule [2] stipulates that the AN is mandatory at least two hours before an aircraft arrives in a particular EAEU territory. If the duration of the flight is, for example, two and a half hours, the AN is carried out within the regulated thirty-minute period after the departure of the ship from the relevant airport.

Meanwhile, we should note some practice-oriented problems associated with the implementation of AN in electronic format, including for the purposes of certain customs operations.

AN is normatively regulated as mandatory for the transactions presented in Table 1.

Table 1 List of customs operations in which advance notification can be used

To perform customs operations related to **the notification of the arrival** of goods in the customs territory of the EAEU	Submission of AN **as a declaration of the means of international transportation**
To confirm compliance with the prohibitions and restrictions on the implementation of sanitary and epidemiological supervision (control) of international transport vehicles and persons	To perform customs operations at the customs border of the EAEU, **requiring a decision by an authorized body** of the member state in the field of sanitary and epidemiological welfare of the population in respect of goods subject to sanitary and epidemiological supervision (control)
To perform customs operations related to **the placement of goods under the customs transit procedure,** including for use as a transit declaration	To perform customs operations related to **the placement of goods for temporary storage**
To perform customs operations related to **customs declaration** of express cargoes	To perform customs operations related to **customs declaration** for the free circulation of goods for personal use, delivered by the carrier as express cargo
To perform customs operations at the customs border of the Union, **requiring a decision by an authorized body** of the member state in the field of veterinary medicine in respect of goods subject to veterinary control (supervision)	To perform customs operations at the customs border of the Union**, requiring a decision by an authorized body** of the member-state on quarantine of plants in respect of regulated products

Source Compiled by the authors

AN can be used not only for the customs arrival operation but also for other operations. However, such an application is possible only if additional information is provided. For example, when using AN to place goods in temporary storage, the authorized body will need to provide information about the place and conditions of storage, as well as other information provided by paragraph 6 of the Decision of the Board of the EEC from April 24, 2018 No. 62.

Stakeholders have the right, but not the obligation, to submit AN to the extent of the listed information. According to the legislation, interested parties are declarants, carriers, persons with authority over goods, and other persons acting on behalf of these persons.

As practice shows, the carrier (representative of the carrier), which, in most cases, submits a package of AN on goods before the arrival of the vehicle at the border crossing point, does not always have the necessary information, which makes it impossible to use the submitted AN for other customs operations.

4 Conclusion

In addition to the problematic aspects described above, it is necessary to separately highlight the issue of liability for failure to submit or untimely submission of AN to the customs authority.

Customs legislation provides for cases where preliminary information on the importation of goods may not be submitted. Legislation has nine categories of such goods, such as goods on the list of military cargoes, onboard supplies, etc.

In other cases, it is clearly defined which persons and to what extent are required or entitled to submit AN on goods imported by air. A failure to submit AN or a violation of the deadline for its submission entails a system of risks and, consequently, the impossibility of optimizing customs operations performed with goods upon arrival. Moreover, it also entails an increase in the time required for customs control.

In paragraph 14 of Article 11, the EAEU Code provides for the possibility of the legislation of member states to establish liability for failure to submit or violation of the deadline for submission of AN to the customs authorities.

If we turn to the Code of the Russian Federation on Administrative Offenses, Chap. 16 "Administrative offenses in the field of customs affairs" does not provide for special responsibility for such an offense.

However, Chap. 19 "Administrative offenses against the order of administration" provides for liability under Article 19.7 "Failure to provide information."

According to the disposition of Article 19.7 of the Code of the Russian Federation on Administrative Offenses, the wrongfulness of a person's deed consists in the failure to submit or untimely submission to a state body of information, the provision of which is stipulated by law and is necessary for a state body to carry out its lawful activity in the form of control or supervision.

Based on a literal meaning of the disposition of Article 19.7 of the Code of the Russian Federation on Administrative Offenses, liability arises for failure to comply with an obligation established by national law rather than an international treaty (the Customs Code of the EAEU) or other international legal acts.

Part 1 of Article 12 of the Federal law "On customs regulations in the Russian Federation" (No. 289-FZ) now grants an individual the right, but not the obligation, to submit AN to the customs authorities. Subparagraph 3 of part 3 of the same article stipulates that the Federal Customs Service of Russia determines the measures to be applied in case of failure to submit or violation of time limits for submission of information, which must be submitted on a mandatory basis. Under these circumstances, failure to submit AN to the customs authority (late submission) does not entail liability under Article 19.7 of the Code of the Russian Federation on Administrative Offenses.

At the same time, the practice of customs authorities has facts of initiation by customs authorities of cases on administrative offenses under Article 19.7. of the Code of the Russian Federation on Administrative Offenses "Failure to provide information," for example, AP Case No. 10702000-001,777/2018 dated September 12, 2018. This is due to the fact that the Federal law "On the free port of Vladivostok" (of July 13, 2015 No. 212-FZ) established the obligation of the carrier to submit AN to customs authorities.

Apparently, there is a need to improve national legislation in the field of customs regulation of the Russian Federation and the EAEU member states.

In a relatively short period of its existence, the AN institute has proven to be an advanced and optimal customs technology that helps speed up the process of customs clearance of goods and vehicles. Timely improvement of legal support for legal decision-making, including in the area of preliminary information, its automation, and optimization of mandatory requirements, will allow solving problems of further modernization of customs administration as clearly and quickly as possible.

References

1. Eurasian Economic Commission Board. (2018). *Decision "On approval of the procedure for providing advance notification about goods intended to be imported into the customs territory of the Eurasian Economic Union by Air (April 24, 2018 No. 62, as amended October 20, 2020 No. 128).* Retrieved October 12, 2021, from https://www.alta.ru/tamdoc/18kr0062/.
2. Eurasian Economic Union (EAEU). (2017). *Annex No. 1 to the Treaty on the Customs Code of the Eurasian Economic Union (April 11, 2017, as amended 29 May 2019).* Moscow, Russia. Retrieved October 12, 2021, from http://www.consultant.ru/document/cons_doc_LAW_215315/c5c6533d9d337bc08b6706133e0b06bdb92b62c0/.
3. Government of the Russian Federation. (2020). *Order "On approval of the development strategy of the customs service of the Russian Federation until 2030" (May 23, 2020 No. 1388-r).* Retrieved October 12, 2021, from https://customs.gov.ru/storage/document/document_file/2020-06/03/2030.pdf.
4. Arestova, Yu. A. (2021). Features applicable to the customs clearance of perishable goods. *Labour and Social Relations Journal, 32*(5), 166–174.
5. Arkhipova, M. V., Bormotova, E. G., Yakushevskaya, E. A., Golovin, Y. O., & Arsentyeva V. S. (2021). International cooperation in the fight against environmental crime: A modern mechanism for combating illegal trade in wildlife. In A. V. Bogoviz, A. E. Suglobov, A. N. Maloletko, O. V. Kaurova, & S. V. Lobova (Eds.), *Frontier information technology and systems research in cooperative economics* (pp. 889–897). Cham, Switzerland: Springer. DOI: https://doi.org/10.1007/978-3-030-57831-2_95.
6. Bolshenkom, S. F., Gorbunova, V. B., & Martynenko, O. V. (2021). A methodological approach to determining the competitive positions of the labor potential in regional consumer cooperation. In A. V. Bogoviz, A. E. Suglobov, A. N. Maloletko, O. V. Kaurova, & S. V. Lobova (Eds.), *Frontier information technology and systems research in cooperative economics* (pp. 661–670). Cham, Switzerland: Springer. DOI: https://doi.org/10.1007/978-3-030-57831-2_71.
7. Bormotova, E. G., Zinovieva, A. L., & Arsentieva, V. S. (2020). Features of customs operations in airport of entry. *Bulletin of the Russian Customs Academy, 3*(52), 55–64.
8. Liu, C.-X. (2021). Consideration of institutional risk factors in the design of international supply chains. In T. G. Shulzhenko (Ed.), *The logistics potential of the 21st century: The youth dimension* (pp. 82–92). St. Petersburg.
9. Permyakova, V. V., & Glushak, O. V. (2021). Analysis of the current state of development and creation of conditions for improving the efficiency of international air transportation in the digital transformation of the economy. In M. V. Davydova (Ed.),, *Antimonopoly policy. Regional practice: Proceedings of the national conference with international participation* (pp. 583–588). Bryansk, Russia: Bryansk State University named after Academician I.G. 4444.

Activities of the Customs Authorities of the Russian Federation for the Protection of Intellectual Property

Nikolay M. Kozhuhanov, Bata I. Ketsba, Alexander O. Pulin, and Sergey V. Shklayev

Abstract The paper analyzes certain aspects of the activities of the customs authorities of the Russian Federation in the context of protecting the rights to intellectual property in relation to goods sent in international postal items. Particular attention is paid to the role of customs authorities in the researched direction of protection. The authors systematize the problems that take place in the studied subject area and propose ways to solve them. The main methods and scientific approaches used in the research include the dialectical method of cognition, the provisions of the philosophy of law, the system of general scientific and particular scientific methods of cognition, and normative and systemic approaches.

Keywords Intellectual property · Activity · Mechanism · Departures · Post · Customs

JEL Classification K23

1 Introduction

The results of a scientific study conducted at the Russian Customs Academy under the code "Intellect-2021" are published within the framework of this publication. In this research, the authors focus on the activities of the customs authorities of the Russian Federation (hereinafter referred to as the customs authorities) to protect the rights to intellectual property objects (IPO) in relation to goods sent in international postal items (IPI). The authors also pay special attention to the role of customs authorities

N. M. Kozhuhanov (✉) · B. I. Ketsba · A. O. Pulin · S. V. Shklayev
Russian Customs Academy, Lyubertsy, Russia
e-mail: n.kozhuhanov@customs-academy.ru

B. I. Ketsba
e-mail: b.ketsba@customs-academy.ru

A. O. Pulin
e-mail: a.pulin@customs-academy.ru

A. V. Bogoviz (ed.), *Big Data in Information Society and Digital Economy*,
Studies in Big Data 124, https://doi.org/10.1007/978-3-031-29489-1_32

in the considered direction of protection, which, in accordance with paragraph 1 of Article 384 of the Customs Code of the EAEU, take measures to protect the rights to intellectual property rights when placing goods under customs procedures, with the exception of placing goods under the customs procedure of customs transit, the customs procedure for destruction, and special customs procedure [1].

In the current geopolitical situation, when foreign manufacturers of goods leave the territory of Russia, and certain countries restrict trade relations with the country by imposing sanctions, the relevance of the issue of protecting intellectual property rights holders by customs authorities does not lose its relevance. Any political conflict will be resolved more quickly, but reputational losses will remain if insufficient attention is paid to ensuring the rule of law when moving goods to IPIs that contain IPO elements.

Assessing the current state of affairs, the authors note that, according to the Federal Customs Service of Russia, customs authorities detected over 38.7 thousand counterfeit items (TCI) sent in 2713 parcels in 2020. In turn, customs authorities detected over 160 TCI sent in 3681 parcels in 2019. There were 226.5 TCI in 1074 parcels in 2018, 37.5 TCI in 1569 parcels in 2017, and 2 TCI in 124 parcels in 2016. The number of counterfeit software is increasing. This trend is objectively related to the overall increase in the number of parcels received through the IPI and the activation of offenders seeking to find a vulnerability in the increased flow to send counterfeit goods.

The need to develop solutions to counteract the growth of counterfeit goods in IPI by the customs authorities and the need to assess their capabilities in this area of work determine the relevance of this research.

2 Materials and Methods

The system approach was applied to study the works of such authors as Agamagomedova [2], Afanasyeva and Dolgikh [4], Bespalko [3], Veretentseva [9], Pulin [7], Tolkachev [8], and others. It allowed the authors to take a broader look at the systematization of individual problems in the direction of analysis.

The authors also used data and materials from the Federal Customs Service of the Russian Federation [5].

Applying the above methods, let us look at the activities of the customs authorities to protect the IPO in relation to goods sent to the IPI as a kind of mechanism.

3 Results

A detailed study of the essence of the issue shows that there are two mechanisms (procedures) for ensuring the protection of rights to IPO in respect of goods sent to IPI, which are implemented by customs authorities in their work. The first mechanism is related to the movement of goods classified as "goods for personal use" (GPU). If the confirmed counterfeit is found, such goods are returned to the sender. The second protection mechanism is associated with the movement of goods that do not fall under the category of GPU, i.e., commercial parties. Administrative offense cases are initiated under Articles 6.33("Circulation of falsified, counterfeit, low-quality, and unregistered medicines and medical devices, as well as circulation of counterfeit dietary supplements"), 7.12 ("Violation of copyright and related rights, invention and patent rights"), and 14.10 ("Illegal use of means of individualization of goods (works and services)") of the Code of Administrative Offenses of the Russian Federation (December 30, 2001) if the confirmed counterfeit is found in a commercial batch. Moreover, there is a subsequent possibility of filing a lawsuit by the copyright holder in the framework of a civil process.

It should be noted that to activate the mechanisms for ensuring the protection of rights to IPO in relation to goods sent to the IPI, used by the customs authorities, the participation of right holders is necessary. In many ways, the result of the work of customs officers depends on the active position of the latter, who, although they have various legal instruments that allow them to quickly, without waiting for a response from the copyright holder to the sent request, respond to the detected counterfeit, still depend on the response of the copyright holder, admits whether he violated his right to IPO.

The main elements of the interaction mechanism are as follows:

1. Copyright holders;
2. Customs authorities;
3. Information and communication technologies (ICT).

In the application form, using ICT, right holders confirm or deny the fact of violation of the right, which allows customs officials to perform appropriate procedural actions. The effectiveness of the actions of the latter largely depends on the desire of the right holder to cooperate with the customs authorities. However, as practice shows, this interaction does not always bring positive results [6].

Based on the use of ICT, the interaction between right holders and customs authorities is regulated by customs legislation. Among the numerous normative legal acts, some of the leading ones are the following:

1 Customs Code of the EAEU. It contains information that must be specified in the DT for protective actions to be taken;
2 Federal Law No. 289-FZ "On Customs regulation in the Russian Federation and on amendments to certain legislative acts of the Russian Federation" (August 3, 2018, as amended April 20, 2021). This legal act reveals the procedural features of the protection in this article;
3 Order of the Federal Customs Service of Russia No. 901 of June 3, 2019. It approves the procedure for using a personal account and organizing the exchange of electronic documents or information between customs authorities, declarants, carriers, persons carrying out activities in the area of customs, authorized economic operators, right holders, and other persons;

4 Order of the Federal Customs Service of Russia No. 131 of January 28, 2019, dedicated to customs register of intellectual property objects (CRIPO).

The existence of regulations governing the interaction in question is often insufficient; law enforcement shows that there are other factors that significantly affect the process.

The authors systematize the problems of interaction.

First, this is the reluctance of the right holder to waste time and resources defending their rights in the process of detaining small consignments of counterfeit goods. In such cases, right holders think that their actions are not economically justified. In fact, the customs authorities have no levers of imperative influence in relation to such right holders. It is necessary to act through explanatory work.

Second, ICT does not allow for full-fledged contact between the considered subjects, in which verbal and non-verbal communication methods could be applied. An electronic message does not often cause the proper emotional reaction in the person reading it. In the absence of liability regulated by law for the non-response of the right holder to information from the customs authority about detected counterfeit goods, there will always be a possibility of failure in the considered protection mechanism.

Third, within the existing mechanism of interaction, the customs authorities act as a subject of law while being a subject of management that depends in their actions (addressing) on the object of management, which violates the basic principles for implementing management mechanisms.

4 Conclusion

Thus, the interaction of customs authorities and right holders in the protection of IPO transferred to IPI has several problems that are complex and not directly related only to the law or technologies used in this process.

The role of the customs authorities is determined by the goals and tasks facing them to ensure the protection of IPO required by law, transferred to the IPI. Despite their managerial position, they are dependent on the right holders. This creates negative prerequisites that make it challenging to apply.

In the author's opinion, to solve the identified problems, the right holder should understand that one-time deliveries of counterfeit goods can result in a poorly controlled flow of counterfeit goods. Therefore, it is extremely important to respond promptly and answer requests from the customs authorities.

The explanatory work of the customs authorities could fit within the framework of the work on the prevention of crimes and administrative offenses within their competence. In general, the authors note that it is advisable for customs authorities to reorient themselves and pay more attention to preventive activities in their work.

It is also important that the information and communication services created and used by the customs authorities be understandable and accessible in operation.

Within the framework of this research, the authors studied the features of interaction between right holders and customs, identified problematic aspects in the current mechanism for protecting IPO sent to IPI, and proposed ways to solve them.

References

1. Eurasian Economic Union. (2017). *Treaty on the Customs Code of the Eurasian Economic Union (adopted April 11, 2017)*. Moscow, Russia. Retrieved September 25, 2021, from http://www.eurasiancommission.org/en/act/tam_sotr/dep_tamoj_zak/SiteAssets/Customs%20Code%20of%20the%20EAEU.pdf.
2. Agamagomedova, S. (2020). Customs control of goods containing IP objects in the development strategy of the customs service of the Russian Federation until 2030. *Intellectual Property and Industrial Property, 10*, 50–55.
3. Bespalko, V. (2008). Criminal-legal protection of intellectual property. In *Improvement of the practice of protection of intellectual property rights by the customs authorities of Russia: Proceedings of the scientific-practical conference* (pp. 99–102). Lyubertsy, Russia: Russian Customs Academy.
4. Dolgikh, M., & Afanasyeva, E. (2010). On the renewal of the exhausted right to a trademark and the legality of parallel import. *Business Law, 1*, 40–44.
5. Federal Customs Service of the Russian Federation. (n.d.). *Official website*. Retrieved September 25, 2021, from http://customs.gov.ru.
6. Nemirova, T. (2017). The activities of customs bodies in protection of intellectual property. *Era of Science, 12*, 52–56.
7. Pulin, A. (2021). Customs protection of intellectual property rights in respect of goods sent by international mail. *Bulletin of the Russian Customs Academy, 2*(55), 131–137.
8. Tolkachev, V. (2005). Counterfeit and unmarked products, goods: Action and retaliation. *Entrepreneur without the Formation of a Legal Entity (PBOUL), 11*. Retrieved October 01, 2021, from https://base.garant.ru/5204628/.
9. Veretentseva, I. (2019). *Administrative and jurisdictional activities of customs authorities in the area of protection of intellectual property rights (Synopsis of Dissertation of Candidate of Legal Sciences)*. Lyubertsy, Russia: Russian Customs Academy.

Analysis of the Practice of Applying Temporary State Price Regulation in the EAEU Member Countries During the Pandemic Period

Dmitriy N. Panteleev, Anastasia A. Sozinova, and Anatolii V. Kholkin

Abstract Based on the study of regulations, data from authorities, and other sources, the paper aims to analyze and summarize the practice of applying temporary state price regulation to identify the results and possible problems of applying temporary state price regulation. This purpose was achieved using analysis, synthesis, deduction, induction, grouping, generalization, comparative law method, tabular method, and graphical method. The normative acts of the EAEU member countries were used as research materials. The originality and scientific novelty of the research lie in the generalization of the practice of applying temporary state price regulation in the EAEU member countries. As a result of the research, the practice of applying temporary state price regulation during the pandemic period was summarized. Similarities and differences in the practice of introducing temporary state price regulation are revealed. The problems of the practice of applying temporary state price regulation due to the gaps in the legislation of the EAEU and its member countries are identified. Measures to overcome these problems include the development of the regulations of the EAEU and the EAEU member countries. Therefore, further studies of this scientific problem should be aimed at determining ways and directions to develop legislative acts of the EAEU member countries and the EAEU as a whole.

Keywords Price regulation · EAEU · Economic regulation · Methods · Analysis of the practice

JEL Classification F55

1 Introduction

Temporary state price regulation is used as a way of influencing the current market situation in the practice of economic management in the EAEU member countries.

D. N. Panteleev · A. A. Sozinova · A. V. Kholkin (✉)
Vyatka State University, Kirov, Russia
e-mail: khav76@mail.ru

A. A. Sozinova
e-mail: aa_sozinova@vyatsu.ru

A. V. Bogoviz (ed.), *Big Data in Information Society and Digital Economy*,
Studies in Big Data 124, https://doi.org/10.1007/978-3-031-29489-1_33

Simultaneously, its most active use falls on the period of spreading COVID-19 and applying various measures to prevent its spread in various countries. In this regard, it is relevant to generalize the practice of applying this method to search for possible problems when applying this method, which is necessary for further development of the management of the national economy.

Simultaneously, not all EAEU member countries applied this method of direct intervention in the economy, allowing for making approximate comparisons and identifying the possible effect of using or refusing to use the studied method of influence.

Based on this, the research topic is relevant and significant.

Based on the study of regulations, data from authorities, and other sources, the paper aims to analyze and summarize the practice of applying temporary state price regulation in decree to identify the results and possible problems of applying temporary state price regulation.

The implementation of this purpose is carried out by solving the following tasks:

1. To study the regulations of the EAEU member countries in the territory of which temporary price regulation was applied;
2. To generalize the practice of applying temporary price regulation and the results achieved;
3. To identify possible problems in the practice of applying temporary state price regulation.

The scientific novelty of this research lies in the analysis and generalization of the practice of applying temporary state price regulation in the EAEU member countries.

The scientific significance of the research is in the study of the experience of application practice, its generalization, and identification of possible problems when applying temporary state price regulation.

The practical significance of the research is that the analysis and generalization of the practice of applying temporary state price regulation will allow avoiding possible problems associated with the use of this method of direct (directive) impact on the economy in the future, which will allow avoiding economic losses.

2 Methodology

The solution to the tasks set in the research was carried out using analysis, synthesis, deduction, induction, grouping, generalization, comparative legal method, tabular method, and graphical method.

The materials and sources of the research were the regulations of the EAEU member countries, information reported by the authorities, and scientific publications of various authors.

The authors studied regulatory acts in the field of application of temporary state price regulation of EAEU member countries. The studied acts and regulations were issued by the authorities authorized to apply the temporary state price regulation procedure. In this case, the specific structure of authority in the part of the consolidation of powers was considered. The composition of these acts is specific to each country. Acts of the Republic of Belarus, the Kyrgyz Republic, and the Republic of Kazakhstan were studied. The authors did not study the acts of the Russian Federation and the Republic of Armenia because the temporary state price regulation was not introduced in these Russian countries.

The following publications were studied as part of the research: R. J. Andrews and K. M. Stange [1]; J. Azarieva and D. Chernichovsky [2]; A. S. Baig, B. M. Blau, and R. J. Whitby [3]; M. Bisceglia, R. Cellini, L. Siciliani, and O. R. Straume [4]; R. Chakraborti and G. Roberts [6]; M. E. Chernew, A. L. Hicks, and S. A. Shah [7]; S. Fischerauer and A. Johnston [8]; C. Jommi, P. Armeni, F. Costa, A. Bertolani, and M. Otto [10]; Y. Liu, X. Chen, and A. N. Rabinowitz [11]; A. Mayorova [12]; S. Poperechny and O. Salamin [15]; V. C. Raimond, W. B. Feldman, B. N. Rome, and A. S. Kesselheim [16]; D. B. Ridley and S. Zhang [17]; H. Yeomans [20]; A. I. Yerchak, I. M. Mikulich, V. A. Gavrilenko, and M. S. Trofimova [21]; A. O. Inshakova, A. A. Sozinova, and T. N. Litvinova [9]; A. Sozinova, N. Savelyeva, and M. Alpidovskaya [19]; A. A. Sozinova, E. V. Sofiina, M. F. Safargaliyev, and A. V. Varlamov [18]; A. A. Sozinova, I. V. Kosyakova, I. G. Kuznetsova, and N. O. Stolyarov [14]; N. K. Savelyeva, A. A. Semenova, L. V. Popova, and L. V. Shabaltina [5].

The study of literary sources showed that, on the one hand, today's science does not pay attention to such an instrument of influence on the market state as temporary state price regulation, and, on the other hand, the issues of price regulation carried out on an ongoing basis are considered sufficiently. Additionally, there are no foreign studies on the practice of applying temporary price regulations in the EAEU countries. Based on this, we can conclude about the relevance and scientific and practical significance of this research.

3 Results

As a result of solving the set tasks, the following results were obtained:

First, the regulations were studied, on the basis of which temporary state price regulation was introduced on the territory of the EAEU member states, according to which the facts of the introduction of temporary state price regulation and key parameters for the introduction of regulation were revealed.

Second, the generalization of the practice of applying temporary state price regulation during the pandemic period was carried out, and the results that were achieved after its introduction were evaluated.

Table 1 was compiled based on the results of the generalization of the practice of application.

Third, the following possible problems in the practice of applying temporary state price regulation were identified. These problems are shown in Fig. 1.

In general, the identified problems require further development of the methodology and regulatory control of such a direct impact tool as temporary state price

Table 1 Generalization of the practice of applying temporary state price regulation by the EAEU member states

Comparison criterion	The Republic of Belarus	The Republic of Kazakhstan	The Kyrgyz Republic	The Republic of Armenia	The Russian Federation
The current regulatory framework was used	Yes	No	Yes	Temporary state price regulation was not introduced	
Period	April 15, 2020 – December 31, 2020	March 16, 2020 – May 11, 2020	March 16, 2020 – June 14, 2020		
Number of goods	19	9	11		
Applied methods of price regulation	– Limit maximum standards of profitability; – Limit maximum additional charges of the importer; – Limit maximum trade additional charges	Introduction of limited retail prices	Establishment of a maximum price level		
Consumer Price Index 2020	101.23	106.7	106.3	103.7	104.9

Source Compiled by the authors

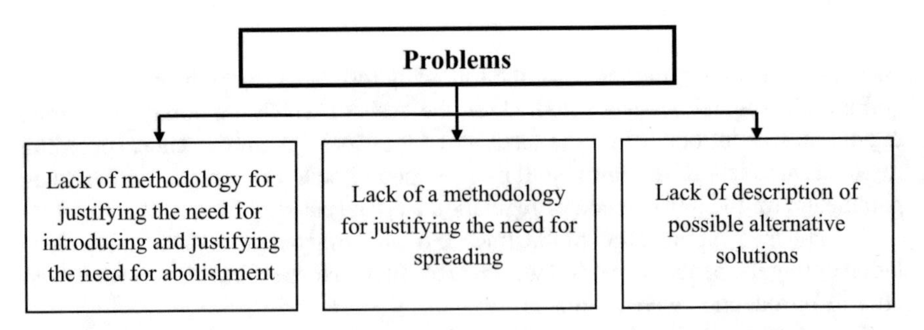

Fig. 1 Problems in the practice of applying temporary state price regulation. *Source* Compiled by the authors

regulation. It will necessarily lead to the need to develop both the legislation of individual member states and the legislation of the EAEU.

4 Discussion

As it was noted above, the application of temporary state price regulation involves regulatory enactment by the authorized body, on the basis of which regulation is introduced on the territory of the EAEU member countries or on the part of its territory. The following parameters are important to generalize regulatory practice:

- The regulatory framework in accordance with which the regulation was introduced;
- The period for which the regulation was introduced;
- The number of items of goods subject to regulation;
- The regulation methods applied;
- The measures taken;
- The result.

It seems appropriate to present the study of the experience of applying temporary state price regulation in the context of the EAEU member countries.

The practice of applying temporary state price regulation in the Republic of Belarus is characterized by the following.

The application of temporary price regulation was carried out based on the existing system of regulations. For these aims, the Ministry of Antimonopoly Regulation and Trade has issued a regulatory act. Thus, the current regulatory framework was used when introducing temporary price regulation in the Republic of Belarus. The regulation was carried out from April 15, 2020, to December 31, 2020. The number of items of goods subject to regulation was 19.

The applied regulation methods were as follows:

- Limiting maximum standards of profitability;
- Limiting maximum additional charges of the importer;
- Limiting the maximum trade markup.

The taken measures were as follows:

- Informing about the need for price regulation;
- Carrying out explanatory activities;
- Taking control measures.

As a result, according to the National Statistical Committee of the Republic of Belarus [14], the price growth index amounted to 101.23% in December 2020. Thus, the effect declared during the introduction of price regulation was achieved.

The application practice of temporary state price regulation developed in the Republic of Belarus in the pre-pandemic period. In particular, in 2017, the regulation was introduced in relation to such a product as white crystalline sugar, which established a ceiling on minimum prices for produced (imported) and sold sugar, as well as a ceiling on maximum trade charges for this product. In 2019, the regulation in relation to buckwheat was introduced, which established a limit on maximum trade charges.

The practice of applying temporary state price regulation in the Republic of Kazakhstan is characterized by the following.

The temporary price regulation in the Republic of Kazakhstan was introduced without relying on previously issued regulations as an emergency measure. The regulation was carried out in the period from March 16, 2020, to May 11, 2020. There were nine items of goods subject to regulation. The applied regulation method was the introduction of limited retail prices.

The taken measures were as follows:

- Informing about the need for price regulation;
- Setting limit market prices;
- Carrying out control measures.

As a result, according to the Bureau of National Statistics of the Agency for Strategic Planning and Reforms of the Republic of Kazakhstan [5], the consumer price index amounted to 106.7% in 2020.

The practice of applying temporary state price regulation in the Kyrgyz Republic is characterized by the following.

The application of temporary price regulation was carried out based on the existing system of regulations. For these aims, the Government of the Kyrgyz Republic has issued a regulatory act. Thus, the current regulatory framework was used when introducing temporary price regulation in the Kyrgyz Republic. The regulation was carried out from March 16, 2020, to June 14, 2020. The number of items of goods subject to regulation was 11. The applied regulation method was the limitation of the price level.

The taken measures were as follows:

- Informing about the need for price regulation;
- Establishing the limit on market prices;
- Carrying out control measures.

In the Kyrgyz Republic, many regulatory bodies were involved in monitoring compliance with the introduced temporary state price regulation: the State Agency for Antimonopoly Regulation, the Financial Police, the State Service for Financial Crimes Enforcement, and local authorities.

The current regulatory acts issued in the pre-pandemic period were used as the regulatory framework.

As a result, according to the National Statistical Committee of the Kyrgyz Republic [13], the consumer price index amounted to 104.9% in 2020.

The generalization of the practice of applying temporary state price regulation allowed identifying the following similarities in the practice of applying temporary state price regulation:

- Achievement of a single goal – prevention or reduction of price increases;
- The regulation on food products was introduced;
- The regulation was based on the previously introduced regulatory framework (except for the Republic of Kazakhstan, where a new regulation procedure was established);
- The information provided by the Treaty to the EEC on the introduction of price regulation was carried out;
- The introduction of measures was accompanied by control over their implementation.

The difference was mainly in the applied methods of regulation. In the Republic of Belarus, the limited standards of profitability and the limited additional chargers were introduced, while limited prices were introduced in the Kyrgyz Republic and the Republic of Kazakhstan.

Differences are also noted in relation to the list of goods. Thus, the widest list was established in the Republic of Belarus, the shortest – in the Republic of Kazakhstan.

The period for which the regulation was introduced also differed. The longest period of regulation was indicated in the Republic of Belarus.

The ambiguity in the obtained results can be noted. During the analyzed period, the consumer price growth index was the lowest in the Republic of Belarus, and the highest—in the Kyrgyz Republic and the Republic of Kazakhstan. Against this background, index values in the countries that did not introduce temporary state price regulation have values that are between those of the countries that introduced the regulation. Thus, the problem of the influence of temporary state regulation of prices on the growth or prevention of price increases requires further research, which consists in identifying and evaluating the influence of causes and factors on the growth of prices and their elimination, which is necessary to assess the impact of temporary regulation of prices.

In general, the practice of introducing temporary state price regulation in the EAEU member countries allowed identifying the following problems of temporary state price regulation as a method of influencing the current market situation, namely:

1. The absence of a methodology for substantiating the need to introduce temporary state price regulation and justifying the need for abolishment in terms of risk assessment;
2. The absence of methodology to justify the need to extend the previously introduced temporary state price regulation in terms of assessing the risk of price increases;
3. The absence of a description of possible alternative ways to solve these problems.

The problem of the absence of a methodology for substantiating the need for introduction or abolishment stems from the following. During the introduction of temporary state price regulation in the pre-pandemic and pandemic period, there were no cases of abolishing the introduced temporary state price regulation by the Eurasian Economic Commission (EEC). This is due to the fact that, as it follows from the regulations, the burden of proving the necessity of deregulation lies with the member countries that have submitted the request to the EEC. Simultaneously, the above normative act does not contain either a reference to the methodology or an indication of the methods and ways of proving the emergence as a result of the introduced regulation and the emergence of a threat (risk) of restricting competition, which makes any proof hardly possible. Any proof based only on the opinion of a participant in the evidentiary process, rather than on a legal standard, may result in other participants in the process not agreeing with either the proof presented or with the ruling on its examination and use in the evidentiary process. This will make it impossible to obtain the result of proof and its legality.

The absence of a methodology for substantiating the need for introduction and extension was also revealed in the analysis of the above regulatory act of the EAEU. The introduction of regulation is fully determined by the legislation of the EAEU member countries. When extending the regulation, which must be coordinated with the EEC, it is possible to decide to refuse the extension, but the above Protocol does not contain the grounds for such a refusal. There is also no indication of possible evidence of the validity of the introduction of temporary state price regulation and the validity of its extension. This also makes it impossible to refuse to agree on an extension, even a purely hypothetical one. Therefore, in practice, there were no cases when the EEC would have decided on the refusal to extend the previously introduced temporary state price regulation during the pandemic and in the pre-pandemic period.

The absence of alternative ways to solve problems is manifested in the fact that one of the reasons for making a decision on the need to abolish the introduced regulation is the proven existence of an alternative way to achieve the goal of introducing regulation, the application of which would lead to less negative consequences for competition conditions. Simultaneously, the EAEU regulations do not indicate possible alternative methods, the use of which has the same goals as temporary price regulation. This makes it impossible to prove the validity and necessity of introduction, as well as the need for abolishment. Additionally, there is no methodology for assessing the impact of the introduced temporary price regulation on the state of competition, the application of which would comply with the requirements of the EAEU regulations.

Therefore, in practice, there were no cases of the Eurasian Economic Commission taking decisions to refuse to renew or abolish the introduced temporary state price regulation.

Simultaneously, the analysis of the values of the consumer price index demonstrated that the introduction of temporary state price regulation does not save from outstripping price growth, which casts doubt on the very feasibility of applying this method of influencing the economic situation.

Undoubtedly, it is necessary to further develop the EAEU regulations and the legislation of the member countries, aimed at developing a methodology for proving the validity of the introduction of regulation and its expediency and necessity.

5 Conclusion

This research summarizes the practice of applying temporary state price regulations based on the regulations of the EAEU member countries during the COVID-19 pandemic.

The object of comparison was the normative acts, which served as the basis for temporary state price regulation in certain EAEU member countries, and the EAEU legislation.

Initially, it was established that the temporary price regulation was not introduced on the territory of the Russian Federation and the Republic of Armenia.

The study of normative acts allowed identifying the following similarities in the practice of introducing temporary state price regulation in the EAEU member countries:

1. When introducing the regulation, the achievement of the purpose in the form of dampening of prices or prevention of price growth was declared;
2. The regulation was introduced mainly on food products;
3. The regulation is based on previously issued normative acts;
4. The introduction of measures was accompanied by strict control of execution and prosecution of bringing violators to responsibility.

Differences were also identified.

Since price regulation was indicated as the main purpose, the authors compared the price growth index in the EAEU countries at the end of 2020. It was discovered that the minimum value was reached only in the Republic of Belarus. In the Republic of Kazakhstan, the value of this index was the highest and exceeded the indices in the countries where regulation was not introduced. Based on this, it was justified that it is necessary to conduct additional research related to identifying the real causes and factors of price growth, which is necessary to confirm the assumption that temporary state price regulation does not affect price growth.

During the analysis of the practice, some problems in the practice of applying temporary state price regulation were also identified, which is due to the imperfection of the EAEU legislation. These problems include the following:

1. The absence of a methodology to justify the need for introduction and the need to abolish the introduced price regulation;
2. The absence of a methodology to justify the need to extend the previously introduced temporary state price;
3. The absence of a description of possible alternative ways of solving the problems underlying the justification for introducing temporary state price regulation.

This necessitates further development of the legislation of the EAEU member countries and the EAEU legislation to overcome the identified problems in the practice of applying temporary state price regulation, which is a necessary condition for increasing the efficiency and effectiveness of this method of state influence on the economy.

The very development of the legislation of the EAEU member countries and the EAEU legislation is an independent topic of scientific research, which can be carried out based on and using the results of this research.

References

1. Andrews, R. J., & Stange, K. M. (2019). Price regulation. Price discrimination and equality of opportunity in higher education: Evidence from Texas. *American Economic Journal: Economic Policy, 11*(4), 31–65. doi: https://doi.org/10.1257/pol.20170306.
2. Azarieva, J., & Chernichovsky, D. (2019). Food prices policy in Israel: A strategic instrument. *Journal of Economic Issues, 53*(4), 1001–1016. https://doi.org/10.1080/00213624.2019.1664236
3. Baig, A. S., Blau, B. M., & Whitby, R. J. (2019). Price clustering and economic freedom: The case of cross-listed securities. *Journal of Multinational Financial Management, 50*, 1–12. https://doi.org/10.1016/j.mulfin.2019.04.002
4. Bisceglia, M., Cellini, R., Siciliani, L., & Straume, O. R. (2021). Optimal dynamic volume-based price regulation. *International Journal of Industrial Organization, 73*, 102675. https://doi.org/10.1016/j.ijindorg.2020.102675
5. Bureau of National Statistics of the Agency for Strategic Planning and Reforms of the Republic of Kazakhstan. (n.d.). *Dynamics of the main socio-economic indicators*. Retrieved February 1, 2022, from https://stat.gov.kz/for_users/dynamic.
6. Chakraborti, R., & Roberts, G. (2021). Learning to hoard: The effects of preexisting and surprise price-gouging regulation during the COVID-19 pandemic. *Journal of Consumer Policy, 44*, 507–529. https://doi.org/10.1007/s10603-021-09493-1
7. Chernew, M. E., Hicks, A. L., & Shah, S. A. (2020). Wide state-level variation in commercial health care prices suggests uneven impact of price regulation. *Health Affairs, 39*(5), 791–799. https://doi.org/10.1377/hlthaff.2019.01377
8. Fischerauer, S., & Johnston, A. (2016). State regulation of retail energy prices: An anachronism in the liberalized EU energy market. *Journal of World Energy Law and Business, 9*(6), 458–474. https://doi.org/10.1093/jwelb/jww032
9. Inshakova, A. O., Sozinova, A. A., & Litvinova, T. N. (2021). Corporate fight against the COVID-19 risks based on technologies of Industry 4.0 as a new direction of social responsibility. *Risks, 9*(12), 212. doi: https://doi.org/10.3390/risks9120212.
10. Jommi, C., Armeni, P., Costa, F., Bertolani, A., & Otto, M. (2020). Implementation of value-based pricing for medicines. *Clinical Therapeutics, 42*(1), 15–24. https://doi.org/10.1016/j.clinthera.2019.11.006
11. Liu, Y., Chen, X., & Rabinowitz, A. N. (2019). The role of retail market power and state regulations in the heterogeneity of farm-retail price transmission of private label and branded products. *Agricultural Economics (United Kingdom), 50*(1), 91–99. https://doi.org/10.1111/agec.12468
12. Mayorova, A. (2019). Baby food prices and their regulation in Russia. In *Proceedings of the 33rd IBIMA Conference: "Education Excellence and Innovation Management through Vision 2020"* (pp. 6524–6528). Granada: Spain.
13. National Statistical Committee of the Kyrgyz Republic. (n.d.). *Prices and tariffs*. Retrieved February 1, 2022, from http://www.stat.kg/ru/statistics/ceny-i-tarify/.
14. National Statistical Committee of the Republic of Belarus. (n.d.). *Consumer price indices for the Republic of Belarus*. Retrieved February 1, 2022, from https://www.belstat.gov.by/ofitsialnaya-statistika/realny-sector-ekonomiki/tseny/potrebitelskie-tseny/operativnye-dannye/indeksy-potrebitelskikh-tsen-po-respublike-belarus/.

15. Poperechny, S., & Salamin, O. (2021). Regulation of prices for agricultural products. *Management Theory and Studies for Rural Business and Infrastructure Development, 42*(3), 323–329. https://doi.org/10.15544/mts.2020.32
16. Raimond, V. C., Feldman, W. B., Rome, B. N., & Kesselheim, A. S. (2021). Why France spends less than the United States on drugs: A comparative study of drug pricing and pricing regulation. *Milbank Quarterly, 99*(1), 240–272. https://doi.org/10.1111/1468-0009.12507
17. Ridley, D. B., & Zhang, S. (2017). Regulation of price increases. *International Journal of Industrial Organization, 50*, 186–213. https://doi.org/10.1016/j.ijindorg.2016.11.004
18. Sozinova, A. A., Sofiina, E. V., Safargaliyev, M. F., & Varlamov, A. V. (2021). Pandemic as a new factor in sustainable economic development in 2020: Scientific analytics and management prospects. In E. G. Popkova, & B. S. Sergi (Eds.), *Modern global economic system: Evolutional development vs. revolutionary leap* (pp 756–763). Cham, Switzerland: Springer. doi: https://doi.org/10.1007/978-3-030-69415-9_86.
19. Sozinova, A., Savelyeva, N., & Alpidovskaya, M. (2021). Post-COVID marketing 2019: Launching a new cycle of digital development. In E. DE LA POZA, & S. E. Barykin (Eds.), *Global challenges of digital transformation of markets* (pp. 419–431). New York, NY: Nova Science Publishers, Inc.
20. Yeomans, H. (2019). Regulating drinking through alcohol taxation and minimum unit pricing: A historical perspective on alcohol pricing interventions. *Regulation & Governance, 13*(1), 3–17. https://doi.org/10.1111/rego.12149
21. Yerchak, A. I., Mikulich, I. M., Gavrilenko, V. A., & Trofimova, M. S. (2019). Peculiarities of management in regulation of minimum prices for strong alcoholic beverages. In V. A. Trifonov (Ed.), *Contemporary issues of economic development of Russia: Challenges and opportunities* (pp. 358–368). London, UK: Future Academies. doi: https://doi.org/10.15405/epsbs.2019.04.40.

Comparison of Temporary State Price Regulation in the EAEU Member Countries

Dmitriy N. Panteleev, Anastasia A. Sozinova, and Anatolii V. Kholkin

Abstract Based on the study of the legislation of each EAEU member country, the paper aims to identify similarities and differences in terms of temporary state price regulation. This aim is achieved by using the following methods: analysis, synthesis, deduction, induction, grouping, generalization, comparison, tabular method, and graphical method. The authors used the normative acts of the EAEU member countries as research materials. The scientific novelty of this research lies in exposing differences in the legislation of the EAEU member countries according to predetermined comparison parameters. This study found no similarity across all comparison parameters, including the period of the introduction of temporary state price regulation. This study did not compare the legislation of temporary regulation of prices of the member countries with the EAEU regulations. Further improvement of the normative legal acts of the EAEU member countries, aimed at unifying the rules of law, can be a measure to overcome these problems. Further study of this scientific problem should be aimed at determining the ways and directions for further development of legislative acts of the EAEU member countries.

Keywords Price regulation · EAEU · Economic regulation · Methods · Comparison

JEL Classification F55

1 Introduction

Temporary state price regulation is one of the ways the government deals with the situation on the market by influencing the price of goods to prevent negative consequences. As the name of this method implies, price regulation is temporary and

D. N. Panteleev · A. A. Sozinova (✉) · A. V. Kholkin
Vyatka State University, Kirov, Russia
e-mail: aa_sozinova@vyatsu.ru

A. V. Kholkin
e-mail: khav76@mail.ru

A. V. Bogoviz (ed.), *Big Data in Information Society and Digital Economy*,
Studies in Big Data 124, https://doi.org/10.1007/978-3-031-29489-1_34

cannot be applied for a long period because the long-term use of this method may negatively affect the state of competition. Therefore, the legislation of almost all EAEU member countries provides for granting the state the right to use this tool. In this case, the only exception is the Republic of Armenia, where legislation does not yet provide for this method of influence.

In addition to national legislation, the application of temporary state price regulation is provided for by supranational legislation of the EAEU.

Since one of the tasks of creating the EAEU is the creation of a single economic space, which involves the unification of legal norms, it is necessary to compare the legal norms governing the application of temporary price regulation in each of the EAEU member countries to identify similarities and differences and eliminate differences as part of the unification process. This determines the relevance of this research.

On this basis, the research aims to identify similarities and differences in terms of temporary state price regulation based on the study of legislation in the temporary price regulation of each EAEU member country.

The implementation of this aim is carried out by solving the following tasks:

1. To study the legislation of the EAEU member countries that form the legal basis for temporary state price regulation;
2. To determine features (parameters) by which the comparison of normative acts is carried out;
3. To identify similarities and differences for each of the comparable parameters and describe and justify them.

The scientific novelty of this research lies in the identification of differences in the legislation of the EAEU member countries in terms of predetermined parameters for comparison.

The scientific significance of this research is in the development and determination of parameters for comparison, which is the development of the method of multidimensional comparison of regulations.

The practical significance of this research is that the comparison of the acts of the EAEU member countries will allow determining further directions and specific measures for developing legislation aimed at achieving unification and ensuring the creation of a common economic space.

2 Methodology

The solution to the tasks set in the research was carried out using the following methods: analysis, synthesis, deduction, induction, grouping, generalization, comparative legal method, tabular method, and graphical method.

The materials and sources of the research were regulatory acts in the field of temporary state price regulation of the EAEU member countries and publications of various researchers.

Legislation acts of the EAEU member countries were investigated considering national specifics because legal systems were different in each member country. This is due to the uniqueness of the organization of authorities, distribution of powers of the authorities, and existent legal regime.

Currently, legislation acts of the Republic of Armenia do not assume the introduction of the temporary state price regulation. For this reason, the legislation of the Republic of Armenia has not been investigated.

The authors also investigated supranational legislation of the EAEU that regulates the temporary state price regulation in the EAEU.

The following publications were studied: Andrews and Stange [1], Azarieva and Chernichovsky [2], Baig et al. [3], Bisceglia et al. [4], Chakraborti and Roberts [5], Chernew et al. [6], Fischerauer and Johnston [7], Jommi et al. [9], Liu et al. [10], Mayorova [11], Poperechny and Salamin [12], Raimond et al. [13], Ridley and Zhang [14], Yeomans [19], Yerchak et al. [20], Inshakova et al. [8], Sozinova et al. [18], Sozinova et al. [17], Sozinova et al. [16], Savelyeva et al. [15].

The study of literary sources showed that, on the one hand, today's science does not pay attention to such an instrument of influence on the state of the market as temporary state price regulation, and, on the other hand, the issues of temporary state price regulation carried out on an ongoing basis are considered in sufficient volumes. Additionally, the study of temporary price regulation in the EAEU countries and its comparison in foreign literature is also not considered. Based on this, we can conclude about the relevance and scientific and practical significance of this research.

3 Results

As a result of solving the set tasks, the following results were obtained.

First, the current regulations on the territory of the EAEU member countries in terms of temporary price regulation by country and the regulations of the EAEU, including the supranational legislation of the EAEU, were studied, as well as the laws of countries and regulations of the authorities authorized to describe the procedure for implementing temporary state price regulation.

Second, the following features (parameters) for comparing normative acts were identified (Fig. 1).

Third, similarities and differences were identified for each feature; their description and justification were given. All similarities and differences are shown in Table 1.

If the feature is common for the member countries of the EAEU on any basis of comparison, then there is the sign "+" in the table; if the feature is not common or there is no similarity, then there is the sign "−" in the table.

Based on the data in Table 1, it can be seen that there is a discrepancy in most of the identified features of comparison, including even such a key parameter as the duration of the period of application of temporary price regulation.

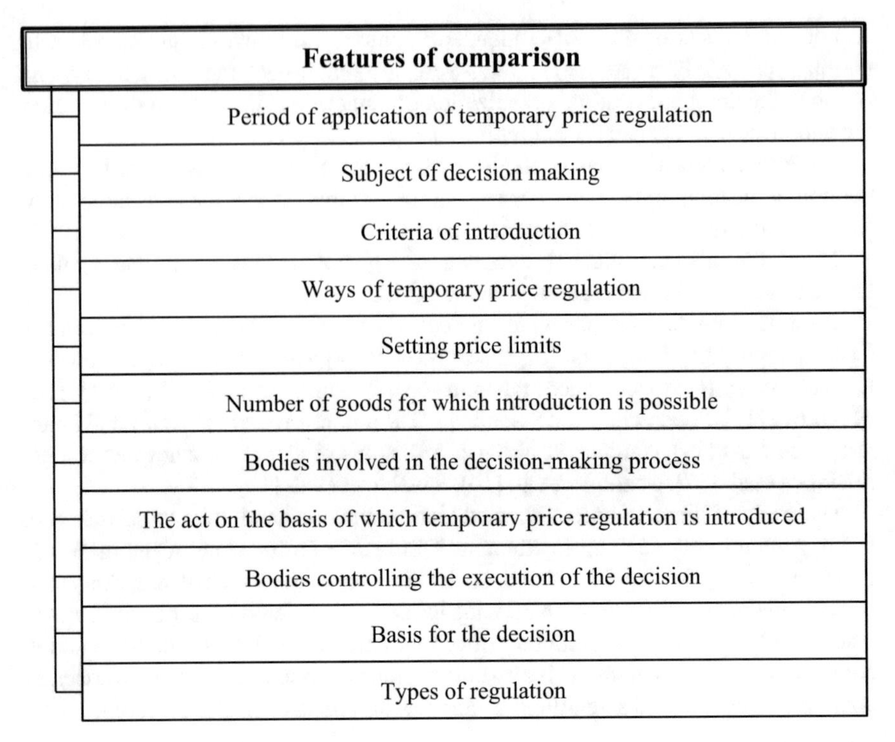

Fig. 1 Features of comparison of temporary state price regulation. *Source* Compiled by the authors

Differences in such parameters as the subject of decision-making and the bodies involved in decision-making and its control are due to the specifics of the structure of the authorities and the distribution of power between the authorities.

In general, the number of differences is too great, which necessitates the unification of the legislation of the member countries in a decree to create a single economic space.

4 Discussion

As it was noted above, the application of temporary state price regulation is based on the legislation of the EAEU member countries. For the detection of similarities and differences, it is required to identify the features by which the comparison of normative acts will be carried out. The following elements of comparison are identified as such features:

Table 1 Generalization of the comparison of the procedure for introducing temporary price regulation in the EAEU member countries

Feature for comparison	The Republic of Belarus	The Republic of Kazakhstan	The Kyrgyz Republic	The Russian Federation
Period of application of temporary price regulation	+	–	+	+
Subject of decision making	–	–	–	–
The criteria of introduction	–	–	–	–
Ways of temporary price regulation	–	+	–	+
Setting price limits	–	+	+	–
Number of goods for which introduction is possible	–	–	–	–
Bodies involved in the decision-making process	–	–	–	–
The act on the basis of which temporary price regulation is introduced	–	–	–	–
Bodies controlling the execution of the decision	–	–	–	–
Basis for the decision	–	–	–	–
Types of regulation	–	–	–	–

Source Compiled by the authors

1. The period of application of temporary price regulation. For the aim of this research, the maximum period established by law is adopted as the period of application of temporary price regulation; during this period, it is possible to introduce temporary state price regulation without considering the possibility of extension.
2. The subject of the decision-making is understood as the executive authority authorized to make a decision on the implementation of temporary state price regulation in accordance with legislative acts of the EAEU member country.

There are differences because each EAEU member state has its own unique system of power. Therefore, similarities will be noted only if these bodies are at the same level of power or the composition of their powers largely coincides.

3. The criteria of introduction are specific numerical indicators that characterize the situation on the commodity market, or facts, events, and circumstances with the occurrence or detection of which the legislation of the member country connects the emergence of the right or obligation to introduce temporary state price regulation.
4. The methods of temporary price regulation are methods allowed by the normative acts of the member countries to change the situation in the commodity market by influencing the price.
5. The establishment of price limits is the method of calculation and fixation of price limits determined by the legislation used in the process of temporary state price regulation.
6. The number of items of goods—the total number of items of goods (product groups) in relation to which the legislation of the member states provides for the possibility of introducing temporary state price regulation.
7. Bodies involved in the decision-making process—the executive authorities, which establish the presence (occurrence) of the criteria for the introduction, prepare and make proposals for the introduction of price regulation, and make the appropriate decision. On this basis, differences may also arise due to the specifics of the structure of authority and the structure of the governmental body created in each member country and the division of powers between authorities.
8. The act on the basis of which the regulation is introduced is a normative act, after the publication and entry into legal force of which temporary state price regulation is introduced and begins its period.
9. The bodies controlling the execution of the decision are the executive authorities of the member country that exercise control over the execution of the decision on the introduction of temporary state price regulation, as well as price control. If non-execution of the decision is found, these authorities fix the offense and bring the violation to justice. In relation to this feature, there may also be a priori discrepancies due to the specifics of the control systems established in each member country.
10. Basis for the decision. The basis is a document that serves as the basis for initiating the procedure for introducing temporary state price regulation. This criterion is inherent only in legislative acts of the Republic of Kazakhstan.
11. Types of regulation. There are varieties of implementation of temporary state regulation in commodity or regional aspects, or their combination. This is peculiar only to the regulation of the Republic of Kazakhstan.

It is necessary to initially research these features of legislative acts in the field of temporary state price regulation in the context of each member country of the EAEU to make a comparison.

The study of temporary state price regulation in the Republic of Belarus (RB) revealed the following. The period of regulation is 90 days. The decision-making subject is the Ministry of Antimonopoly Regulation and Trade of the Republic of Belarus (MART RB).

The introduction criteria are not established in the normative acts that determine the possibility of introducing price regulation. The possible criteria—ensuring price stability—are given in other regulations.

The ways of temporary price regulation involve the determination of the following:

- Fixed prices (tariffs);
- Limit prices (tariffs);
- Limit charges (discounts and extra charges);
- Limit standards of profitability;
- The procedure for establishing and applying prices (tariffs);
- Indexation of prices (tariffs);
- Declaration of prices (tariffs).

National regulations and legislation do not set price caps.

The government body that participates in the decision-making process is the MART RB.

The act on the basis of which the regulation is initiated is the order of the MART RB.

The enforcement authorities include two bodies; the body that introduced price regulation and the body that exercises general control over economic activity (State Control Committee of the Republic of Belarus).

In general, the following conclusions are made with respect to the Republic of Belarus:

1. A wide range of temporary price regulation measures was established;
2. The subject of decision-making is the executive authority;
3. Regarding the number of subjects involved in the decision-making process on the initiation, it could be argued that the procedure is simple, which also leads to the promptness of decision-making and the concentration of responsibility for its consequences.

The study of temporary state price regulation in the Republic of Kazakhstan (RK) revealed the following. The period of regulation is 180 days. The decision-making subject is the Agency for protection and development of competition of the Republic of Kazakhstan (AOPDC RK).

The introduction criteria are as follows:

- Emergencies,
- Natural disasters, as a national security measure.

The condition of introduction is the lack of other ways to solve the problem, the application of which would lead to less damage to competition conditions.

The way of temporary price regulation is by setting the price limit.

The antimonopoly authority sets price limits according to the following:

1. When introducing the regulation in relation to a product:

    1.1. If official statistics are available, based on statistics;
    1.2. In the absence of statistical data—based on responses to requests;

    The price is set as the arithmetic average for three years.

2. When introducing regulation in relation to a market entity: based on data provided by the market entity. The price is determined as the arithmetic average for three years.

Government bodies involved in the decision-making process are as follows:

1. The President of the RK gives orders on the initiation of temporary price regulation;
2. The Prime Minister of the RK gives instructions on the introduction of temporary price regulation;
3. AOPDC RK introduces temporary price regulation on its own initiative based on applications from legal entities and individuals.

The number of goods items includes 19 names of goods.

The act, on the basis of which the temporary regulation is initiated, is an order of the Agency for Protection and Development of Competition of the Republic of Kazakhstan.

The body that initiates the temporary state regulation of prices is the body that controls the execution of the decision.

The basis for the decision to initiate the temporary state price regulation may be as follows:

- Instructions of the President of the RK;
- Instructions of the Prime Minister of the RK;
- Initiative of the antimonopoly body and applications from individuals or legal entities.

Additionally, the following intermediate conclusions can be drawn:

1. The subject of decision-making is the authorized body of executive power.
2. In terms of the number of subjects involved in decision-making, the procedure is confusing and complex; it reduces efficiency and requires the response of the country's top leadership to apply this method of influence.
3. The possibility to apply various types of regulation allows flexibly responding to emerging threats from the market situation. However, the introduction of regulation in relation to certain specific subjects can be perceived as discrimination.
4. Various reasons, including the reaction of the supreme authority, can create a delay effect: the subordinate body will wait for the reaction of the supreme authority and not do anything that could reduce the speed of decision-making.

The study of temporary state price regulation in the Kyrgyz Republic revealed the following. The period of regulation is 90 days. The decision-making subject is the Government of the Kyrgyz Republic (KR).

Introduction criteria are as follows:

1. The increase in prices for socially important goods by 20% or more within a month;
2. The threat of a crisis.

The choice of the method of temporary price regulation depends on the specifics of the goods. One of the following methods of regulation is applied:

- Maximum prices;
- Maximum or minimum price mark-ups.

Price limits are established by the state antimonopoly authority (State Agency for Antimonopoly Regulation under the Government of the Kyrgyz Republic, SAAR KR) for each type of product based on calculations submitted by the manufacturer or importer.

The number of goods is 25.

Government bodies involved in the decision-making process are as follows:

1. The Ministry of Agriculture and Land Reclamation of the KR and the Ministry of Economy of the KR monitor prices for socially significant goods.
2. Authorized representatives of the Government of the KR in the regions, local state administrations, and mayors of the cities of Bishkek and Osh monitor the situation in the markets for the sale of basic socially significant goods and prepare and submit proposals to the Ministry of Economy of the KR on price regulation.
3. The Ministry of Economy of the KR submits a proposal to introduce temporary price regulation.
4. The Government of the KR makes a decision on the initiation of temporary state price regulation.

The act, which serves as the basis for the temporary state regulation of prices, is the Decree of the Government of the Kyrgyz Republic.

The bodies controlling the implementation of the decision include the financial police, tax authorities, and SAAR KR.

Additionally, the following intermediate conclusions can be drawn:

1. The decision-making body is the highest executive authority, which reduces the efficiency of the decision.
2. The presence of a large number of bodies involved in the decision-making procedure also does not contribute to increasing the efficiency of decision-making on the introduction of regulation.
3. The presence of a large number of controlling bodies can lead to a violation of the principle of single punishment.

The study of temporary state price regulation in the Russian Federation (RF) revealed the following. The period of regulation is 90 days. The subject of decision-making is the Government of the RF. An initiation criterion is an increase in average retail prices by 10% or more within 60 consecutive calendar days. The way of temporary price regulation is the fixation of limited retail prices. The price value is set by the Decree of the Government of the RF.

The number of goods is 24.

Government bodies involved in the decision-making are as follows:

1. The Ministry of Economic Development and Trade (MEDT) of the Russian Federation carries out an operational analysis of prices, determines the procedure for assessing the influence of the seasonal factor, and prepares, coordinates, and submits a draft of the relevant Decree to the Government of the Russian Federation.
2. The Federal Antimonopoly Service of Russia (FAS) controls the implementation of the Decree.
3. The Ministry of Agriculture of the RF establishes the procedure for assessing the seasonal factor and coordinates the draft of the Decree. The Ministry of Industry and Trade of the RF coordinates the draft Decree.
4. The Government of the RF adopts the Decree and initiates the temporary state price regulation.
5. The MEDT monitors prices.
6. Federal executive authorities carry out a comprehensive analysis of price dynamics and factors of their formation and develop and submit proposals for the application of economic regulation measures (together with the FAS, the MEDT, and other interested bodies).
7. Deputy Prime Ministers of the RF coordinate the draft Decree of the Government of the RF.
8. The Government of the RF accepts or rejects the draft Decree.

The act on the basis of which the regulation is initiated is the Decree of the Government of the RF. The authority controlling the execution of the decision is the FAS.

Additionally, the following intermediate conclusions can be drawn:

1. The decision-making body is the Government of the RF—the highest body of executive power.
2. The presence of a huge number of bodies involved in the decision-making procedure makes it impossible to apply this method of influence in practice.

In general, comparing the procedure for temporary state price regulation established in different countries, the following conclusions can be drawn:

1. In most countries, the period for introducing temporary state price regulation coincides—90 days.
2. Differences in the composition of the bodies participating in the decision-making procedure and controlling the execution of the decision can be qualified as

insignificant; it is determined by national specifics. The elimination of these differences at this stage of the historical development of the EAEU is seen as redundant and inappropriate.
3. Differences in the number of commodity items for which the initiation of temporary state price regulation is possible are deeper because, despite the possible similarities in the names of goods, it does not mean the similarity of their qualitative characteristics, which makes it challenging to conduct a comparative analysis and develop measures to unify legislation.

Based on the analysis results, it was revealed that there were no similarities between all EAEU participants in all comparable features, even in at least one feature.

5 Conclusion

This research compares the procedure for temporary state price regulation. The research was based on legislation acts in the field of the temporary state price regulation of the EAEU member countries.

The object of comparison was the regulations governing the introduction of temporary state price regulation in the country.

For comparison, the following features (parameters) of comparison were determined:

1. Period of application of temporary price regulation;
2. Subject of decision making;
3. Introduction criteria;
4. Ways of temporary price regulation;
5. Setting price limits;
6. Number of goods items;
7. Bodies involved in the decision-making process;
8. The act on the basis of which the regulation is introduced;
9. Bodies controlling the execution of the decision;
10. Basis for the decision;
11. Types of regulation.

A comparative analysis was carried out for these parameters, during which it was found that there were no similarities in the legislation in terms of the identified parameters. Some of the differences are due to the specifics of the created system of power, and some are due to different legislation.

This necessitates further development of legislative acts of the EAEU member countries to achieve the unity of the regulations underlying the organization of the process of temporary state price regulation.

The very development of legislative acts of the EAEU member countries is an independent topic of scientific research, which can be carried out based on and using the results of this research.

References

1. Andrews, R. J., & Stange, K. M. (2019). Price regulation, price discrimination, and equality of opportunity in higher education: Evidence from Texas. *American Economic Journal-Economic Policy, 11*(4), 31–65. https://doi.org/10.1257/pol.20170306
2. Azarieva, J., & Chernichovsky, D. (2019). Food prices policy in Israel: A strategic instrument. *Journal of Economic Issues, 53*(4), 1001–1016. https://doi.org/10.1080/00213624.2019.1664236
3. Baig, A. S., Blau, B. M., & Whitby, R. J. (2019). Price clustering and economic freedom: The case of cross-listed securities. *Journal of Multinational Financial Management, 50*, 1–12. https://doi.org/10.1016/j.mulfin.2019.04.002
4. Bisceglia, M., Cellini, R., Siciliani, L., & Straume, O. R. (2021). Optimal dynamic volume-based price regulation. *International Journal of Industrial Organization, 73*, 102675. https://doi.org/10.1016/j.ijindorg.2020.102675
5. Chakraborti, R., & Roberts, G. (2021). Learning to hoard: The effects of preexisting and surprise price-gouging regulation during the COVID-19 pandemic. *Journal of Consumer Policy, 44*, 507–529. https://doi.org/10.1007/s10603-021-09493-1
6. Chernew, M. E., Hicks, A. L., & Shah, S. A. (2020). Wide state-level variation in commercial health care prices suggests uneven impact of price regulation. *Health Affairs, 39*(5), 791–799. https://doi.org/10.1377/hlthaff.2019.01377
7. Fischerauer, S., & Johnston, A. (2016). State regulation of retail energy prices: An anachronism in the liberalized EU energy market. *Journal of World Energy Law and Business, 9*(6), 458–474. https://doi.org/10.1093/jwelb/jww032
8. Inshakova, A. O., Sozinova, A. A., & Litvinova, T. N. (2021). Corporate fight against the COVID-19 risks based on technologies of Industry 4.0 as a new direction of social responsibility. *Risks, 9*(12), 212. https://doi.org/10.3390/risks9120212.
9. Jommi, C., Armeni, P., Costa, F., Bertolani, A., & Otto, M. (2020). Implementation of value-based pricing for medicines. *Clinical Therapeutics, 42*(1), 15–24. https://doi.org/10.1016/j.clinthera.2019.11.006
10. Liu, Y., Chen, X., & Rabinowitz, A. N. (2019). The role of retail market power and state regulations in the heterogeneity of farm-retail price transmission of private label and branded products. *Agricultural Economics, 50*(1), 91–99. https://doi.org/10.1111/agec.12468
11. Mayorova, A. (2019). Baby food prices and their regulation in Russia. In *Proceedings of the 33rd IBIMA Conference: "Education Excellence and Innovation Management through Vision 2020"* (pp. 6524–6528). Granada, Spain.
12. Poperechny, S., & Salamin, O. (2021). Regulation of prices for agricultural products. *Management Theory and Studies for Rural Business and Infrastructure Development, 42*(3), 323–329. https://doi.org/10.15544/mts.2020.32
13. Raimond, V. C., Feldman, W. B., Rome, B. N., & Kesselheim, A. S. (2021). Why France spends less than the United States on drugs: A comparative study of drug pricing and pricing regulation. *Milbank Quarterly, 99*(1), 240–272. https://doi.org/10.1111/1468-0009.12507
14. Ridley, D. B., & Zhang, S. (2017). Regulation of price increases. *International Journal of Industrial Organization, 50*, 186–213. https://doi.org/10.1016/j.ijindorg.2016.11.004
15. Savelyeva, N. K., Semenova, A. A., Popova, L. V., & Shabaltina, L. V. (2022). Smart technologies in agriculture as the Basis of Its innovative development: AI, ubiquitous computing, IoT, robotization, and blockchain. In E. G. Popkova, & B. S. Sergi (Eds.), *Smart innovation in agriculture* (pp. 29–35). Springer. https://doi.org/10.1007/978-981-16-7633-8_4.
16. Sozinova, A. A., Kosyakova, I. V., Kuznetsova, I. G., & Stolyarov, N. O. (2021). Corporate social responsibility in the context of the 2020 economic crisis and its contribution to sustainable development. In E. G. Popkova & B. S. Sergi (Eds.), *Modern global economic system: Evolutional development vs. revolutionary leap* (pp. 83–90). Springer. https://doi.org/10.1007/978-3-030-69415-9_10.

17. Sozinova, A. A., Sofiina, E. V., Safargaliyev, M. F., & Varlamov, A. V. (2021). Pandemic as a new factor in sustainable economic development in 2020: Scientific Analytics and management prospects. In E. G. Popkova & B. S. Sergi (Eds.), *Modern global economic system: Evolutional development vs. revolutionary leap* (pp. 756–763). Springer. https://doi.org/10.1007/978-3-030-69415-9_86.
18. Sozinova, A., Savelyeva, N., & Alpidovskaya, M. (2021). Post-COVID marketing 2019: Launching a new cycle of digital development. In E. De La Poza & S. E. Barykin (Eds.), *Global challenges of digital transformation of markets* (pp. 419–431). Nova Science Publishers Inc.
19. Yeomans, H. (2019). Regulating drinking through alcohol taxation and minimum unit pricing: A historical perspective on alcohol pricing interventions. *Regulation & Governance, 13*(1), 3–17. https://doi.org/10.1111/rego.12149
20. Yerchak, A. I., Mikulich, I. M., Gavrilenko, V. A., & Trofimova, M. S. (2019). Peculiarities of management in regulation of minimum prices for strong alcoholic beverages. In V. A. Trifonov (Ed.), *Contemporary issues of economic development of Russia: Challenges and opportunities* (pp. 358–368). Future Academy. https://doi.org/10.15405/epsbs.2019.04.40.

Problems of Temporary State Price Regulation in the EAEU

Anatolii V. Kholkin, Elena V. Sofiina, Nadezhda K. Savelyeva, and Anastasia A. Sozinova

Abstract The research aims to identify the problems of temporary state price regulation within the EAEU member countries based on the regulations governing the activities of the EAEU and their practical application. The authors used the following methods to achieve this goal: analysis, synthesis, deduction, induction, grouping, generalization, experiment, tabular method, and graphical method. The research materials are the normative acts of the Eurasian Economic Union. The originality and scientific novelty of the research are in identifying possible problems while applying temporary state price regulation due to the imperfect legislation of the EAEU, including acts adopted by the EAEU and its bodies. The research revealed a considerable number of problems. The authors grouped them into classes based on the cause of occurrence: the lack of consistency, the presence of gaps, and the lack of synchronization of processes. Further development of the EAEU regulations will help overcome these problems. Further study of this scientific problem should aim to determine the ways and directions for further development of the EAEU legislative acts.

Keywords Price regulation · EAEU · Economic regulation · Methods · Problems

JEL Classfication F55

A. V. Kholkin (✉) · N. K. Savelyeva · A. A. Sozinova
Vyatka State University, Kirov, Russia
e-mail: khav76@mail.ru

N. K. Savelyeva
e-mail: nk_savelyeva@vyatsu.ru

A. A. Sozinova
e-mail: aa_sozinova@vyatsu.ru

E. V. Sofiina
Federal Research Center of Agricultural Economics and Social Development of Rural Areas—All-Russian Research Institute of Agricultural Economics, Moscow, Russia

Kirov Agricultural Sector Advanced Training Institution, Kirov, Russia

A. V. Bogoviz (ed.), *Big Data in Information Society and Digital Economy*, Studies in Big Data 124, https://doi.org/10.1007/978-3-031-29489-1_35

1 Introduction

Temporary state price regulation is one of the tools to stabilize the current market situation, mainly in the markets of essential goods and food products. This instrument is provided by Section VII of the protocol "On general principles and rules of competition" (Appendix No. 19 to the Treaty on the Eurasian Economic Union (Signed in Astana on May 29, 2014)) [1] (the Protocol), and by the national legislation of the EAEU member countries. During the COVID-19 pandemic, many EAEU member countries applied this tool, in particular the Republic of Belarus, the Kyrgyz Republic, and the Republic of Kazakhstan. Even before the COVID-19 pandemic, some participating countries, especially the Republic of Belarus, actively applied this method. As for the Russian Federation, it did not apply this method, even during the COVID-19 pandemic. The Republic of Belarus actively used temporary price regulation, particularly for such goods as sugar. The government set a minimum price below which it was forbidden to sell goods.

However, this method has several conditions and limitations because the introduction of temporary state price regulation is a restriction on freedom of trade and competition. In particular, the restrictions set by paragraphs 81–89 of the Protocol include the temporary nature of introducing this measure and the possibility of challenging the introduced regulation by another member country or refusing to extend the previously introduced regulation.

However, this method is associated with some problems due to the imperfect legislation of the EAEU and the national legislation of the member countries. This imperfection is especially evident in the attempts of other member countries to challenge the price regulation introduced by another country, even if it affects the interests of market entities located on the territory of the member country that introduced the regulation. This leads to the fact that it is impossible in practice to cancel the introduced regulation by external influence through the institutions of the EAEU. This problem exists in the EAEU.

Therefore, the research aims to identify the problems of temporary state price regulation within the EAEU member countries based on the regulations governing the activities of the EAEU and their practical application.

We should solve the following tasks to achieve the goal set:

1. To study the regulations underlying the temporary state price regulation in the EAEU and the practice of using this tool.
2. To identify possible problems in the application of this tool for regulating the market state;
3. To describe and justify these problems.

The scientific novelty of the research lies in the identification of possible problems while applying temporary state price regulation due to the imperfect legislation of the EAEU.

The research is of scientific significance because problems of temporary state price regulation will contribute to the development of methods and techniques for substantiating the introduction, extension, and abolition of temporary price regulation and the processes of regulating the economy as a whole.

The research has practical significance because the identified problems will allow determining further directions and specific measures for developing the EAEU regulations and the legislation of the EAEU member countries in terms of temporary state price regulation.

2 Methodology

To solve the research tasks, we applied the methods of analysis, synthesis, deduction, induction, grouping, generalization, thought experiment, tabular method, and graphical method.

The materials and sources of the research were the EAEU regulations and scientific works, including:

1. The protocol "On general principles and rules of competition (Appendix No. 19 to the Treaty on the Eurasian Economic Union" (Signed in Astana on May 29, 2014) (as amended on October 1, 2019)) (the Protocol) [1].
2. The procedure for submitting applications of the Member States of the Eurasian Economic Union to the Eurasian Economic Commission on the facts of the introduction of state price regulation, their consideration by the Eurasian Economic Commission, and consultations (approved by the Decision of the Board of the Eurasian Economic Commission on December 25, 2018 No. 221) (the Procedure) [2].

We studied the works of the following researchers: Andrews and Stange [3], Azarieva and Chernichovsky [4], Baig et al. [5], Bisceglia et al. [6], Chakraborti and Roberts [7], Chernew et al. [8], Fischerauer and Johnston [9], Jommi et al. [11], Liu et al. [12], Mayorova [13], Poperechny and Salamin [14], Raimond et al. [15], Ridley and Zhang [16], Yeomans [21], Yerchak et al. [22], Inshakova et al. [10], Sozinova et al. [20], Sozinova et al. [19], Sozinova et al. [18], Savelyeva et al. [17].

The study of these works showed that today's science does not pay attention to such an instrument of market regulation as temporary price regulation, but the issues of price regulation have been considered properly. Therefore, we can conclude that our research has its relevance and scientific and practical significance.

3 Results

We have obtained the following results.

First, we studied the regulations in force on the territory of the EAEU. In particular, the greatest attention was paid to:

1. The protocol "On general principles and rules of competition" (Appendix No. 19 to the Treaty on the Eurasian Economic Union (Signed in Astana on May 29, 2014) (as amended on October 1, 2019)) (the Protocol) [1].
2. The procedure for submitting applications of the Member States of the Eurasian Economic Union to the Eurasian Economic Commission on the facts of the introduction of state price regulation, their consideration by the Eurasian Economic Commission, and consultations (approved by the Decision of the Board of the Eurasian Economic Commission dated December 25, 2018 No. 221) (the Procedure) [2].

Second, we identified problems, grouped them, and generalized them into classes (Fig. 1).

Table 1 shows classes of problems.

Third, we have described in detail and substantiated all problems.

Thus, we have found many problems preventing the adequate use of such an important tool as temporary state price regulation. Simultaneously, these problems are objective due to the imperfect regulatory acts of the EAEU, and their elimination is possible only through the improvement of the EAEU acts.

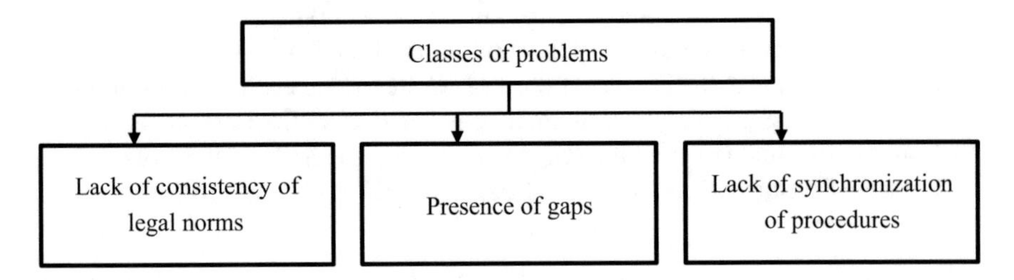

Fig. 1 Classes of problems of temporary state price regulation. *Source* Compiled by the authors

Table 1 Composition of the problems

Class	Problem
Lack of consistency in legal norms	1. The Procedure does not specify which body or official of the EEC sets a date for consultations, determines the attendance; in case of failure of representatives of the member countries to appear, it sets a date for new consultations 2. The presence of a return of documents from the responsible structural unit of the Commission to the member of the Board and vice versa 3. Conflict between actions and rights (duties) of a Member of the Board of the Commission 4. The decision to terminate the consideration is made by a person who is not authorized to consider the application 5. Internal conflict between the grounds for the abolition of temporary state price regulation 6. Inadequacy of the grounds for the abolition of temporary state price regulation 7. Assigning the burden of proving the need to abolish price regulation solely to the member country that has not introduced this regulation

(continued)

Table 1 (continued)

Class	Problem
Presence of gaps	1. No requirement to indicate and justify the reasons for cancellation or extension in the application 2. Absence of requirements for the composition of evidence substantiating the need to extend or cancel price regulation 3. No requirements for the documents attached to the application 4. The sequence and methodology for checking the received application for completeness are not defined 5. Verification of the submitted evidence in the application is not provided 6. Lack of the right and opportunity to protest the abandonment of the application without consideration 7. The procedure for leaving the application without consideration is not defined 8. Lack of indication to extend the term of the regulation when submitting the application 9. No procedure was established for submitting the application drawn up in a language other than Russian 10. The procedure for confirming receipt of the application is not defined 11. No deadline was set for postponing consultations in case of the absence of a representative of the country that introduced price regulation 12. In terms of evidence of an existing alternative way to achieve the goals of the introduced price regulation, neither the composition of the evidence nor the procedure for their research and evaluation is defined 13. The consultation process is not fully regulated 14. The possible outcome of the consultations is not determined 15. Incomplete regulation of the process of registration of the application 16. The scope of actions performed by a Member of the Board upon receipt of an application is not defined 17. The deadline for notification of the receipt of the application is not defined 18. The procedure for confirmation and verification of the implementation of measures to resolve the issue at the level of authorized bodies of the Member States is not defined 19. Incomplete regulation of the procedure for considering the application 20. The composition of evidence confirming the compliance (non-compliance) of the decision on price regulation with paragraphs 81 and 82 of the Protocol, the procedure for their study and evaluation are not specified 21. The procedure and methods for conducting an analysis for the existence of grounds for an extension of the term are not established because the grounds for the extension or refusing to extend are not established 22. There is no right and opportunity to get acquainted with the draft decision 23. There is no time limit and procedure for sending requests to the EEC 24. The obligation to notify the termination of the consideration of the application is not assigned either to a specific structural unit or to an official 25. The method of notices and notifications is not defined 26. The consequences of filing applications in relation to the running of the term of the regulation are not fully established 27. The concept of a barrier to entering the industry and at least signs of the emergence or creation of a barrier are not established 28. Absence of a methodology for proving building barriers, as well as the presence of competition restriction 29. An alternative way to achieve the goals set during the introduction of temporary state price regulation is not determined

(continued)

Table 1 (continued)

Class	Problem
Lack of synchronization of procedures	1. Lack of deadlines for checking the submitted application for completeness 2. Lack of synchronization of the deadlines for submitting the application and the timing of consideration of the application 3. It is not determined whether the actions are carried out by the Member of the Board of the Commission in the process of consideration or outside its framework 4. The place of consultations in the process of considering the application is not determined

Source Compiled by the authors

4 Discussion

As noted above, the application of temporary state price regulation at the EAEU level is currently regulated by:

1. Section VII of the protocol "On general principles and rules of competition" (Appendix No. 19 to the Treaty on the Eurasian Economic Union) [1].
2. The Procedure for submitting applications of the Member States of the Eurasian Economic Union to the Eurasian Economic Commission on the facts of the introduction of state price regulation, their consideration by the Eurasian Economic Commission, and consultations (Approved by the Decision of the Board of the Eurasian Economic Commission dated December 25, 2018 No. 221) [2].

Simultaneously, this process is regulated by the Protocol in general terms and by the Procedure in more detail. Therefore, in our research, we analyze the Procedure in more depth to identify unsettled aspects, deadlines, conflicts of rights and obligations, and a possible conflict of participants' interests in the process.

Based on this, the Procedure determines that the basis for the extension of the previously introduced regulation is the submission to the Commission by the member country that previously introduced the regulation of an application for the extension of price regulation. The Commission considers this appeal and, if necessary, consults. Based on the consideration of the application, the Commission makes a decision. There are also a number of requirements for the content of the appeal.

However, the Procedure does not provide for the following:

1. Requirements to justify the need to extend or cancel the previously introduced regulation in terms of indicating and justifying the reasons for the cancellation or extension.
2. Requirements for the composition of evidence justifying the need for extension or cancellation, at least at the level of an indicative list or examples.
3. Requirements for the documents attached to the applications and their composition, content, and design.

Moreover, the Procedure forms the basis for leaving the application without consideration (paragraph 13), namely a design defect in terms of the lack of documents (materials and information) attached to the application. However, the Procedure does not define such moments and actions that can make the abandonment without consideration from a legal point of view, namely:

1. The sequence and methodology for checking the received application for completeness.
2. The sequence and methodology for verifying the submitted documents (materials and information), which are evidence of admissibility, relevance, reliability, and sufficiency to confirm the need to extend or cancel price regulation.
3. Since the verification is associated with implementing a number of procedures, its implementation requires time. Therefore, it is necessary to establish the timing of such verification separately. The Procedure establishes only a general period for sending a notice of abandonment without consideration—five working days from the date of receipt of the application (clause 13).
4. Due to the absence of clearly defined grounds for leaving the application without consideration, except for the established fact of a design defect, there can be a dispute due to unreasonable leaving without consideration. However, the Procedure does not establish the possibility and procedure for challenging the abandonment of the application without consideration.

The procedure for possible actions in case the application leaves without consideration, which the member country can take, is also not defined, namely: eliminate the design defect that caused the refusal to consider it, submit the missing documents (materials, information), or re-submit the application. In this regard, another possible problem is the uncertainty of the course of the period established for filing the application in paragraph 5 of the Procedure, because it is not determined that when submitting the application, the period will continue, stop, or be suspended.

Additionally, paragraph 12 of the Procedure introduces the requirement to draw up and submit documents in Russian. However, the text of the studied normative act does not define the procedure for identifying the fact of drawing up and submitting documents in another language.

Further analysis of the Procedure has revealed that the procedure for confirming the receipt of the application has not been regulated normatively, and the procedure for establishing the moment (time) of receiving the application and notification of receipt has not been established. According to the authors, this issue is mandatory. The need to regulate it is due to the following reasons: the obligation to send the application is urgent (although there are no sanctions for non-compliance with the deadline and extension of the deadline), the EAEU member countries are in different time zones, and an application sent on one date may be received on another date.

One of the most important methods of considering the application and, accordingly, an element of the process of considering the application, is consultations. Therefore, in the Procedure, Section III is devoted specifically to the regulation of the consultation process. However, there can be several problems in terms of the regulation of this element, namely:

1. No period has been set for postponing consultations if a representative of the country that introduced price regulation has not come. The general period for holding consultations is one month.
2. The Procedure does not specify which body or official of the EEC sets a date for consultations, determines the attendance, and, in case representatives of the member countries are absent, sets a date for new consultations.
3. Clause 19 of the Procedure gives the participant the right to provide evidence of the existence of an alternative way to achieve the goal of introducing temporary state price regulation. Simultaneously, neither the composition of the evidence nor the order of their examination during consultations nor their evaluation is determined.
4. The consultation process has not been regulated in terms of:

- Identifying the person in charge of the consultation process;
- Maintaining a transcript (or audio, video recording), its availability for the participants of the consultation;
- The right of participants to make changes to the Protocol is not established;
- The official signing of the Protocol has not been identified;
- The form of consultations (in person or remotely) is not defined.

5. The possible outcome (result) of the consultations, which may include the achievement of an agreement or its absence, has not been determined. The same is with the composition and sequence of further actions taken by the member countries and the EEC if an agreement has been reached and claims have been waivered or if an agreement is not reached.

Having analyzed paragraph 20 of the Procedure and compared it with other paragraphs of the Procedure (in particular, paragraph 5), we underline the following fact. The established general term for consideration of the application is two months (from 59 to 61 days), while the deadline for submitting the application is 35 days before the expiration of the period for introducing price regulation. These dates do not match. This discrepancy may lead to the possible occurrence of a period when the previously introduced regulation should formally be terminated due to the expiration of the introduction period, but the extension of the regulation period did not happen because the application was not considered. This problem is also due to the fact that the Procedure does not indicate the continuation of the previously introduced regulation when submitting the application. This desynchronization may also occur if a representative of a member country does not attend consultations within 35 days prior to the extension. Such a suspension may adversely affect the economic situation of the EAEU member country and the EAEU as a whole.

Additionally, having analyzed the behavioral analysis of paragraph 20 of the Procedure, we point out that there is the problem of incomplete regulation regarding the registration of the application because it does not state the following:

- Which unit or official of the EEC registers the application;
- How the notice of registration is made, that is, confirmation of the beginning of the period for considering the Application.

A joint study of paragraphs 21 and 22 of the Procedure has revealed the following. Clause 21 of the Procedure indicates the mandatory sending of the application to the member of the Board of the Commission in charge of competition and antimonopoly regulation (the Board). Simultaneously, paragraph 22 of the Procedure introduces the obligation of a member of the Board to transfer the application to the responsible structural unit of the Commission within two days from the date of receipt (as indicated in paragraph 22 of the Procedure, this period is two working days). However, the scope of actions taken by a member of the Board with the received application is not established. This action regarding the transfer of the application and the subsequent return is a return movement of the document and, in fact, is not necessary. It only complicates the process of considering the application.

Clause 23 of the Procedure introduces the obligation for the responsible structural unit of the Commission to notify the member countries within five working days of the receipt of the application. However, the notification method is not specified.

The study of paragraph 24 of the Procedure, which establishes the scope of actions carried out by the responsible structural unit of the Commission within ten working days from the date of receipt of the application, showed the following unsettled points:

- The procedure for confirming and verifying the implementation of measures to resolve the issue at the level of the authorized bodies of the member countries is not defined in terms of the composition of evidence confirming the actions taken to resolve the issue, the result of the actions, and the procedure for verifying and evaluating this evidence;
- The procedure for considering the application on the merits has not been established in terms of the composition of the participants, the composition of their actions, the assessment of actions, the result and its execution, and the review process and its form (in person, in absentia, remotely);
- The composition of evidence confirming the compliance (non-compliance) of the decision on price regulation with paragraphs 81 and 82 of the Protocol, the procedure for their study and evaluation is not indicated;
- The procedure and methods for conducting an analysis for the existence of grounds for an extension of the term have not been determined because the grounds for the extension and refusing to extend are not established.

The study of paragraph 25 of the Procedure contains the scope of actions carried out by a member of the Board during the review. Simultaneously, the Procedure does not establish the participation of a member of the Board in the consideration process. That is, the right or obligation to participate in the consideration is not established, and the scope of actions that an official can carry out is normatively defined. This is a contradiction (conflict) between rights (duties) and prescribed actions.

Also, the Procedure does not define the following issue: these actions are carried out during the consideration of the Application or in parallel with the consideration because there are only ten days for consideration of the Application, and a draft decision is prepared based on the results of the consideration.

When analyzing the Procedure as a whole, we have found the following shortcomings:

1. Lack of opportunity and right to familiarize with the draft decision of the member country in relation to which this decision should be taken.
2. Deadlines for sending the EEC request for materials and the deadline for their submission are not defined.
3. The place of consultations in the process of considering the Application has not been determined. They are carried out in parallel or are included in the review process because a period of up to one month is allotted for consultations, and ten working days are allotted for considering the Application, excluding the time spent (up to two working days) at a member of the Board.

The study of paragraph 26 of the Procedure allowed us to identify the following problems:

1. In case of cancellation or termination of temporary price regulation by a member country, a member of the Board of the Commission terminates consideration of the Application. Simultaneously, as follows from the remaining paragraphs of the Procedure, the responsible structural unit of the Commission has to consider the Application. Thus, the decision to terminate the consideration is made by a person who is not authorized to consider the Application. This also indicates a conflict between the rights, duties, and powers of the Commission's units and officials.
2. The obligation to notify the EAEU member countries is established about the termination of consideration, while the Procedure does not impose this obligation on either a specific official or a structural subdivision of the Commission. That is, this obligation is not assigned to a specific subject.

The analysis of paragraphs 27–33 of the Procedure has not revealed significant problems and shortcomings.

However, in general, the analysis of the entire Procedure has revealed the following problems:

1. The Procedure does not define the methods of notices and notifications.
2. There is no indication that if an Application for Extension is not submitted, temporary price regulation is automatically extended.

Additionally, a joint study of the Procedure and the Protocol allowed us to identify the following significant problems of temporary state price regulation:

1. The presence of an internal conflict between the grounds for the abolition of temporary state price regulation.
2. The concept of a barrier to entering the industry has not been defined, and signs of the emergence or creation of a barrier have not been established.
3. Inadequacy of the grounds for the abolition of temporary state price regulation.
4. Absence of a methodology for proving the creation of barriers; the presence of signs of restriction of competition.

5. Assigning the burden of proving the need to abolish price regulation solely to the member country that has not introduced this regulation.
6. An alternative way to achieve the goals set during the introduction of temporary state price regulation has not been determined.

As for the presence of an internal conflict, we should note the following. The basis for the abolition of temporary state price regulation, as indicated in paragraph 87 of the Protocol, is the established fact that the introduced regulation led to the restriction of competition or it is a proven increase in the risk of restricting competition. The creation of market entry barriers or a reduction in the number of economic entities in the market is indicated as a restriction of competition.

Simultaneously, subparagraph 15 of paragraph 2 of the Protocol establishes the following signs of restriction of competition:

- Reduction in the number of economic entities;
- Increase or decrease in the price of goods that is not associated with changes in the general conditions for the circulation of goods;
- Refusal of market entities from independent actions in the commodity market;
- Determination of the general conditions for the circulation of goods on the market through agreements between subjects, either as a result of the execution of mandatory instructions or through coordination;
- Other circumstances leading to the emergence of an opportunity for one subject (several subjects) to unilaterally influence the general conditions for the circulation of goods.

Thus, paragraph 87 sets out grounds that are different from the signs of restriction of competition in paragraph 2 of the Protocol, which indicates the existence of an internal conflict. Such a conflict makes it difficult to apply the Protocol in terms of proving the need to abolish price regulation or its absence.

In terms of the absence of the concept and signs of the creation or emergence of a barrier, we should underline the following. Neither the text of the Protocol nor the text of the Treaty nor the text of the Procedure has no definition of a barrier, any signs of a barrier, and examples of barriers. It also makes it difficult to prove the need to abolish price regulation.

The inadequacy of the grounds for the abolition of the introduced price regulation is manifested in the following.

First, one of the grounds for cancellation is the creation of barriers mentioned above.

Second, another reason is the reduction in the number of economic entities in the market. On this basis, we need to point out that it is not normatively fixed what a reduction in the market of economic entities means. Therefore, it is impossible to adequately determine such a reduction, eliminating it from the action of other factors that led to the termination of activities, bankruptcy or liquidation of an economic entity. Additionally, it is impossible to trace the reduction of subjects in the market adequately due to the following reasons:

1. A possible source of information about the reduction in the number of entities may be the data of momentary observations conducted by statistical authorities in relation to retail outlets. However, the subject does not always have access to such data.
2. Another possible source of information is statistical reporting. The subject does not always have access to it. Additionally, statistical reporting is not always compiled on a monthly basis and the period for introducing price regulation is up to 90 days, which may not fall within the corresponding reporting period. Another case is due to the time spent on reporting and processing: the data is received when the period for introducing the regulation has expired, and there is no need to cancel it. The data compiled based on statistical reports of statistical collections may not be sufficiently detailed because they may contain information on the whole industry or for a period of time, including time periods before the introduction of price regulation. They also cannot be suitable for proof.
3. Data on the liquidation of economic entities can become an alternative source. However, with regard to this source, there are also problems associated with the inertia of the economy, as well as the fact that the liquidation process will be delayed in time. A similar conclusion can be drawn with respect to information about bankruptcy.
4. A consequence of the introduced price regulation can be changing the type of activity of the subject, such as leaving the market and reducing the number of subjects by leaving to another area of activity. However, according to this indicator, it is impossible to establish the number of left entities due to the following reasons:

    - The process of changing the type of activity code is extended in time and may go beyond the period of regulation;
    - A change in the type of activity does not always require a change in the code. In particular, if initially, when creating an economic entity, the main, additional, and backup codes were received. Therefore, if there is an actual change in the type of activity, there will be no need for state registration of such changes in the form of a code change.

In terms of the lack of a methodology to prove the existence or absence of grounds for the abolition or extension of temporary state price regulation, we should point out the following. Normative acts do not indicate the following:

- The composition of indicators for the created barriers or a reduction in the number of economic entities, the procedure for their calculation and interpretation of the obtained values;
- The composition of indicators characterizing the risk of a reduction in the number of market entities, or the introduction (creation) of barriers, or another procedure for assessing risk;
- The composition of the sources of information necessary for the calculation of indicators;

- The composition of the evidence, the procedure for their evaluation, and the requirements for them.

The imposition of the burden to prove the need for repeal on the member country that did not introduce the regulation arises from the Protocol, paragraph 4 of paragraph 87. This presumption makes the possible deregulation practically unrealistic because another member country intending to file an Application for the need to cancel the introduced price regulation is obliged to collect all necessary evidence of the existence of grounds, including the possibility of achieving the goal of introducing regulation in an alternative way. This is not possible due to the following reasons:

- Uncertainties in the composition of evidence;
- The lack (or difficulty) of access to information sources because the main source of information is statistical data, and, within the EAEU, there are no unified terms for collecting and processing statistical reporting, and statistical reporting data do not always contain up-to-date information
- The inability to request the necessary information through the EEC because the acts of the EAEU do not provide for the right to such a request and the response time to the request is very long.

In terms of proving the existence of an alternative method, we should underline the following. The obligation to prove the possibility of achieving the goal in a way other than price regulation is assigned to the member country by the Protocol, paragraph 4 of paragraph 87. Simultaneously, the EAEU acts do not explain what methods of regulation or other impacts on the economy can be used as alternatives when proving the need to cancel the introduced regulation, even at the level of examples.

This makes it difficult to challenge the introduced price regulation and may lead to abuse of the right of member countries in an attempt to "protect their own market," which does not contribute to further integration within the EAEU.

The reason for these problems is the imperfect regulations that determine the procedure for the introduction, extension, and cancellation of temporary state price regulations.

5 Conclusion

This research identifies and describes possible problems that may arise in implementing temporary state price regulation within the framework of the EAEU.

To identify problems, the authors have studied the regulations in force in the EAEU, determining the procedure for introducing, prolonging, and abolishing temporary state price regulations.

They have grouped all problems into three classes due to the common causes of problems. These classes of problems are as follows:

1. Problems of the systemic nature of the rules of law, which determine the order of the processes and the functionality of departments and officials.
2. The presence of gaps in terms of the composition of evidence, the procedure and methods for their research and evaluation, rights and obligations of participants in the process, and the procedure for implementing individual procedures.
3. Problems of lack of synchronization of procedures and stages of their implementation.

The authors described all problems and established the cause of their occurrence: the imperfection of the regulations in force within the framework of the EAEU. All problems are defined as objectively existing at the level of subjects participating in the process of introducing, prolonging, and abolishing temporary state price regulation.

Since the causes of these problems are the existing defects of normative acts, the elimination of all identified problems is possible only by amending the existing normative acts.

Therefore, the identified problems necessitate further development of the EAEU regulations to improve the processes that arise as a result of the temporary state price regulation introduced by one of the member countries, carried out by the EEC. This should lead to an increase in the efficiency of the activities of the EEC and a reduction in risks and uncertainties in the activities of economic entities participating in the market.

The direct development of the EAEU regulations is an independent topic and objective of the research, which can be carried out further.

References

1. The protocol on General Principles and Rules of Competition (Appendix No. 19 to the Treaty on the Eurasian Economic Union (Signed in Astana on May 29, 2014)). Retrieved February 1, 2022, from http://www.consultant.ru/document/cons_doc_LAW_163855/fb5a3b34bcd16abfc4df4e0f0c26c9d184a3b34d/.
2. The procedure for submitting applications of the Member States of the Eurasian Economic Union to the Eurasian Economic Commission on the facts of the introduction of state price regulation, their consideration by the Eurasian Economic Commission, and consultations (approved by the Decision of the Board of the Eurasian Economic Commission dated December 25, 2018 No. 221). Retrieved February 1, 2022, from http://www.consultant.ru/document/cons_doc_LAW_315019/170632957b42a6d9eb25a7fb73ecce141c7e0a63/.
3. Andrews, R. J., & Stange, K. M. (2019). Price regulation, price discrimination, and equality of opportunity in higher education: Evidence from Texas. *American Economic Journal-Economic Policy, 11*(4), 31–65. https://doi.org/10.1257/pol.20170306.
4. Azarieva, J., & Chernichovsky, D. (2019). Food prices policy in Israel: A strategic instrument. *Journal of Economic Issues, 53*(4), 1001–1016. https://doi.org/10.1080/00213624.2019.1664236.
5. Baig, A. S., Blau, B. M., & Whitby, R. J. (2019). Price clustering and economic freedom: The case of cross-listed securities. *Journal of Multinational Financial Management, 50*, 1–12. https://doi.org/10.1016/j.mulfin.2019.04.002.

6. Bisceglia, M., Cellini, R., Siciliani, L., & Straume, O. R. (2021). Optimal dynamic volume-based price regulation. *International Journal of Industrial Organization, 73*, 102675. https://doi.org/10.1016/j.ijindorg.2020.102675.
7. Chakraborti, R., & Roberts, G. (2021). Learning to hoard: The effects of preexisting and surprise price-gouging regulation during the COVID-19 pandemic. *Journal of Consumer Policy, 44*, 507–529. https://doi.org/10.1007/s10603-021-09493-1.
8. Chernew, M. E., Hicks, A. L., & Shah, S. A. (2020). Wide state-level variation in commercial health care prices suggests uneven impact of price regulation. *Health Affairs, 39*(5), 791–799. https://doi.org/10.1377/hlthaff.2019.01377.
9. Fischerauer, S., & Johnston, A. (2016). State regulation of retail energy prices: An anachronism in the liberalized EU energy market. *Journal of World Energy Law and Business, 9*(6), 458–474. https://doi.org/10.1093/jwelb/jww032.
10. Inshakova, A. O., Sozinova, A. A., & Litvinova, T. N. (2021). Corporate fight against the COVID-19 risks based on technologies of industry 4.0 as a new direction of social responsibility. *Risks, 9*(12), 212. https://doi.org/10.3390/risks9120212.
11. Jommi, C., Armeni, P., Costa, F., Bertolani, A., & Otto, M. (2020). Implementation of value-based pricing for medicines. *Clinical Therapeutics, 42*(1), 15–24. https://doi.org/10.1016/j.clinthera.2019.11.006.
12. Liu, Y., Chen, X., & Rabinowitz, A. N. (2019). The role of retail market power and state regulations in the heterogeneity of farm-retail price transmission of private label and branded products. *Agricultural Economics, 50*(1), 91–99. https://doi.org/10.1111/agec.12468.
13. Mayorova, A. (2019). Baby food prices and their regulation in Russia. In *Proceedings of the 33rd IBIMA Conference: "Education Excellence and Innovation Management through Vision 2020"*. (pp. 6524–6528). Granada, Spain.
14. Poperechny, S., & Salamin, O. (2021). Regulation of prices for agricultural products. *Management Theory and Studies for Rural Business and Infrastructure Development, 42*(3), 323–329. https://doi.org/10.15544/mts.2020.32.
15. Raimond, V. C., Feldman, W. B., Rome, B. N., & Kesselheim, A. S. (2021). Why France spends less than the United States on drugs: A comparative study of drug pricing and pricing regulation. *Milbank Quarterly, 99*(1), 240–272. https://doi.org/10.1111/1468-0009.12507.
16. Ridley, D. B., & Zhang, S. (2017). Regulation of price increases. *International Journal of Industrial Organization, 50*, 186–213. https://doi.org/10.1016/j.ijindorg.2016.11.004.
17. Savelyeva, N. K., Semenova, A. A., Popova, L. V., & Shabaltina, L. V. (2022). Smart technologies in agriculture as the basis of its innovative development: AI, ubiquitous computing, IoT, robotization, and blockchain. In E. G. Popkova, & B. S. Sergi (Eds.), *Smart innovation in agriculture* (pp. 29–35). Springer. https://doi.org/10.1007/978-981-16-7633-8_4.
18. Sozinova, A. A., Kosyakova, I. V., Kuznetsova, I. G., & Stolyarov, N. O. (2021). Corporate social responsibility in the context of the 2020 economic crisis and its contribution to sustainable development. In E. G. Popkova, & B. S. Sergi (Eds.), *Modern global economic system: Evolutional development vs. revolutionary leap* (pp. 83–90). Springer. https://doi.org/10.1007/978-3-030-69415-9_10.
19. Sozinova, A. A., Sofiina, E. V., Safargaliyev, M. F., & Varlamov, A. V. (2021). Pandemic as a new factor in sustainable economic development in 2020: Scientific analytics and management prospects. In E. G. Popkova, & B. S. Sergi (Eds.), *Modern global economic system: Evolutional development vs. revolutionary leap* (pp. 756–763). Springer. https://doi.org/10.1007/978-3-030-69415-9_86.

20. Sozinova, A., Savelyeva, N., & Alpidovskaya, M. (2021). Post-COVID marketing 2019: Launching a new cycle of digital development. In E. De La Poza & S. E. Barykin (Eds.), *Global challenges of digital transformation of markets* (pp. 419–431). Nova Science Publishers Inc.
21. Yeomans, H. (2019). Regulating drinking through alcohol taxation and minimum unit pricing: A historical perspective on alcohol pricing interventions. *Regulation & Governance, 13*(1), 3–17. https://doi.org/10.1111/rego.12149.
22. Yerchak, A. I., Mikulich, I. M., Gavrilenko, V. A., & Trofimova, M. S. (2019). Peculiarities of management in regulation of minimum prices for strong alcoholic beverages. In V. A. Trifonov (Ed.), *Contemporary issues of economic development of Russia: Challenges and opportunities* (pp. 358–368). Future Academy. https://doi.org/10.15405/epsbs.2019.04.40.

Analysis of the Market for Cryogenic Blasting in Russia

Dmitriy N. Panteleev, Nadezhda K. Savelyeva, and Olga V. Fokina

Abstract The paper aims to analyze the cryogenic blasting market in Russia and determine the prospects for its development. The authors use general scientific and experimental-theoretical methods. The supply of equipment for cryopreservation is mainly carried out by foreign companies, although the presence of individual Russian enterprises in the market is becoming more active. Cryogenic blasting services are more expensive than conventional cleaning methods. Prices for cryogenic blasters vary widely depending on the brand, design, performance, and specifications. The main competitive advantages of Russian cryogenic blasters are the price and efficiency of equipment operation. Despite the active promotion of cryogenic blasting by interested companies, this technology is becoming increasingly popular and it is likely to be present in the Russian market along with other high-pressure cleaning technologies. Cryogenic blasting is increasingly used in various industries, where the largest market share is occupied by the industrial cleaning segment, which is expected to continue to dominate. This issue is actively discussed in journalistic and industrial circles, but it is insufficiently reflected in science.

Keywords Cryogenic blasting · Cryogenic blasters · Dry ice · Market · Development prospects · Prices · Consumer requirements

JEL Classifications O1 · L6

1 Introduction

The history of the development of cryogenic blasting began in the 1930s. However, due to technological difficulties, cryoblasters appeared on industrial markets only at

D. N. Panteleev (✉) · N. K. Savelyeva · O. V. Fokina
Vyatka State Univeristy, Kirov, Russia
e-mail: cazador.nur@gmail.com

N. K. Savelyeva
e-mail: nk_savelyeva@vyatsu.ru

A. V. Bogoviz (ed.), *Big Data in Information Society and Digital Economy*,
Studies in Big Data 124, https://doi.org/10.1007/978-3-031-29489-1_36

the end of the 1980s. The technology of cryogenic blasting began to be introduced more actively in the late 20th and early twenty-first centuries.

Dry ice cleaning appeared on the Russian market relatively recently—about 20 years ago. The main problem was the lack of information and the lack of confidence of potential users in the new technology. In 2011, cryoblasting was called "virgin soil in cleaning" [12]. Nowadays, "thanks to the active promotional activities of companies, the stage of familiarizing the market with dry ice cleaning technology ended [3]. However, there are still technological and financial problems.

In this regard, the analysis of the Russian market of cryogenic blasting to determine the prospects for its development is of particular interest; it involves studying the market for dry ice cleaning services and the equipment market, as well as analyzing competitors' and consumers' requirements.

2 Materials and Method

The problem of market analysis is covered in sufficient detail in existing studies and publications, including studies by Kataeva et al. [19], Lysova et al. [21], Saveleva et al. [28–30, 32], Sozinova [5, 33–35], and Fokina [10].

Problems and prospects for the development of cryogenic blasting were considered in scientific, public, and industrial circles in Russia and abroad. The researchers note that dry ice blasting as a cooling medium has already proven its potential [24]. The undoubted advantages of cryogenic cleaning over traditional methods are noted [18, 20]. Nevertheless, it is also noted that the use of ice particles as a substitute for mineral abrasives has been studied by only a few research groups [17].

Specialized publications mainly present information aimed at potential consumers and reveal the prospects for the introduction of cryogenic blasting and call cryoblasting the "future of cleaning." In recent years, the situation concerning the introduction of cryoblasting in the industry has noticeably changed for the better, and it has clear superiority over all other industrial cleaning methods [38]. The publications also suggest that the method of cryoblasting using carbon dioxide is currently considered one of the most advanced methods for cleaning contaminated surfaces. In addition to the main ones, additional advantages of this cleaning method are revealed: the presence of a disinfecting effect, saving time, and the possibility of applying to bases made of materials of any kind [36, 39].

To interest potential customers, foreign and Russian industrialists-manufacturers of installations and enterprises providing cryogenic cleaning services highlight characteristics of the used equipment, features of the technology, variety of its applications, and the advantages of cryogenic blasting [1, 2, 4, 11, 16, 23, 25] etc.

3 Data Availability

Areas of application of cryoblasting and consumer requirements for the design of cryoblasters, which are in the subsequent section of the study, are available at https://figshare.com/ with the identification number https://doi.org/10.6084/m9.figshare.19410233.v2.

4 Results

Currently, Russian enterprises increasingly apply cryogenic blasting. The largest market share is held by the industrial cleaning segment, which is expected to continue to dominate.

Experts note that especially promising areas for the implementation of cryoblasting are the electric power industry, ferrous and non-ferrous metals industry, mechanical engineering, aircraft and shipbuilding, the chemical industries, food production, enterprises and sites for the production of plastic parts; when solving a wide range of repair work [38].

The main material used in cryoblasting is "dry ice." The change in the production of "dry ice" in Russia is shown in Fig. 1.

Experts predict further growth in market volumes due to a deeper introduction of dry ice cleaning technology into the practice of cleaning companies, utility providers, and specialized divisions of large industrial enterprises. As a result, equipment for cryoblasting will be more in demand [5].

The largest companies providing cryogenic cleaning services in the Russian market are "NTK Soltek" (Moscow), "Dr. Sauber" (Moscow), "Polus Holoda" (Moscow), "Vekfort" (Moscow), "Tip-Top Cleaning Service" (Moscow), "Irbistech"

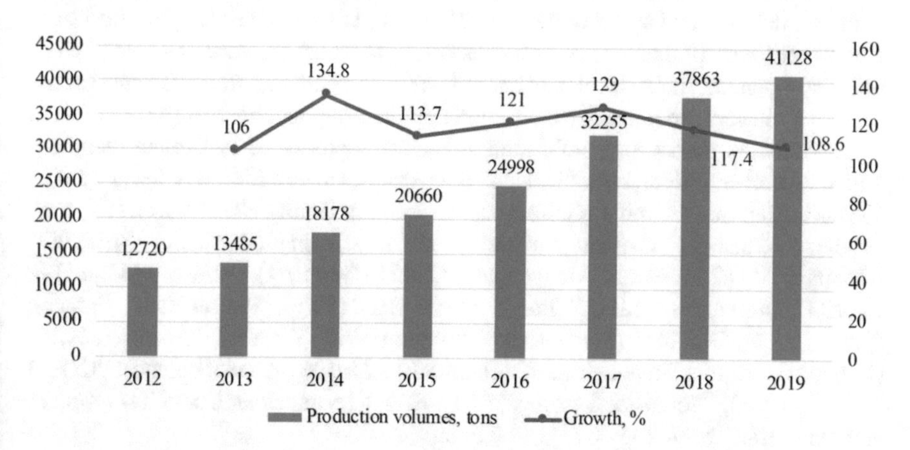

Fig. 1 Production volumes of "dry ice" in Russia. [7, 8, 27, 37]. *Source* Compiled by the authors based on

(Moscow), "Iceventek" (Moscow), "Carblast-Industry" (the Moscow Region), "SK Blasting-Service" (the Moscow Region), LLC "K2-URAL" (Ekaterinburg), "Real-Invest + " (Kazan), "Volga Cleaning Company" (Tolyatti), "Dry Ice" (Perm), and "Sila Kholoda" (Tomsk).

Almost all companies providing cryogenic blasting services showed revenue growth in 2019. However, due to the COVID-19 pandemic, there was a fall in 2020. Companies in Moscow and the Moscow Region suffered from the crisis more than others.

Also, the following foreign companies offering cryogenic cleaning services are presented on Russian Internet portals: "Cryoserv'ice" (France), "Cryo'geni" (France), "SM3I" (France), "Cryo'tech—La Technologie Cryogenique" (France), "NIDD" (France), "CNH" (France), "Lance Verte Décapage" (France), "Delta Diffusion" (France), "Agro Clean" (Belgium), "S.B.M.I." (Belgium), "ACP Belgium" (Belgium), "European Cleaning Et Maintenance" (Belgium), "Ecovert Facility Services" (Switzerland), "A&G Chemical Production S.R.L." (Italy), and "ASPEL" (Morocco) [9].

Cryogenic blasting services are more expensive than conventional cleaning methods. This is determined by higher costs for work, materials, and equipment. Although, many experts note that "… in industrial cleaning, the main thing is competent decisions based on efficiency and not on low price" [13, 21]. Companies located in Moscow offer cryoblasting services from 4000–5000 rubles/hour, regional companies—from 3500 rubles/hour, and some (e.g., "Dry Ice") declare the price from 2000 rubles/hour. The price of the offered service by intermediate parties is on average 7–10%, sometimes 50% or more, higher.

Consumers mainly pay attention to such characteristics of cryogenic blasting as a high degree of cleaning, a possibility of saving time and labor costs, removing stubborn dirt, and cleaning hard-to-reach surfaces. The main is to reduce the cost of cleaning.

The determining parameters of cryoblasters are price, power, reliability, operating costs, availability, and speed of delivery of repair parts and consumables. Additionally, weight, ease of maintenance, the possibility to adjust pressure, the cost of additional components, noise level, and ease of use are considered. Requirements for the design of blasters are primarily determined by their application area [22].

The Russian market of cryoblasters is mainly represented by foreign manufacturers: "Karcher" (Germany), "Icetech" (Germany), "White Lion Dry Ice and Laser Cleaning Germany" GmbH (Germany), "CryoSnow GmbH" (Germany), "Mycon" GmbH (Germany), "Carblast Fahrzeugetechnik" GmbH (Germany), "EisTEC" (Germany), "DCA Deckert Anlagenbau" GmbH (Germany), "Südstrahl GmbH & Co. KG" (Germany), "M.E.C." (Italy), "C.P.S. SRL" (Italy), "Ryoserv'ice" (France), "Asco Carbon Dioxide" (Switzerland), "Icesonic Int" (Croatia), "Artimpex N.V." (Belgium), "Aquila Triventek A/S" (Denmark), "Dry Ice Zone" (Poland), "COLD Jet" (the USA), "SodaBlastSystems" (the USA), "Technoclean Co." LTD (Korea), and "Sinocean Technology Co." Ltd (China).

Among Russian manufacturers, "Irbistech" (Kazan) is actively developing, which began development in 2016 [15]. Another Russian company producing equipment for dry ice cleaning is "SibTIM Group of Companies" (Omsk), which started development in 2003.

In 2019, supplies of cryoblasters from abroad decreased by 28% compared to the previous year. The decrease in supply volumes is due to the stagnation of the Russian economy and the activities of "Irbistech," which is implementing an active communication campaign to promote products. In 2018, the most active regions-consumers of cryoblasters were Moscow and the subjects of the Volga Federal District. In 2019, consumption shifted to Moscow and the Moscow Region (Table 1).

Prices for cryoblasters vary widely depending on the brand, design, performance, and other characteristics. Prices for individual models are presented in Table 2.

The difference in prices for equipment offered by intermediate parties and manufacturers/representatives of manufacturers is up to 25–30%.

The cost of the "Minicryoblaster EisTEC" (Germany) on the company's website is 768000 rubles [6]. The cost of minicryoblasters produced by "Irbistekh" (Russia) for MiniCryoBlaster GT with the capacity of 30 kg/h is 340000 rubles, the

Table 1 Regions-consumers of cryoblasters

Region	2018		2019		2019 to 2018, ±
	USD	%	USD	%	
Moscow	348,393	79.8	200,885	87.4	%−147,508
the Moscow Region	–	–	5322	2.3	5322
the Republic of Tatarstan	28,807	6.6	–	–	−28,807
the Mari El Republic	29 273	6.7	–	–	−29,273
the Nizhny Novgorod Region	6 163	1.4	–	–	−6163
Other regions	24,120	5.5	23,717	10.3	−403

Source Compiled by the authors

Table 2 Prices for cryoblasters with the capacity of 100 kg/h

No	Cryoblaster	Country of origin	Price, rub
1	Karcher IB 15/80	Germany	1 500 000
2	TRIVENTEK Triblast T2	Denmark	1 200 000
3	EisTEC	Germany	950 000
4	IceTech Xtreme	Germany	1 700 000
5	White Lion 3000	Germany	1 600 000
6	Cold Jet Aero 40FP	The USA	1 600 000
7	ATX-25 E	India	1 000 000
8	Cryonomic COB 71	Belgium	1 500 000
9	YGAX-750	China	700 000
10	ASCO 1701	the USA	2 000 000

Source Compiled by the authors using[26]

cost of MiniCryoBlaster BM1000 with the capacity of 25 kg/hour is 220000 rubles [14]. Another Russian company, "SibTIM Group of Companies," sells the STIM.066499.001 unit with the capacity of up to 80 kg/h for 649,000 rubles [31]. The main competitive advantages of Russian cryoblasters targeted at the Russian market are the price and operating efficiency.

Most equipment suppliers also offer cryocleaning services. Some companies (e.g., "Karcher," "Carlblst," "Crionomic," and "Krio-Klining") offer cryoblaster rental services.

It is expected that in the context of the recovery of the Russian economy, the cryoblasting market will show steady growth due to an increase in demand (primarily in the automotive sector). Many cleaning companies are also planning to select and purchase their own dry ice cleaning equipment and create business units capable of doing this work on an ongoing basis. The use of full-time personnel and own technical means will save significant funds, and it is an expedient solution from the point of view of the production organization.

5 Conclusion

Thus, it can be noted that cryogenic blasting in Russia is gaining popularity in many areas. It is supported by developing markets for dry ice and cryogenic equipment. Particularly promising areas are the cleaning of industrial equipment and machine parts. The efficiency of cryoblasting is primarily determined by the technical characteristics of the blasters. The cryoblaster market is mainly represented by foreign models, but the gradual introduction of new models of domestic production to the market is expected thanks to the scientific and industrial developments of Russian companies focused on the requirements of the Russian consumer.

Despite the fact that cryogenic blasting service providers and equipment manufacturers actively promote the benefits of cryogenic blasting and predict that this technology will replace many traditional cleaning methods in the near future, cryogenic blasting is likely to be present in the Russian market along with other technologies of high pressure—hydraulic, steam-jet, and sandblast cleaning.

References

1. Carblast industry. (n.d.). Cryoblasting – dry ice cleaning. Retrieved February 9, 2022, from http://carblast.ru/item/crio_blast
2. Cryonomic. (n.d.). How does dry ice or cryogenic blasting technology work? Retrieved February 9, 2022, from https://cryonomic.ru/content/kak-rabotaet-tehnologiya-ochistki-suhim-ldom-ili-kriogennogo-blastinga
3. Dr. Sauber. (n.d.). Selection of dry ice cleaning equipment. Retrieved February 8, 2022, from https://dr-sauber.ru/vybor-oborudovaniya-dlya-ochistki-suhim-ldom

4. Dr. Sauber. (n.d.). Industrial cleaning. Retrieved February 9, 2022, from https://dr-sauber.ru/promyshlenniy-klining.
5. Dr. Sauber. (n.d.). Dry ice. Retrieved February 10, 2022, from https://dr-sauber.ru/news-by-tag/suhoy-lyod
6. Eistec GmbH. (n.d.). Cryoblaster mini. Retrieved February 15, 2022, from https://eistec.ru/eistec-mini/
7. Emiss. (n.d.). Production of main types of products in kind from 2010 to 2016. Retrieved February 12, 2022, from https://www.fedstat.ru/indicator/40557
8. Emiss. (n.d.). Production of main types of products in kind since 2017. Retrieved February 12, 2022, from https://www.fedstat.ru/indicator/58636
9. Europges. (n.d.). Cryogenic cleaning. Retrieved February 10, 2022, from https://www.europages.com.ru/предприятия/Очистка%20производственная/чистка%20криогенная.html
10. Fokina, O. V., Sozinova, A. A., Tyufyakova, E. S., Gurova, E. S., & Lysova, E. A. (2018). Marketing tools for increasing effectiveness of entrepreneurial structures' activities in the conditions of import substitution. *Espacios, 39*(28). Retrieved February 8, 2022, from http://www.revistaespacios.com/a18v39n28/a18v39n28p08.pdf
11. IceVentek. (n.d.). CO_2 Purification and production: Useful information. Retrieved February 16, 2022, from http://xn--80aejarjt3am.xn--p1ai/dryice/info/
12. Infoclen. (2011, October 3). Cryoblasting - the untrodden cherry in clinking. Retrieved February 9, 2022, from https://infoclean.su/stati-analitika-intervyu/krioblasting-nepodnjataja-celina-v-klininge.html
13. Infoclen. (2011, October 7). Cryogenic blasting. What do the old-timers think of the market? Retrieved February 16, 2022, from https://infoclean.su/stati-analitika-intervyu/kriogennyi-blasting-chto-dumayut-starozhily-o-rynke.html
14. IrbisTech. (n.d.). IrbisTech presents efficient solutions for dry ice production and cleaning. Retrieved February 15, 2022, from https://new.irbistech.com/ru/#
15. IrbisTech. (n.d.). About IrbisTech. Retrieved February 13, 2022, from https://new.irbistech.com/ru/о-ирбистех/
16. IrbisTech. (n.d.). About cryogenic blasting technology. Retrieved February 9, 2022, from https://new.irbistech.com/ru/о-криогенном-бластинге/
17. Jerman, M., Zeleňák, M., Lebara, A., Foldyna, V., Foldyna, J., & Valentincic, Y. (2021). Observation of cryogenically cooled ice particles inside the high-speed water jet. *Journal of Materials Processing Technology, 289*, 116947. https://doi.org/10.1016/j.jmatprotec.2020.116947
18. Karpuschewski, B., Emmer, T., Schmidt, K., & Petzel, M. (2013). Cryogenic wet-ice blasting – Process conditions and possibilities. *CIRP Annals – Manufacturing Technology, 62*(1), 319–322. https://doi.org/10.1016/j.cirp.2013.03.102
19. Kataeva, N. N., Marakulina, I. V., Sanovich, M. A., Sozinova, A. A., & Vasilyuk, N. (2017) Transformation of approach to market segmentation within crisis management of global entrepreneurship. In E. G. Popkova (Ed.), *Overcoming uncertainty of institutional environment as a tool of global crisis management* (pp 491–496). Springer. https://doi.org/10.1007/978-3-319-60696-5_62
20. Komarova, N. G. (2018). Abrasive treatment – Cryoblasting. In S. Yu. Shirokova (Ed.), *Current issues in engineering and technology* (pp. 280-282). Bashkir State University.
21. Lysova, E. A., Bratukhina, E. A., Sozinova, A. A., & Matushkina, Y. N. (2020). Digital modernization of the region's educational market and its influence on quality of education. *International Journal for Quality Research, 14*(1), 253–270. https://doi.org/10.24874/IJQR14.01-16
22. MEC CO_2. (n.d.). Dry ice blasting technologies. Retrieved February 12, 2022, from https://www.ecotech24.ru/images/doc/cb.pdf
23. MEWO. (n.d.). Cryogenic media-blasting deburring. Medi-blasying-, drum deburring- and washing systems for non-ferrous metls, rubber and plastic components. Retrieved February 9, 2022, from https://www.maplan.co.uk/files/ww/MEWO%20Product%20Information.pdf
24. Muhammad, J., Ning, H., Xiang, H., Wei, Z., Manish, K. G., & Akib, M. K. (2022). Measurement of machining characteristics under novel dry ice blasting cooling assisted milling of AISI 52100 tool steel. *Measurement, 191*, 110821. https://doi.org/10.1016/j.measurement.2022.110821

25. Nitrofreeze. (n.d.). Dry ice blasting during planned shutdowns. Retrieved February 9, 2022, from https://www.nitrofreeze.com/dry-ice-blasting-during-planned-shutdowns/
26. Polyus holoda. (n.d.). Cryoblaster 2021: Overview. Retrieved February 15, 2022, from http://polus-holoda.ru/krioblastery-obzor-2018g/
27. Rosproizvoditel. Reference Information System. (n.d.). Analysis of the market of solid carbon dioxide ("dry ice") in Russia. Retrieved February 12, 2022, from https://rosproizvoditel.ru/services/24877-analiz-rynka-tverdogo-dioksida-ugleroda-suhogo-lda-rossii
28. Saveleva, N. K., Nagovitsyna, E. V., Lapteva, I. P., Shchinova, R. A., & Koikova, T. L. (2018). The necessity for developing the integrated system of economic information on the bank services market in the conditions of globalization. *Espacios, 39*(28). Retrieved February 9, 2022, from http://www.revistaespacios.com/a18v39n28/a18v39n28p13.pdf
29. Saveleva, N., Lapteva, I., Araslanova, O., Matushkina Y., Koykova, T., & Graboyy, K. (2017). Integrated assessment technology in management of competition efficiency in business corporations. *MATEC Web of Conferences, 106*, 08086. https://doi.org/10.1051/matecconf/201710608086
30. Savelyeva, N. K. (2017). Method of determining the efficiency of price and non-price competition in service sector. *MATEC Web of Conferences, 106*, 08084. https://doi.org/10.1051/matecconf/201710608084
31. SibTIM (2021, December 31). Price list. Retrieved February 15, 2022, from http://sibtim.ru/index.php/price
32. Sozinova, A. A. (2018). Marketing concept of managing the reorganization of entrepreneurial structures using the latest information technologies. *Quality – Access to Success, 19*(S2), 118–122. Retrieved February 10, 2022, from https://tinyurl.com/yckvpzau
33. Sozinova, A. A., Fokina, O. V., & Shchinova, R. A. (2017). Marketing tools for increasing company's reorganization effectiveness. In E. Popkova (Ed.), *Overcoming uncertainty of institutional environment as a tool of global crisis management* (pp. 315–320). Springer. https://doi.org/10.1007/978-3-319-60696-5_40
34. Sozinova, A. A., Malysheva, N. V., Zonova, A. V., Fokina, O. V., & Shchinova, R. A. (2018). Effectiveness of marketing of business structures as a basis of import substitution in Russia: The role of information technologies. *Espacios, 39*(28). Retrieved February 12, 2022, from https://www.revistaespacios.com/a18v39n28/a18v39n28p03.pdf
35. Sozinova, A., Savelyeva, N., & Alpidovskaya, M. (2021). Post-COVID marketing 2019: Launching a new cycle of digital development. In E. DE LA POZA, & S. E. Barykin (Eds.), *Global challenges of digital transformation of markets* (pp. 419–431). Nova Science Publishers, Inc.
36. Stroy Podskazka. (n.d.). What is cryoblasting and how is it performed? Retrieved February 9, 2022, from https://stroy-podskazka.ru/specializirovannoe-oborudovanie/chto-takoe-krioblasting-i-kak-ego-delayut/
37. Tekhgaz-TK. (2020, February 4). Domestic carbon dioxide market 2019. Retrieved February 12, 2022, from https://www.techgas.ru/2020/02/rynok-uglekisloty/
38. UVAO. (2018, July 27). Prospects for introduction of cryogenic blasting technology at the enterprises of the Russian Federation. Retrieved February 9, 2022, from https://uvao.ru/finances/8370-perspektivy-vnedreniya-tehnologii-kriogennogo-blastinga-na-predpriyatiyah-rf.html
39. WikiMetall.ru. (n.d.). Soft and cryogenic blasting technology. Retrieved February 9, 2022, from https://wikimetall.ru/metalloobrabotka/blasting.html

Modeling the Dynamics of Development of Russian Retail Trade in the Modern Economy

Yekaterina A. Golubeva, Natalia S. Somenkova, Marina V. Smagina, Yekaterina V. Bulanova, and Natalia A. Yagunova

Abstract Retail trade affects the development of all socio-economic processes in the country. Being an important source of cash flow, trade forms the basis for the economic stability of the state. The share of retail trade is 4.8% of Russia's GDP and about 12 million people are employed there. In this regard, retail trade as an important sector of the economy is the object of close attention. Currently, due to anti-Russian sanctions and restrictions, imposed due to the current difficult geopolitical situation, retail trade is in the zone of influence of many negative factors leading to a slowdown in the development of retail chains and trade turnover. The purpose of this article is to analyze and forecast the retail trade turnover in the Russian Federation. A system of statistical indicators affecting the turnover of retail trade in Russia has been formed. Comparative and regression analysis were used to collect and process the initial information. The result of the study is the construction of a factor model of the dynamics of retail trade turnover in the Russian Federation and a multiplicative model in which a polynomial of the third degree describes the trend, and a model with fictitious variables describes changes in the seasonal component. It has been established that retail trade turnover depends on the standard of living of the population, consumer price indices and the index of tariffs for freight transportation.

Keywords Retail trade · Trading business · Factor model · Modeling · Economic development

JEL Classifications C1 · F17 · O1

1 Introduction

Today, retail trade is one of the fastest growing sectors of the economy of the Russian Federation. Retail is a significant industry, its contribution to GDP was 13.4% in 2020, and the share of people employed in trade reached 15.4% of the employed

Y. A. Golubeva (✉) · N. S. Somenkova · M. V. Smagina · Y. V. Bulanova · N. A. Yagunova
National Research Lobachevsky State University of Nizhny Novgorod, Nizhny Novgorod, Russia

A. V. Bogoviz (ed.), *Big Data in Information Society and Digital Economy*,
Studies in Big Data 124, https://doi.org/10.1007/978-3-031-29489-1_37

population. In 2020, the contribution of trade to the tax revenues of budgets of all levels amounted to more than 10%.

But at the same time, there are many different problems in this industry related to the introduction of anti-Russian sanctions and restrictions imposed due to the current difficult geopolitical situation. In addition, the consumer market is very dependent on the income of the population and rising prices for products, goods and services.

The purpose of this publication is to analyze and forecast retail trade turnover in the Russian Federation. The goal of the study determined the solution of the following main tasks: to analyze the indicators that affect the retail trade turnover; build factorial and multiplicative models of retail turnover dynamics in the Russian Federation.

2 Methodology

The object of study is the dynamics of retail trade turnover in Russia as one of the most important budget-forming areas. The study of the development of retail trade in the modern economy has found sufficient reflection in the works of Russian authors such as Vasilchuk [8], Voitkevich [9], Zhukovskaya, Mityakov [11], Zavyalov, Zavyalova [10], Iosipenko, Reshetnikova [2], Magomedov [3], Reprintseva [4], Somenkova [6]. The issues of forecasting indicators of the development of retail trade are studied in the works of Golubeva, Smagina [1], Shchepakin, Oblogin, Mikhailova [5].

At the same time, the issues of retail trade turnover modeling remain insufficiently studied. In this regard, the relevance and practical importance of developing a model for predicting the retail trade turnover in the Russian Federation is increasing.

3 Results

The study of the dynamics of change in retail trade is carried out in order to identify the factors influencing this process and, in the future, to develop adequate measures leading to the development of this area of trade.

Using the Statistica package, a model of factor analysis of the retail trade turnover in the Russian Federation was built in the work. The statistical base for the study was the data of the Federal State Statistics Service of the Russian Federation for 2010–2020 [7].

To describe the dynamics of retail trade turnover, 8 indicators were considered:

x_1– average nominal wage;

x_2– real money incomes (in % to the period of the last year);

x_3– average per capita monetary income of the population;

x_4– index of tariffs for freight transportation (in % of the previous period);

x_5– consumer price index;

x_6– consumer price indices for food products;

x_7 – consumer price indices for non-food products;

x_8 – the cost of a minimum set of food.

Factor analysis will begin with the building of a matrix of pairwise correlation coefficients between variables (available at https://figshare.com/ with the identifier https://doi.org/10.6084/m9.figshare.19368557.v1), which reflects the degree of influence of features on each other. Variables that are highly correlated can be combined into a single factor.

Next, using the Statistica package, we will build a table that contains the eigenvalues and cumulative dispersions (available at https://figshare.com/ with ID https://doi.org/10.6084/m9.figshare.19368614.v1). In the first column of the constructed table, the dispersions of the features under consideration are calculated; they are used to determine the number of factors in the model. The Cumulative column contains the cumulative dispersion. In our case, the first factor explains 39.53% of the total variance, the second—29.61%, the third—13.68%, etc.

To answer the question of how many factors should be distinguished in the factorial model, consider the Kaiser criterion and the scree criterion.

The Kaiser criterion states that factors whose eigenvalues are greater than 1 must be included in the factor model. Factors that do not extract a dispersion equivalent to the dispersion of at least one variable are omitted. According to this criterion, in our case only the first three factors are necessarily should be singled out. The remaining factors do not satisfy the condition imposed in the criterion on the eigenvalues.

A graphical method for determining the required number of factors is the scree criterion. When using it, the eigenvalues are displayed in the form of a graph (Fig. 1). The place on the graph corresponding to the maximum deceleration of the decrease in eigenvalues serves as the desired point.

It can be seen from the constructed graph that it is necessary to single out 3 factors in our case.

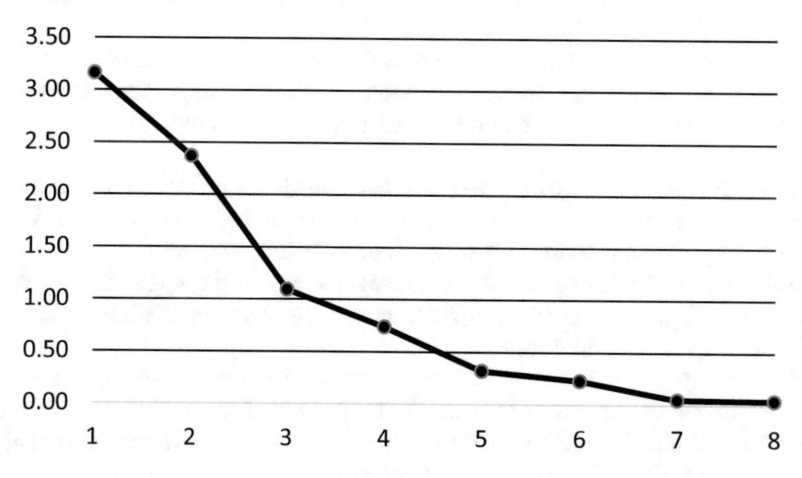

Fig. 1 Scree criterion. *Source* Compiled by the authors

Table 1 Parameters of the factor model of retail trade turnover

$N = 120$	Regression summary for dependent variable: y $R = 0.95738922$, $R^2 = 0.91659413$, Adjuster $R^2 = 0.91443708$ $F(3,116) = 424.93$, $p < 0.0000$, Std. error of estimate: 1434E2					
	b^*	Std. Err. of b^*	b	Std. Err. of b	$t(116)$	p-Value
Intercept	–	–	2,148,083	13,089.06	164.1129	0.000000
f_1	0.860461	0.026814	421,781	13,143.94	32.0894	0.000000
f_2	−0.401248	0.026814	−196,684	13,143.94	−14.9639	0.000000
f_3	−0.123292	0.026814	−60,436	13,143.94	−4.5980	0.000011

Note A definition for the significance of * is coefficients of the multiple regression equation without an intercept term
Source Compiled by the authors

To classify the declared variables into factors, we determine the correlations between the variables and the selected factors, i.e. factor loadings (available at https://figshare.com/ with ID https://doi.org/10.6084/m9.figshare.19368923.v1). From the resulting table, we conclude that 3 variables correspond to the first and second factors, one variable corresponds to the third. In this case, the first and third factors are characterized by large values of correlation coefficients compared to the second factor.

The selected components of the factorial model must be named based on the names of the variables included in them. In this example, the factors are named as follows:

Factor 1—the standard of living of the population;

Factor 2—change in consumer price indices;

Factor 3—change in the index of tariffs for freight transportation.

Let us construct a model of multivariate dependence of changes in retail trade turnover on the selected factors (Table 1).

The quality of the multidimensional model is judged by the value of the adjusted coefficient of determination. In the example under consideration, it is equal to 0.9144, which indicates a fairly good quality of fitting the regression model to the observed data.

In addition, it is necessary to evaluate the statistical significance of the model parameters and the regression equation as a whole, as well as to check it for the presence of multicollinearity, heteroscedasticity and autocorrelation.

Fisher's F-statistic is used to assess the significance of the equation as a whole. In our case $F_{\text{factual}} = 424.93$, and $F_{\text{table}} = (3; 116; 0.05) = 2.68$, therefore, the regression equation is significant.

The values of t-statistics required to assess the statistical significance of the parameters of the regression model turned out (Table 1) in modulus to be greater than the tabular value of Student's t-test ($t_{\text{table}} = 1.66$), which indicates the statistical significance of the coefficients and the free term.

Based on the obtained data, we get the following regression equation:

$$y = 2148083 + 421781 f_1 - 196684 f_2 - 60436 f_3.$$

The matrix of pair correlations of factors f_1, f_2, f_3 has the form:

$$\begin{pmatrix} 1 & 6.93936 \cdot 10^{-15} & -6.012 \cdot 10^{-15} \\ 6.93936 \cdot 10^{-15} & 1 & 2.11875 \cdot 10^{-15} \\ -6.012 \cdot 10^{-15} & 2.11875 \cdot 10^{-15} & 1 \end{pmatrix}$$

The matrix determinant is 1, which indicates the absence of multicollinearity in the model. Also, the absence of multicollinearity is evidenced by the statistical significance of the parameters of the regression model.

To detect heteroscedasticity, consider the Goldfeld-Quandt test. The results obtained for $k = 44$ (available at https://figshare.com/ with ID https://doi.org/10.6084/m9.figshare.19368959.v1) indicate the absence of heteroscedasticity in the considered multidimensional dependence.

To study the constructed factorial model for autocorrelation, we apply the Durbin-Watson criterion. In our case DW = 0.681696. The tabular values of the Durbin–Watson statistics are respectively equal to $d_l = 1.653$—upper, $d_u = 1.755$—bottom. Thus, $0 \le 0.681696 \le d_l$, which indicates the existence of a positive autocorrelation.

In order to eliminate autocorrelation, we use an autoregressive transformation. Here, the new levels of the series $\tilde{y}_t$, $\tilde{f}_{1t}$, $\tilde{f}_{2t}$ are calculated by the formulas.

$$\tilde{y}_t = y_t - \rho \cdot y_{t-1}, \tilde{f}_{1t} = f_{1t} - \rho \cdot f_{1t-1}, \tilde{f}_{2t} = f_{2t} - \rho \cdot f_{2t-1}, \tilde{f}_{3t} = f_{3t} - \rho \cdot f_{3t-1},$$

where the coefficient $r \approx 1 - \frac{\mathrm{DW}}{2}$, $t = 2, ..., n$ can be taken as an estimate of the coefficient ρ.

The first time series observations are calculated using the Price–Winsten correction:

$$\tilde{y}_1 = y_1 \cdot \sqrt{1-\rho^2}, \tilde{f}_1 = f_1 \cdot \sqrt{1-\rho^2}, \tilde{f}_2 = f_2 \cdot \sqrt{1-\rho^2}, \tilde{f}_3 = f_3 \cdot \sqrt{1-\rho^2}$$

After transformations, we get that DW = 1.946247. Therefore, $d_u \le 1.946247 \le 4 - d_u$, this means that there is no autocorrelation, and the transformed model has the form:

$$y = 2178241 + 365398 f_1 - 166601 f_2 - 83440 f_3.$$

The resulting model describes 91.67% of the variation in retail trade turnover.

The factor model carries very important information about the causes of this or that behavior of the studied indicator, so the forecast obtained on its basis is very useful in studying the dynamics.

Evaluation of the quality of the built model includes checking its predictive abilities. One of the parameters of such an assessment is the coefficient of fluctuation, calculated by the formula.

$$V = \frac{S}{\overline{y}},$$

where $S = \sqrt{\frac{\sum e_i^2}{n-m-1}}$—the value of the standard error of the regression,

$\sum e_i^2 = \sum (y - \hat{y}_x)^2$—the sum of the squared deviations of the actual values of the levels of the series from the calculated ones,

n—number of observations,

m—number of explanatory variables in the model,

$\overline{y}$—mean value of the dependent variable.

The coefficient of fluctuation is used to determine the intensity of fluctuations in the general value or forecasts. The predictive quality of the model is high if the value V is very small. At the same time, the smaller the value of the regression standard error, the more likely this model adequately describes the behavior of the real object under study.

In the built model $S = \sqrt{\frac{3.04294 \cdot 10^{12}}{120-3-1}} = 161963.761$, $\overline{y} = 2148082.91$, therefore, $V = 0.0754$, which is a good enough indicator.

To have a general judgment about the quality of the model from the relative deviations for each observation, determine the average approximation error:

$$\overline{A} = \frac{1}{n} \cdot \sum \left| \frac{y - y_x}{y} \right| \cdot 100\%$$

.

In our case, $A = 6.7\%$. These values of the average approximation error are quite acceptable, therefore, the built model of the dynamics of retail trade turnover is of high quality and can be used to predict future values of the indicator under study.

Thus, the model under study describes the variation in retail trade turnover by 91.67% and has good predictive qualities. The resulting model showed that the retail trade turnover depends on the standard of living of the population, consumer price indices and the index of tariffs for freight transportation. The indicator under consideration increases with an increase in the standard of living of the population and with a fall in consumer prices and tariffs for freight transportation.

To determine the type of trend in the time series under consideration, the Statistica software package was used. Several different dependences of retail trade turnover on time were constructed [1] (Table 2).

An analysis of the calculated coefficients of determination shows that the best equation is a polynomial of the third degree:

$$y_t = 1165253 + 27115.97 \cdot t - 219.863 \cdot t^2 + 0.916375 \cdot t^3$$

.

The coefficient of determination of the model takes a fairly high value $R^2 = 0.85524$. Therefore, the regression equation explains 85.524% of the dispersion of the resulting attribute (retail trade turnover). This fact is well illustrated by the mutual

Table 2 Selecting the type of trend

	Type of trend	R^2
Linear trend	$y_t = 1385532 + 12426.2 \cdot t$	0.84062
Exponential trend	$y_t = e^{14.22493+0.005467 \cdot t}$	0.81831
Hyperbola	$y_t = 2310104 - 2373111/t$	0.22577
Power-law trend	$y_t = 13.65275 \cdot t^{0.016787}$	0.84176
Quadratic parabola	$y_t = 1275491 + 17353.37 \cdot t - 37.0464 \cdot t^2$	0.84929
Cube parabola	$y_t = 1165253 + 27115.97 \cdot t - 219.863 \cdot t^2 + 0.916375 \cdot t^3$	0.85524

Source Compiled by the authors

arrangement of the graphs of the actual values of the retail trade turnover and the levels of the series calculated according to the trend (Fig. 2).

To describe the dynamics of seasonal fluctuations in retail trade turnover, we use the method of fictitious variables. Due to the fact that the amplitude of seasonal fluctuations increases, we apply the multiplicative model. After eliminating the trend, we introduce 11 fictitious variables $D_i, i = 1, ..., 11$ corresponding to the months from January to November. In this case, the variable corresponding to the month of December is used as a reference variable. Each fictitious variable takes two values, 0 and 1.

After estimating the parameters, the model describing seasonal fluctuations in retail trade turnover takes the form:

$$S_t = 1.226318 - 0.30658 \cdot D_1 - 0.32429 \cdot D_2 - 0.25172 \cdot D_3 - 0.27693 \cdot D_4 - \\ -0.26704 \cdot D_5 - 0.25003 \cdot D_6 - 0.22365 \cdot D_7 - 0.19714 \cdot D_8 - 0.21588 \cdot D_9 - \\ -0.19828 \cdot D_{10} - 0.20314 \cdot D_{11}.$$

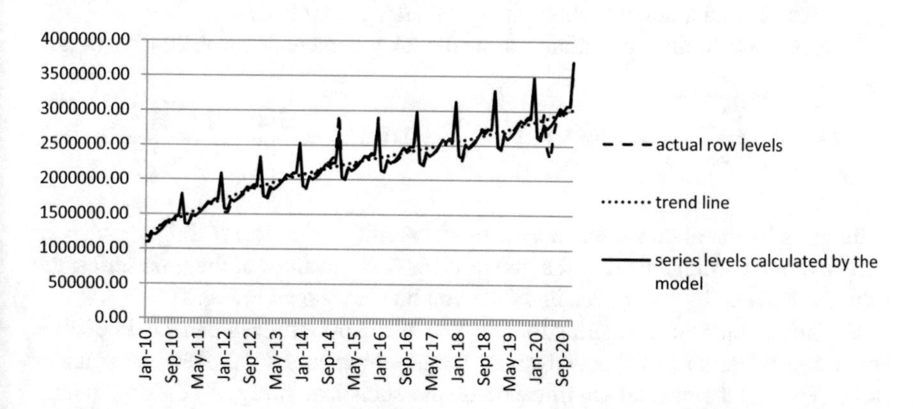

Fig. 2 Dynamics of retail trade turnover in the period 2010–2020. *Source* Compiled by the authors

All parameters of the obtained regression model are statistically significant. The results of the study are shown in a table available at https://figshare.com/ with ID https://doi.org/10.6084/m9.figshare.19369061.v1.

The significance of the regression equation is evaluated using the Fisher F-test, with $F_{\text{factual}} = 82.2156$, $F_{\text{table}}(120, 11) = 1.86$, which indicates the significance of the equation as a whole.

The obtained estimates of the model parameters are the difference between the average levels of the resulting attribute for the corresponding group and the reference group. Since the month of December was chosen as the reference variable, the values of the parameters are easily explained. In December, the highest value of the retail trade turnover for the cycle (year) is observed, then in January there is a sharp drop, which somewhat intensifies in February. In March of each year, a gradual increase in the studied indicator begins, which continues throughout the rest of the year, ending with a sharp jump in December.

The adjusted multiple correlation coefficient in this model is $R^2_{\text{adjust}} = 0.87$, which indicates that the resulting equation describes 87% of retail trade turnover fluctuations.

The general model containing the description of the trend and seasonal components is as follows:

$$
\begin{gathered}
y_t = \left(1165253 + 27115.97 \cdot t - 219.863 \cdot t^2 + 0.916375 \cdot t^3\right) \cdot \\
(1.226318 - 0.30658 \cdot D_1 - 0.32429 \cdot D_2 - 0.25172 \cdot D_3 - 0.27693 \cdot D_4 - \\
-0.26704 \cdot D_5 - 0.25003 \cdot D_6 - 0.22365 \cdot D_7 - 0.19714 \cdot D_8 - \\
-0.21588 \cdot D_9 - 0.19828 \cdot D_{10} - 0.20314 \cdot D_{11}) \cdot
\end{gathered}
$$

In the resulting model, the multiple correlation coefficient and the adjusted multiple determination coefficient, despite the large number of parameters, differ insignificantly and amount to more than 0.98 ($R^2_{\text{adjust}} = 0.9807$).

To assess the predictive qualities of the model, we calculate the following values:

$$
S = \sqrt{\frac{\sum e_i^2}{n - m - 1}} = 72085.5; \; V = \frac{S}{\overline{y}} = 0.033; \; A = \frac{100}{\overline{y}} \cdot \sqrt{\frac{\sum e^2}{n}} = 3.1
$$

.

Judging by the obtained results, the model is suitable for forecasting.

The mutual arrangement of the graphs of the initial values of the time series, the values calculated by the model, and the trend line are shown in Fig. 2.

Quantification and accounting for seasonal fluctuations reflected in the time series are an important part of the analysis of the development of the phenomena under study. The great practical significance of the statistical study of seasonal fluctuations lies in the fact that the quantitative characteristics obtained by analyzing the series of intra-annual dynamics reflect the specifics of the development of the studied phenomena by months and quarters of the annual cycle. This is necessary for understanding the patterns of development of socio-economic phenomena in intra-annual dynamics, forecasting and developing operational measures for the qualified management of their development over time.

4 Conclusions

To describe the behavior of the retail trade turnover, two econometric models were constructed in which the structure of the time series was studied by various methods.

The resulting factor analysis model showed that retail trade turnover depends on the standard of living of the population and consumer price indices and tariffs for freight transportation. The indicator under consideration increases with an increase in the standard of living of the population and with a fall above the indicated indices. With the help of factor analysis, it was possible to describe 91.67% of the variation in the studied time series, mainly because indicators that caused seasonal fluctuations in retail trade turnover were correctly and completely identified. The forecast obtained by this model is quite accurate, since the relative approximation error turned out to be 6.7%.

The multiplicative model, in which a polynomial of the third degree describes the trend, and the dynamics of the seasonal component is described by the model with fictitious variables approximates the time series of retail trade turnover with an accuracy of more than 98%. The coefficient of fluctuation, the standard error of the regression, and the relative approximation error take on low values, which indicates the good qualities of the model, including predictive ones.

Thus, the multiplicative model most accurately describes the initial data and gives the correct forecast. The factor model allowed us to identify the variables that affect the dynamics of the development of the retail trade turnover in the Russian Federation.

References

1. Golubeva, E. A., & Smagina, M. V. (2021). Retail turnover modeling. *Science of Krasnoyarye, 10*(6–1), 139–146.
2. Iosipenko, V. D., & Reshetnikova, N. V. (2020). Modern trends in the development of retail trade in food products. *Patterns of Development of Regional Agro-Food Systems, 1*, 53–57.
3. Magomedov, A. M. (2020). Problems of development of distance trading during the COVID-19 pandemic. *Economics and Management: Problems, Solutions, 8*(104), 59–68.
4. Reprintseva, E. V. (2021). Trends in the development of trade in the regions in the context of the coronavirus pandemic. *Bulletin of the Kursk State Agricultural Academy, 6*, 181–186.
5. Shchepakin, M. B., Oblogin, M. V., & Mikhailova, V. M. (2020). Factor model for managing the development of the wholesale and retail trade market in the national economy. *Economics, Entrepreneurship and Law, 10*(4), 1095–1122.
6. Somenkova, N. S. (2021). Problems and prospects for the development of foreign economic activity in the region. *Management Accounting, 11–1*, 113–120.
7. Trade in Russia (2021). stat. comp. Rosstat. M., 2021. Retreived January 20, 2022, from https://rosstat.gov.ru/storage/mediabank/Torgov_2021.pdf.

8. Vasilchuk, E. S. (2021). Features of the development of the regional retail market in a pandemic. *Innovative Economy: Prospects for Development and Improvement, 3*(53), 19–26.
9. Voitkevich, N. I. (2020). Current trends in Russian retail trade. *Bulletin of Samara State University of Economics, 8*(190), 13–20.
10. Zavyalov, D. V., & Zavyalova, N. B. (2020). Problems and prospects for the development of the sphere of commodity circulation in the digital economy. *Economics, Business and Law, 10*(6), 1701–1720.
11. Zhukovskaya, I. F., & Mityakov, D. A. (2020). Retail trade in the context of digitalization of the economy: Development trends in Russia. *Financial Economics, 3*, 243–247.

The Future of Big Data in the Information Society and Digital Economy on the Eve of the Fifth Industrial Revolution (Conclusion)

Aleksei V. Bogoviz

Doctor of Economics, Professor, Independent Researcher, Moscow, Russia

Due to the fourth industrial revolution, big data has gained traction and practical application in the information society and digital economy. As demonstrated by the best practices of the Eurasian Economic Union (EAEU), cyber-physical systems, the elements of which are integrated with big data, are widespread. The hands-on experience with big data highlighted and scientifically interpreted in this book has built a bridge between theory and practice of big data, contributing to the scientific concepts of Industry 4.0, the information society, and the digital economy.

While generating new scientific knowledge, this book also raised new research questions. In particular, today's world is on the threshold of the Fifth Industrial Revolution, the rapid arrival of which is due to the acceleration of scientific and technological progress. Big data will also play an essential but new role in the cyber-social systems of Industry 4.0. All aspects of big data (from collection to processing and use) must be rethought so that the institution of big data does not become an "institutional trap," but rather evolves with the information society and digital economy and supports the transition to Industry 5.0.

This raises the question of how to achieve the socialization of big data—in the collection (integrate automatically collected data and data received from users), processing (make it easy to combine manual and automated analysis of big data), and use (achieve a user-friendly big data interface with intelligent decision support). It is suggested that further research in the continuation of this book be devoted to finding an answer to this question.

A. V. Bogoviz (ed.), *Big Data in Information Society and Digital Economy*, Studies in Big Data 124, https://doi.org/10.1007/978-3-031-29489-1

Printed in the United States
by Baker & Taylor Publisher Services